The Advancement of Science

SCIENCE WITHOUT LEGEND,
OBJECTIVITY WITHOUT ILLUSIONS

PHILIP KITCHER

New York   Oxford
OXFORD UNIVERSITY PRESS
1993

Oxford University Press

Oxford New York Toronto
Delhi Bombay Calcutta Madras Karachi
Kuala Lumpur Singapore Hong Kong Tokyo
Nairobi Dar es Salaam Cape Town
Melbourne Auckland Madrid

and associated companies in
Berlin Ibadan Madrid

Copyright © 1993 by Philip Kitcher

Published by Oxford University Press, Inc.
200 Madison Avenue, New York, New York 10016

Oxford is a registered trademark of Oxford University Press

Parts of Chapter 2 previously appeared in "Darwin's Achievement" (in Nicholas Rescher [ed.] *Reason and Rationality in Science*, Washington, D.C.: University Press of America, 1975, 127–190). Some short discussions in Chapters 3 and 4 are taken from "Theories, Theorists, and Theoretical Change" (*Philosophical Review*, 87, 1978, 519–547). Chapter 8 contains sections drawn from "The Division of Cognitive Labor" (*Journal of Philosophy*, 87, 1990, 5–21), and from "Authority, Deference, and the Role of Individual Reason" (in Ernan McMullin [ed.] *The Social Dimension of Science*, Notre Dame: University of Notre Dame Press, 1992, 247–270). I am grateful to the editors of these books and journals for their kind permission to reprint this material.

Library of Congress Cataloging-in-Publication Data
Kitcher, Philip, 1947–
The advancement of science : science without legend, objectivity
without illusions / Philip Kitcher.
p. cm.   Includes bibliographical references and index.
ISBN 0-19-504628-5
1. Science—Philosophy. 2. Science—History. I. Title.
Q175.K533 1993   500—dc20   92–19532

2 4 6 8 9 7 5 3 1

Printed in the United States of America
on acid-free paper

For Andrew and Charles

The Advancement of Science

Preface

This is the book I have wanted to write ever since I began studying the history and philosophy of science. More exactly, it is concerned with the topics that have interested me the most throughout the past twenty-odd years. The versions of it that I would have produced at different times during that period would have diverged, sometimes quite dramatically, from what I have now written. I suspect, furthermore, that my ideas about the growth of scientific knowledge will continue to evolve. These are not, I hope, my last thoughts about the issues that occupy me. However, I also hope that their formulation will help to stimulate improved thoughts, both for me and for others.

I owe an enormous intellectual debt to two people who taught me when I was a graduate student at Princeton in the early 1970s: C. G. Hempel and Thomas Kuhn. Although my treatment of virtually all the questions I address differs from theirs, I could not have arrived at my own conclusions without their deep insights. I have also been greatly influenced by the writings of Alvin Goldman, Hilary Putnam, and W. V. Quine, all of whom have shaped my ideas in important ways.

I began work on the writing of this book during 1988–89, when I had leave from teaching at the University of California at San Diego. That leave was largely made possible by a Fellowship from the John Simon Guggenheim Foundation, and I am most grateful for the foundation's support. Earlier, the development of my ideas had been aided by the opportunity to serve as co-director of an institute to investigate new consensus in the philosophy of science (held at the University of Minnesota), during which I had the chance to listen to presentations made by many of the world's leading philosophers of science. I would also like to thank the Office of Graduate Studies at the University of California at San Diego for supporting a workshop on naturalizing the philosophy of science.

Many people have read large parts of this book and offered valuable comments on it. I am grateful to John Dupre and Elliott Sober for many philosophical insights, and to Larry and Rachel Laudan for illuminating discussions. Michael Rothschild (of the UCSD Economics Department) provided extensive comments on Chapter 8, which have enabled me to make considerable improvements. From my colleagues Martin Rudwick and Robert Westman I have learned much about the craft of history and about the more specific aspects of the history of science for which they are well known. Martin Rudwick's careful reading of Chapter 6 was particularly helpful. To the many others who have listened to lectures in which I presented rough-hewn fragments of the material of this book, and who have offered insightful suggestions, my belated (and inadequate) thanks.

In recent years I have been fortunate to work with a group of extremely talented graduate students. Their questions, comments, criticisms, and suggestions have improved many parts of this book. In particular, I would like to thank Gillian Barker, Michael Bishop, Jeffrey Brown, Sylvia Culp, Michael Dietrich, Bruce Glymour, Peter Godfrey-Smith, Gary Hardcastle, Todd Jones, Alex Levine, Sam Mitchell, Eric Palmer, Neelam Sethi, and Andrew Wayne, all of whom have left their mark on the pages that follow. I am also grateful to Joe Ramsey for his preparation of the index.

During the past six years, my thinking about epistemology and the history and philosophy of science has been greatly helped by discussions with Stephen Stich and Steven Shapin. Neither is likely to agree with the conclusions of this book, but both can pride themselves on having diverted me from even sillier things that I might have said. I am also extremely grateful to an anonymous reader for Oxford University Press, who not only gave me numerous specific suggestions for improvement but also convinced me that the bloated penultimate version needed to be put on a diet.

Above all, I would like to thank three exceptionally perceptive philosophers and generous friends, who sent me detailed written comments on the entire penultimate draft. Small tokens of my thanks to Wesley Salmon, Kim Sterelny, and John Worrall are scattered throughout the following pages, but there is no way to convey in footnotes the more global suggestions they have made that have helped me, again and again, to bring my ideas and arguments into clearer focus. I am immensely grateful to them.

Finally, as always, Patricia Kitcher has provided sage advice, generous encouragement, and warm support, without which this project would never have been completed. Our two sons, Andrew and Charles, have offered both distraction and delight. As I reflect on the writing of this book, I recall the many baseball games, soccer practices, attacks on computer games, piano festivals, tennis foursomes, and mountain hikes from which I have returned to philosophy, refreshed. The joy of their conversation and their many enthusiasms deserve a better return, but this is the book I wrote and, with all its flaws, it is theirs.

P. K.

La Jolla, Calif
June 1992

Contents

The Advancement of Science

My mistress' eyes are nothing like the sun;
Coral is far more red than her lips' red:
If snow be white, why then her breasts are dun;
If hairs be wires, black wires grow on her head.
I have seen roses damask'd, red and white,
But no such roses see I in her cheeks;
And in some perfumes is there more delight
Than in the breath that from my mistress reeks.
I love to hear her speak, yet well I know
That music hath a far more pleasing sound:
I grant I never saw a goddess go,—
My mistress, when she walks, treads on the ground:
  And yet, by heaven, I think my love as rare
  As any she belied with false compare.

William Shakespeare
Sonnet CXXX

1

Legend's Legacy

Once, in those dear dead days, almost, but not quite beyond recall, there was a view of science that commanded widespread popular and academic assent. That view deserves a name. I shall call it "Legend."

Legend celebrated science. Depicting the sciences as directed at noble goals, it maintained that those goals have been ever more successfully realized. For explanations of the successes, we need look no further than the exemplary intellectual and moral qualities of the heroes of Legend, the great contributors to the great advances. Legend celebrated scientists, as well as science.

The noble goals of science have something to do with the attainment of truth. Here, however, there were differences among the versions of Legend. Some thought in ambitious terms: ultimately science aims at discovering the truth, the whole truth, and nothing but the truth about the world. Others preferred to be more modest, viewing science as directed at discovering truth about those aspects of nature that impinge most directly upon us, those that we can observe (and, perhaps, hope to control). On either construal, discovery of truth was valued both for its own sake and for the power that discovery would confer upon us.

According to Legend, science has been very successful in attaining these goals. Successive generations of scientists have filled in more and more parts of the COMPLETE TRUE STORY OF THE WORLD (or, perhaps, of the COMPLETE TRUE STORY OF THE OBSERVABLE PART OF THE WORLD). Champions of Legend acknowledged that there have been mistakes and false steps here and there, but they saw an overall trend toward accumulation of truth, or, at the very least, of better and better approximations to truth. Moreover, they offered an explanation both for the occasional mistakes and for the dominant progressive trend: scientists have achieved so much through the use of SCIENTIFIC METHOD.

Variants of Legend often disagreed, sometimes passionately, on details of method, but all concurred on some essential points. There are objective canons of evaluation of scientific claims; by and large, scientists (at least since the seventeenth century) have been tacitly aware of these canons and have applied them in assessing novel or controversial ideas; methodologists should articulate the canons, thus helping to forestall possible misapplications and to extend the scope of scientific method into areas where human inquiry typically falters; in short, science is a "clearing of rationality in a jungle of muddle, prejudice, and superstition."[1] Indeed, many advocates of Legend would maintain that science is the pinnacle of human achievement not because of its actual successes but in virtue of the fact that its practice, both in attaining truth and in lapsing into error, is thoroughly informed by reason. Even those whose historical situations lead into making mistakes still do the best they can in the interests of truth, judging reasonably in the light of the available evidence and bowing to whatever new findings expose their errors.

Who were—and are—these advocates of Legend? Many reflective people, if asked for an assessment of science, would respond with a version of Legend. Practising scientists have sometimes been more ambivalent, proclaiming Legend on high days and holidays, sometimes unconsciously using it in formulating their plans, sometimes, in their cups, confessing that the realities of science do not reflect Legend's rosy glow.[2] However, the most detailed articulations of Legend have been provided not by the practitioners but by their amanuenses in history of science, philosophy of science, and sociology of science.

To find confident testimonials to the progressiveness of science and the rationality of individual scientists, we need look no further than some widely read writings of the 1940s and 1950s. James Bryant Conant's volumes of *Case Studies in Experimental Science,* designed to convey to general readers "the variety of methods by which [the physical] sciences have advanced" (Conant 1957 vii), attempted to recreate the evidence and reasoning that led scientists to support novel ideas. Objectivity and rationality are the order of the day. Similarly, Bernard Barber, reflecting on the cultural values of the "modern world," recognized a "belief in and an approval of "progress" in this world, a progress which is not necessarily of a unilinear evolutionary kind, *but which is somehow cumulative in the way in which science and rational knowledge are cumulative"* (Barber 1952 66; italics mine).[3] Although the truth about nature may be difficult to attain, Barber insisted that the cornerstone of "scientific morality" is "devotion to the 'truth' " and that "it is worth endless human striving to attain those provisional and approximate statements of truth that make up the substance of science at any given historical moment in its course of development" (Barber 1952 87).

Philosophical articulations of Legend are even more rich and impressive. The failure of simple attempts to demarcate the cognitively meaningful (science) from the cognitively meaningless (post-Kantian German metaphysics) inspired the original members of the Vienna Circle and their philosophical kin to investigate the characteristics of good science.[4] Rudolf Carnap, Carl

1. The phrase is Bruno Latour's, who uses it for ironical effect. See (Latour 1988 6) and also (Latour 1987 chapter 5).

2. James Watson's celebrated (1966) is a striking confessional. The reaction of many reviewers is interesting and instructive. See, in particular, (Lewontin 1968).

3. Barber's reference to evolution appears to embody a common misunderstanding. See (Williams 1966 34–55) and (Gould 1989).

4. In my judgment it is important to separate two movements with an overlapping cast of characters. The earlier, logical positivism, took as its central philosophical problem that

Gustav Hempel, Ernest Nagel, Hans Reichenbach, and Karl Popper endeavored to analyze good science by focusing on questions about the confirmation of hypotheses by evidence, the nature of scientific laws and scientific theories, and the features of scientific explanation. They differed on many points of detail, but some major articles of doctrine united them. Among these was the conviction that the succession of theories in the physical sciences constituted a progression and that the achievements of earlier theories (confined, in some versions, to the deliverance of statements about the observables) were retained in later theories. Another was the understanding that, while there is no systematic way to *generate* new hypotheses, once hypotheses have been proposed there are principles for their proper assessment in light of statements of evidence. Inspired by the work of Gottlob Frege and Bertrand Russell, the architects of modern mathematical logic, logical empiricist philosophers of science proposed to uncover the logic of confirmation, the logical structure of theories and the logic of explanation, thus formulating with precision those canons and criteria that they took to be tacitly employed by scientists in their everyday work. References to logic reverberate like drumrolls through the classic works of logical empiricist philosophy of science, works that, because of their clarity, rigor, and attention to a wide range of considerations, belong among the greatest accomplishments of philosophy in our century.

So much for the dear dead days. Since the late 1950s the mists have begun to fall. Legend's lustre is dimmed. While it may continue to figure in textbooks and journalistic expositions, numerous intelligent critics now view Legend as smug, uninformed, unhistorical, and analytically shallow. Some of these critics, the science bashers, regard the failure of science to live up to Legend's advertising as reason enough to question the hegemony of science in contemporary society. I shall not be concerned with them, but with the critiques of the Legend bashers, those who believe that Legend offered an unreal image of a worthy enterprise.[5]

---

of presenting a criterion of cognitive significance. Although logical positivists sometimes offered analyses of particular parts of particular sciences, their involvement in issues of general methodology was rather limited. With the demise of the demarcation problem, attention shifted to trying to understand the features of science that make it cognitively valuable. Thus, in the 1940s and 1950s, former positivists and sympathizers began intensive studies of scientific confirmation, explanation, and the structure of theories. These studies belonged to a new movement, logical empiricism, which no longer insisted on positivist doctrines about meaning and meaningfulness, and no longer viewed the positivist demarcation problem as either central or solvable. Ironically, the movement included, prominently, one outspoken former critic of positivism, Karl Popper, whose important (1934) had begun to tackle many of the problems that later became central to logical empiricism. I want to emphasize that the detailed articulations of Legend are the fruits of logical empiricism (although the logical positivists were surely devotees of Legend). For a classic account of the demise of the demarcation problem see (Hempel 1950, reprinted in Hempel 1965), whose final section outlines the agenda for logical empiricist philosophy of science.

5. Ultimately philosophy of science ought to attend to the most sweeping criticisms of the scientific enterprise and its role in our lives. However, before any such examination

My aim in this book is to probe the notions of progress and rationality, dear to Legend's champions but often trimmed, defaced, or discarded by detractors. Contemporary scholars have raised penetrating questions about truth and scientific progress. Can we legitimately view truth as a goal of science? Is it meaningful to talk of approximations to the truth or to see science as "converging" on truth about nature? Can we expect to attain even part of the truth about nature? Does the historical record show even the accumulation of truths about observable phenomena?

These questions raise venerable worries (some as old as the Greeks). Some can be posed with renewed force in light of contemporary accounts of change in science. The famous opening sentence of (Kuhn 1962)—"History, if viewed as a repository for more than anecdote or chronology, could produce a decisive transformation in the image of science by which we are now possessed"— began the enterprise of comparing the actual course of science with Legend's ideal of accumulation of truth. Disentangling two strands in the skeptical questions of the last paragraph, we might pick out issues of the *coherence* of certain putative aims, issues that have been addressed many times in the history of Western philosophy, and issues of the *attainability* of those aims. In the wake of Kuhn's work, the practice of science, past and present, is seen as raising doubts about whether the aims beloved of Legend have *actually* been attained, and these doubts then serve as the basis for wondering whether they are *attainable*. These latter concerns can also be motivated by adding the perspectives of the biologist, the cognitive scientist, and the sociologist. If scientists, like other people, are recognized as biological entities, who have evolved under selection pressures that are (at least prima facie) quite remote from the demands of the quest for the truth about quarks or about jumping genes, who are, in consequence, cognitive systems with identifiable limitations and deficiencies, and who are embedded in complex networks of social relations, then the chances that their activities will result in the attainment of truth can seem distressingly low.

Attacks on Legend's conception of the rationality of science accompany the critique of scientific progress. Philosophers, inspired by the view of scientists as paradigmatic reasoners, undertook the task of exposing the rules of good inference that their subjects were (unconsciously) following. Ideally, the result of their labors was to be a system of logic, as lucid and precise as that which has been provided for deduction. Criticisms of the rules specified by Carnap, Reichenbach, Popper, and others, on the grounds either that those rules did not endorse the right inferences or that they could not be articulated to reconstruct the actual inferences made by scientists, led to the more general claim that scientific inference should not be viewed as an algorithmic activity, that there is no set of rules that underlies the appraisal of hypotheses and theories in science. Here again (Kuhn 1962) is a seminal work. Kuhn did not simply propose that existing accounts of good reasoning

---

can go forward, we need an accurate picture of science, its goals, and its achievements. This book can thus be seen as a prelude to a more wide-ranging study.

in science were wrong, that the idealizations they employed were too simple, or that their vocabulary needed enrichment. Instead, he suggested that the entire project of finding a "calculus of scientific reasoning" ought to be abandoned.

One way of developing that conclusion would be to claim that there are no standards of good reasoning in science that are binding on all scientists at all times—or at least that such standards would be insufficiently powerful to endorse all the major decisions made in the history of science. In a celebrated discussion, Paul Feyerabend (1970, 1975) maintained that Galileo's defense of Copernicanism violated philosophers' favorite rules of good reasoning, and he concluded that the only maxim that can be applied across all historical instances is "Anything goes." Norwood Russell Hanson (1958) campaigned for the theory-ladenness of observation, suggesting that there is no theory-neutral body of evidence to which scientific theories must conform. So, by emphasizing the lack of shared methodological rules and the inefficacy of those few canons that are shared, by insisting on the variability of what is taken as evidence, some critics of Legend made it seem appropriate to talk of major scientific decisions as "conversion experiences."[6] Because massive underdetermination of belief by "objective" factors came to seem omnipresent, there opened up a vacuum into which social explanations of scientific behavior could be inserted. Instead of an ordered abode of reason, science came to figure as the smoke-filled back rooms of political brokering. This was not the way in which Kuhn intended to go, and, to the disappointment of some who have drawn inspiration from him, he has continued to insist on a set of commitments that scientists share, that cannot be articulated as rules but that function in distinguishing good reasoning from bad.[7]

Critics of scientific rationality maintain not only that those who resisted past decisions that we endorse as "correct" may have been just as reasonable as their opponents but also that our endorsement itself is crucially dependent on that initial decision. We should not comfort ourselves with the idea that the choice among rival theories was *temporarily* underdetermined and that those who prevailed were—luckily—subsequently vindicated by the exposure of decisive evidence. Instead, we should recognize that either the choice

6. The phrase is Kuhn's (Kuhn 1962 151). In the later chapters of his book, Kuhn assembles many of the principal arguments that others have deployed to reach far more radical conclusions. He contends that observation is theory-laden (1962 chapter X) and that shared methodological canons are too weak to determine scientific choices (1962 chapter XII). Yet the penultimate paragraph of the latter chapter suggests a reshaping of traditional ideas about scientific rationality, not a decisive break. "Because scientists are reasonable men, one or another argument will ultimately persuade many of them. But there is no single argument that can or should persuade them all. Rather than a single group conversion, what occurs is an increasing shift in the distribution of professional allegiances" (1962 158). Chapters 6 and 8 later will develop some of the ideas at which Kuhn seems to hint here.

7. See, in particular, chapter 13 of (Kuhn 1977). In Chapter 6, I shall develop a position that has many affinities with Kuhn's attempt to steer a middle course between Legend and relativism.

actually made *or its rival* would have been self-sustaining: had matters gone differently, our counterparts would have seen a history of success, congratulating themselves on the fact that subsequent evidence had supported the initial decision. From this perspective, claims about the success and rationality of science are a wonderful illusion, fostered by our ability to rewrite history and to lose sight of the possibilities of discarded points of view. Legend's praise of science is only the kind of boosterism that any proponent of any "form of life" might engage in. Contemporary societies—at least those of the industrialized world—have made science part of their "forms of life," but those are, in the end, no better or worse than the practices of the Azande, the recommendations of homeopathy, or the promises of parapsychology. They are simply ours.[8]

Despite the efforts of a few philosophers,[9] little headway has been made in finding a successor for Legend. If anything, recent work in the history of science and in the sociology of science has offered even more sweeping versions of the original critiques. Philosophy of science, meanwhile, has enjoyed a renaissance of studies of particular sciences.[10] However, the contemporary attention to a wide range of examples is typically eclectic: practitioners borrow what general philosophical categories they need for the purposes of their individual studies, attempting only to rely on concepts and distinctions that seem to have survived Legend's demise.

Perhaps these individual studies (largely ignored by those who condemn philosophy of science as defunct) represent the proper role for the discipline. Perhaps there is no general picture of the sciences that philosophers can give. Perhaps individual scientific disciplines and achievements are just that—individual, bound together by nothing more than "family resemblance."[11] Although these suggestions may offer sound advice, I am not yet ready to abandon the search for generality. For without *some* substitute for Legend, minimal, eclectic, and tacit though it may be, it is not clear that individual studies of particular sciences can be pursued. Moreover, the arguments of the Legend bashers deserve answers or acknowledgment of their correctness.

The goal of this book is to draw a picture of how science advances, using

8. For a strong version of this type of relativism, see (Barnes and Bloor 1982) and (Collins 1985). The case of parapsychology is discussed in some detail in (Collins and Pinch 1982). Similar ideas ares advanced in (Feyerabend 1975, 1979, 1987). The notion of a "form of life," now much in vogue among some sociologists of science, descends from (Wittgenstein 1953). A locus classicus for the application of this notion to the history of science and for many of the themes noted in the text is (Shapin and Schaffer 1985).

9. Most notably Imre Lakatos (1970), Larry Laudan (1977, 1984, 1990), and Dudley Shapere (1984).

10. Not only in philosophy of physics, to which Legend's main elaborators made major contributions, but increasingly in philosophy of psychology and philosophy of biology as well.

11. (Wittgenstein 1953 §67). One might abandon the attempt to offer a general philosophical vision of science for Wittgensteinian reasons, holding that there is no essence to the enterprise, while commending the activity of studying and characterizing particular scientific contributions or debates. Arthur Fine has sometimes advocated this kind of view in lectures.

the commonsensical ideas that underlie Legend, the insights of Legend's critics, as well as the contributions of contemporary philosophers, historians, sociologists, and cognitive scientists. The chief questions I shall address are those that Legend took as central: What is scientific progress? How is science pursued rationally? Problems about progress occupy Chapters 4 and 5. The goal of Chapter 4 is to provide a definition of scientific progress, or, more exactly, to characterize some varieties of scientific progress and to connect them to aims that I take to be worth satisfying. Chapter 5 attempts to defend my characterizations against criticisms to the effect that the aims I specify are incoherent or unattainable, and I try to rebut charges that science does not make progress in the senses I pick out. In Chapter 6 I turn to questions about scientific rationality, attempting first to specify what an ideal—or ideals—of rationality ought to be. The traditional project of articulating canons of rationality for the epistemic projects of individuals is taken up in Chapter 7. Finally, in Chapter 8, I turn to the important, though largely neglected, issue of how various combinations of individual epistemic strategies might advance or retard the community's epistemic enterprise.[12]

Before these accounts can be developed, some preliminary work is necessary. My attempts to combine what I take to be important (and currently unappreciated) insights of logical empiricism with the *aperçus* of historians and sociologists will rest on a novel way of idealizing the phenomena. Instead of thinking of science as a sequence of theories and of theories as sets of statements, I shall offer a multi-faceted description of the state of a science at a time. Moreover, I shall be concerned to treat the growth of science as a process in which cognitively limited biological entities combine their efforts in a social context. Placing the knowing subject firmly back into the discussion of epistemological problems seems to me to be the hallmark of naturalistic epistemology.[13] Chapter 3 is my attempt to construct a framework for naturalism in the philosophy of science.

The constructions of Chapter 3 will be guided by a review of some phenomena. In Chapter 2 I shall offer an extended illustrative example, on which subsequent discussions will be able to draw.[14] I shall examine some facets of

12. The idea of understanding how social institutions might promote knowledge has been largely ignored in traditional epistemology. *Parts* of the problem begin to emerge in (Kuhn 1962), in (Sarkar 1982), and in (Hardwig 1985). Some suggestive points are made in (Fuller 1988), and Alvin Goldman (1987) offers a general overview of the issues. Some highly pertinent source material with interesting interpretations can be found in (Ghiselin 1989) and (Hull 1988).

13. This is a common feature of the enterprises of (Goldman 1986) and (Quine 1970a), although they apparently disagree on the possibility of undertaking normative inquiries in epistemology. The present book exemplifies my kinship on this issue with Goldman (see also my 1983). Other accounts of what makes for a naturalistic epistemology can be found in the literature. Laudan (for example) conceives of naturalistic epistemology as committed to the notion that methodological rules are empirical (see Laudan 1989). General discussions of the character of naturalistic epistemology can be found in (Kornblith 1985), (Giere 1985), (Kim 1988), and (Kitcher 1992).

14. Other examples from past and present science will be discussed more briefly in later chapters, but it seems to me to be impossible to elaborate a convincing picture without providing one, fairly detailed, illustration. Ultimately, philosophical pictures should be

the history of evolutionary ideas, from the midnineteenth century to the present, with a view to fixing ideas about goals, methods, progress, rationality, individual scientific behavior, and the social structure of science.

All the discussions that follow are suffused with a vision of philosophy of science that should be made explicit. In my judgment, philosophical reflections about science stand in relation to the complex practice of science much as economic theory does to the complicated and messy world of transactions of work, money, and goods. Much traditional philosophy of science, in the style of some economic modeling, neglects grubby details and ascends to heights of abstraction at which considerable precision and elegance can be achieved. We should value the precision and elegance, for its own sake, for its establishing a standard against which other efforts can be judged, and for the possibility that extreme idealizations may lay bare large and important features of the phenomena. But, like ventures in microeconomics, formal philosophy of science inevitably attracts the criticism that it is entirely unrealistic, an aesthetically pleasing irrelevancy. To rebut such charges—or to concede them and to do better service to philosophy's legitimate normative project— we need to idealize the phenomena but to include in our treatment the features that critics emphasize.

Legend's legacy is the task of recognizing the general features of the scientific enterprise, most importantly by scrutinizing the apparent progressiveness of science, the seeming individual rationality of individual scientists, and the collective rationality of the scientific community. In light of the serious challenges to Legend, there is no doubt that the task is difficult. However, recurrent pronouncements of the death of philosophy notwithstanding, it is too important to be discarded.

---

tested against a range of cases, historical and contemporary. Articulating the methodology underlying this test procedure is itself a formidable task—and, moreover, one that cannot be detached from our philosophical picture of science.

2

Darwin's Achievement

1. A Triumph for Heresy

Between 1844, when Charles Darwin first confessed to Joseph Hooker his unorthodox ideas on "the species question," and 1871, when Thomas Henry Huxley was prepared to declare that "in a dozen years the 'Origin of Species' has worked as complete a revolution in Biological Science as the 'Principia' did in Astronomy," biology underwent the most important transformation in its history.[1] The *Origin* has continued to dominate the biological sciences, and, as we look back over the hundred and thirty years since its publication, the initial impression is of a steady, cumulative building on Darwin's original achievement. Scrutinizing the initial phase of the Darwinian revolution and its aftermath offers us the opportunity to examine in a concrete context some of the ideas about progress and rationality that are at the heart of Legend.

Legend might view Darwin as combatting superstition, achieving a surprising victory because of his ability to marshall the evidence. But part of the secret of Darwin's quick success surely lay in his political skill. We should not be beguiled by the picture of the unworldly invalid of Down, whose quiet walks in his beloved garden were the occasion only for lofty musings on points of natural philosophy. Darwin's study was the headquarters of a brilliant campaign (which he sometimes saw in explicitly military terms),[2] directed with energy and insight. His letters are beautifully designed to make each of his eminent correspondents—Hooker and Huxley, Charles Lyell, Alfred Russell Wallace, and Asa Gray—feel that he is the crucial lieutenant, the man on whose talents and dedication the cause depends.[3] Morale is kept up and the troops are deployed with skill.

1. (Huxley 1896 120). The passage is quoted in (F. Darwin 1888 III 132). Three years before it was written (i.e., in 1868), Darwin was prepared to talk of "the almost universal belief in the evolution (somehow) of species" (F. Darwin 1903 I 304). For Darwin's initial confession to Hooker, see (F. Darwin 1888 II 23).

2. See, for example, (F. Darwin 1903 I 304). Michael Ghiselin has provided a concise analysis of Darwin's tactics. See his (1969) especially chapter 6.

3. See (F. Darwin 1888 II 165, 216 ff., 232, 273, 302–303, 308, 330; III 11). Two of these passages provide an interesting comparison. In November 1859, Darwin wrote to Huxley that, in advance, he had "fixed in [his] mind three judges on whose decision [he] determined mentally to abide" (F. Darwin 1888 II 232). By early 1860, the trio had expanded

Darwin earned a sympathetic hearing from his friends, in part by working his way up the English scientific hierarchy in the years after the *Beagle* voyage, in part by intertwining his scientific exchanges with social occasions.[4] Darwin was, undeniably, well connected, it was important to him to be well connected, and his connections were evident in the discussions that followed the publication of the *Origin*. Yet, as I shall argue in what follows, his exploitation of the social structure of British (and American) science was not the entire secret of his success. Those who fought on his behalf were recruited to the campaign through his patient presentation of elaborate arguments about biological issues, and, in their public defenses of Darwin's ideas, they explained and amplified the reasoning distilled in the *Origin*.[5]

Assessing the role of reason in the initial triumph of Darwin's heresy is only one of the projects of this chapter. If we are to advance a clear account of scientific progress, it will be useful to examine the career of Darwin's ideas, attempting to identify the extent to which they have endured and been extended within subsequent biology. Contemporary discussions of alleged foibles in our understanding of evolution often complain that the initial controversies about natural selection have never been satisfactorily resolved, so that the history of biology reveals an oscillation among competing rivals rather than a steady advance toward truth.[6]

What follows is a case study, an approach to an episode in the history of biology that abstracts, sometimes in extreme fashion, from many important historical details in the hope of illuminating philosophical issues. Because case studies are in bad odor among some historians of science, I should make plain what I intend to achieve. I hope to satisfy two adequacy conditions: (i) if something is attributed to a past figure then the attribution is correct, (ii) nothing is omitted which, if introduced into the account, would undermine

---

to a quartet: Huxley, Hooker, and Lyell had been joined by Asa Gray. Darwin concluded a letter to Gray, "It is the highest possible gratification to me to think that you have found my book worth reading and reflection; for you and three others I put down in my own mind as the judges whose opinions I should value most" (F. Darwin 1888 II 272). There is no inconsistency here, but the juxtaposition of the passages reveals Darwin's tact.

4. For Darwin's securing of a place in the British scientific establishment, see (Moore 1985 450–452, 460–462) and (Manier 1978). The mixed intellectual and personal character of Darwin's relations with his colleagues is beautifully captured in Hooker's account of visits to Down (see F. Darwin 1888 I 387–388, Ospovat 1981 91–94).

5. Huxley offers a clear account of how he came to espouse Darwin's theory—see his (1896 246). He also attributes the rapidity of the Darwinian revolution to the arguments of the *Origin* (1896 286). Of course, Huxley's stress on reason and argument may be self-deception. It is interesting to note that, in the 1850s, before he became a personal friend of Darwin's, Huxley offered a ranking of the leading British biologists of the day. Darwin figures in the second tier: he is described as "one who might be anything if he had good health" (L. Huxley 1913 I 94).

6. The most explicit formulations of this type of critique are given in (Ho and Saunders 1984) and in several essays in (Ho and Fox 1988). (See, in particular, the introduction and the article by Brian Goodwin in the latter work.) But there are less strident attacks on neo-Darwinism that question cumulativist views of the history of evolutionary biology: examples are (Gould 1982, Eldredge 1985, Levins and Lewontin 1985, Stanley 1984).

the point being made. My guiding analogy is with a sketch that lays down the main lines around which the detailed picture is subsequently drawn. Philosophically oriented history should not "reconstruct" in the sense of drawing lines that would have to be altered in a more detailed presentation.[7]

2. Before Darwin

To understand the context in which Darwin's ideas were received and debated, we need to identify some major themes in early nineteenth century thinking about the history of earth and of life. From the close of the eighteenth century on, it had become increasingly evident to the cognoscenti that the time span for life on earth was far longer than had traditionally been assumed. In the early decades of the nineteenth century Georges Cuvier established the phenomenon of extinction, and, in the wake of his painstaking paleontological work, natural historians formulated an increasingly detailed picture of a succession of organisms.[8] But the problem of identifying the process(es) driving the succession of organisms—Darwin's "mystery of mysteries" (1859 1)—was bypassed in favor of other biological projects.

The tradition of natural theology, especially strong in Britain,[9] was dedicated to the idea of exhibiting the exquisite fit of animal to environment. The utilitarian conception of life, favored by the British who conceived of the Creator as engineer extraordinaire, contrasts with the approach of the German *Naturphilosophen*. Influenced by the ideas of the poet Johann Wolfgang Goethe, the Romantic morphologists sought archetypes (*Urbilde*) for the major groups of organisms. The principal task of comparative biology, as they conceived it, was not to discern the intricate ways in which organic structures contributed to the well-being of their bearers but to expose the underlying communities of structure. Ultimately, all organisms were to be organized by considering the ways in which each diverged from the corresponding archetype, so that the plan of creation would be exhibited by un-

7. Historians frequently (often legitimately) complain that the character of historical events is misrepresented in attempts to grind a philosophical axe. My own case study is not intended as a "reconstruction" in the famous (notorious?) sense of Imre Lakatos (1971), for I am interested here not in what Darwin et al. ought to have said and done, but in what they actually said and did. Moreover, the present version of the chapter was condensed from a draft more than twice as long in which the historical detail was far more elaborate—but the philosophical points less perspicuous. Both versions told the same story. Hence my confidence that appreciating the richness of the episode does not invalidate the uses which I shall make of it.

8. For a clear account of the expansion of the time scale, see (Rudwick 1972/85). For much more detail on Cuvier, see (Coleman 1964).

9. Toby Appel (1987) provides a good survey of attitudes toward natural theology in the early nineteenth century, hypothesizing that France was less interested in the exhibition of design in nature because of the dominance of Catholicism. Cuvier, whose commitment to teleology was strong and thus atypical of his countrymen, was a Protestant.

derstanding the unities of type and the principles governing the divergences from the archetypes.[10]

Cuvier's insistence on the existence of a functional role for every structural element collided with the program of the *Naturphilosophen* in a public debate before the Paris Academy of Sciences in 1830, where Cuvier demolished his colleague Geoffroy Saint-Hilaire.[11] Despite Cuvier's success in exposing weaknesses in Geoffroy's arguments for homologies (deep structural affinities) between vertebrates and invertebrates, those who reflected on the debate in the 1830s were usually convinced that Geoffroy's notion of homology, while in need of clarification, was valuable. Between panfunctionalism and mystical appeals to unity of type, biologists sought a compromise.

One form of compromise, which loomed large on the British intellectual scene in 1859, was proposed by Richard Owen. In 1831, long before he had become the dean of comparative biology in Britain, Owen visited Paris. Initially persuaded by Cuvier, he modified his ideas during the 1830s, and, in a series of memoirs on various structures in various animal groups, he developed the language in which comparative anatomy was to be described. Central to Owen's program was a conception of homology, a descendant of Geoffroy's seminal idea, which he enunciated in his anatomical lectures and published in his (1848).

> *Analogue*—A part or organ in one animal which has the same function as a part or organ in a different animal.
>
> *Homologue*—The same organ in different animals under every variety of form and function (1848 7).

What is most significant about this double introduction of terms is the implicit recognition of two types of affinity in animal structures. Some similarities (those on which Cuvier insisted) represent commonality of function. Others (those Geoffroy tried to discern everywhere) indicate community of type.

As his ideas developed in the 1830s and early 1840s, Owen revived the concept of the archetype, regarding it as a formal distillation of the homologies revealed in his specimens. The task of the comparative anatomist was to include more than the simple tracing of individual homologies. Ultimately,

10. In the works of Lorenz Oken and Karl von Baer, this approach was elaborated with considerable imagination and some illumination. Embryological similarities were interpreted as signs of sameness of archetype, and, indeed, the process of development could be construed in terms of divergences from archetypal form. As Ernst Mayr points out, Oken's fanciful ideas were frequently erroneous, although his approach "was actually very productive in arthropod morphology, helping to homologize mouth parts and other cephalic appendages with extremities" (1982 458). A sympathetic and insightful account of the ideas of the *Naturphilosophen* is provided by Gould in his (1977).

11. Geoffroy advocated a "philosophical anatomy," an approach to comparative morphology that had obvious affinities with the ideas of the *Naturphilosophen* but that included important contributions of his own. In particular, he introduced a version of the homology/analogy distinction.

Owen hoped to display the archetypes—structural plans or Platonic Ideas—that underlie the diversity of organisms. Functional biology was to investigate the ways in which the plan is modified in its embodiment, so that organisms are adapted to their environments.

Owen's method of fusing the ideas of some of his predecessors could be extended to a view of the history of life that articulated the tradition of natural theology in a novel way. Paleontology reveals a succession of organisms, each of which embodies one of a limited stock of archetypes. To uncover the plan of Creation is to trace the ways in which the sequence of living things through time articulates the timeless Ideas, how each organism is exquisitely adapted to its environment and yet how each is fundamentally structured on one of a small number of plans. To appreciate both the economy of archetypes and the harmonies of adaptation is to recognize the beneficence and wisdom of the Creator.[12]

In the early 1830s Charles Lyell published his magisterial *Principles of Geology* (which Darwin was to read aboard the *Beagle*), ordering the fossils of the Cenozoic and offering, contra Cuvier, a picture of the history of the earth in which changes in environment result from the action of presently observable forces over immense periods of time. Within the next fifteen years, several outstanding puzzles of the fossil record were resolved, and the paleontologists of the 1840s and 1850s could, with some confidence, pronounce on the relative ordering of strata, assigning fossils to their temporal ranges. Commitment to the idea of extinction and renewal of the earth's fauna entailed, of course, commitment to the existence of *some* process through which new species were produced. There were, apparently, two options. Either the new species were descended from those now extinct (the *transmutationist* or *evolutionary* hypothesis) or they had been called forth in a new act of creation. The sophisticated drew back from speculation. Lyell declined to pronounce on the processes by means of which novel species have appeared. Similarly, Owen maintained a studied agnosticism. It was enough for both of them to declare that the types of organisms that emerge are determined by "secondary causes."

From a post-Darwinian point of view this failure to attempt a resolution of the issue appears strange. Surely one *must* ask how new species have originated! To dissolve the mystery, we need to note two aspects of the pre-Darwinian situation. First, with transmutationism in disrepute, partly as a result of Lamarck's formulation of it, partly because of the force of the familiar objections that Cuvier had marshaled so effectively, inquiry into the process(es?) of species origination threatened to degenerate quickly into irresponsible speculation. Second, both Lyell and Owen could pose for themselves and their colleagues manageable and apparently significant research problems—the replacement of appeals to catastrophes with "uniformitarian" ex-

---

12. See (Owen 1849). Both Rudwick and Michael Ruse (1979) offer clear accounts of Owen's ideas and testify to their power and subtlety. See also (Russell 1916).

planations, the continued description of the fossil record, the display of homologies and, ultimately, the delineation of the unfolding of the archetypes in
the history of life.[13]

But, even before Darwin, there were some areas within biology in which
considerations of pattern seemed insufficient, in which considerations of history became evident, and in which speculations about creation surfaced. Just
as the early nineteenth century witnessed a great increase in understanding
of the distributions of animals (and plants) through time, so too the voyages
of exploration (and the reports of the naturalists who sailed on them) yielded
a greatly enriched stock of biogeographical phenomena. The discoveries of
very different faunas on different continents—especially the Australian marsupials—made it very difficult to maintain that there had been a single center
of creation from which (by some mysterious principle of distribution) the
animals had dispersed. For early nineteenth century biogeography, the principal questions were to understand what the spatial distributions of plants
and animals actually are and why they are that way. In its latter project,
biogeography became historical and the history involved facing the topic of
the origins of species.

The leading biogeographers of the early nineteenth century invoked the
notion of "centers of creation," a concept that had been proposed in the
eighteenth century and had quietly been accepted as a way of modifying the
Biblical story of creation.[14] Thinking of species as immutable, pre-Darwinian
biogeographers were severely constrained in their historical analyses. Similar
contemporary species have similar, but distinct, populations of founders, and,
after one has reconstructed their histories of dispersal, one is forced to conclude that they were separately created, perhaps at the same center. What
principles underlie the origination of some groups of similar species in certain
places, and the origination of different groups of similar species in different
places, even when the centers of creation have, apparently, quite similar

13. Similarly, in both the French and German research traditions, one can find the
posing of questions about *pattern* in the history of life that are independent of problems
about the *process(es)* through which new species have emerged. See, for example, the
influential essay on classification by Henri Milne-Edwards (1844).

14. For a concise and informative discussion of this concept and its employment in pre–
nineteenth century biogeography, see (Mayr 1982 439–443). Alphonse de Candolle, whose
discussion of the geographical distribution of plants was recognized by Darwin for its
comprehensiveness, proposed that there are more than twenty separate botanical regions.
Similarly, Edward Forbes offered an explanation of the distributions of British plants and
animals by positing separate centers of creation. Both Forbes and de Candolle had begun
to naturalize biogeography by seeking historical accounts of current distributions. Both
rejected, with good reason, the once popular idea that the distributions of species could
be understood teleologically by noting that each species range covered just those environments to which the species was best suited. By the 1840s and 1850s it was clear from
abundant examples that species sometimes were absent from regions in which they might
thrive and present in places that were less than ideal for them. De Candolle and Forbes
(and Lyell before them) saw the present distributions of plants and animals as resulting
from processes of dispersal. Yet, after the history of dispersal had been retraced, the
biogeographer inevitably came to the point of origin, the center of creation.

environmental conditions?[15] Biogeography, by the 1850s, was teetering on the edge of those issues that the program of comparative morphology (at least, in the influential version advanced by Owen) studiously avoided.

In the early decades of the century, appeals to the transmutation of species were hardly respectable.[16] But in 1844, the year of Darwin's first acknowledgment, in a private letter, of his heretical views, an anonymous author published *Vestiges of the Natural History of Creation.* The author, who was later discovered to be Robert Chambers, called down on his head a vast number of critical reviews—or, more exactly, thunderous denunciations—which assured the popular success of his book.[17]

Perhaps Chambers drew the fire that would inevitably be directed at the first effort to introduce evolutionary ideas into British discussions of natural history. At any rate, by the mid–1850s there were signs of a softening in attitude. In 1855, the Reverend Baden Powell, Savilian Professor of Geometry at Oxford and a well respected natural philosopher, published his *Essays on the Spirit of the Inductive Philosophy, the Unity of Worlds, and the Philosophy of Creation.* In the third part of this work, Baden Powell offered a measured defense of the possibility of the transmutation of species.[18]

Even my brief review of the background against which Darwin's work was set should make it evident that the *Origin* was not pitched into a vacuum of biological ideas and that its competition was not some dominant, dogmatic fundamentalism. Darwin's contemporaries and predecessors were practicing biology in sophisticated ways. Above all, they were practicing biology in *different* ways. In 1859, there were remnants of the natural theology tradition, there were serious projects of displaying the functional significance of traits and structures, ventures in biogeography and in paleolontological reconstruction. Perhaps most importantly, in the work of Richard Owen, arguably the leading British biologist of the 1850s, there was a serious and successful program in comparative morphology. Owen had fused important elements of the ideas of the *Naturphilosophen,* of Geoffroy, of Cuvier, and of Lyell. His research suggested that it might be possible to attain significant scientific goals (including the natural theologians' ideal of fathoming the plan of Creation) without delving into murky questions about modes of species origination.

15. Lyell was convinced that the answer depended on unnoticed features of the ecological and geological conditions, that there are laws relating the characteristics of the species produced to the conditions of their places of origin. Alphonse de Candolle, by contrast, maintained that appeal to climatic and physical conditions alone would prove insufficient. See (De Candolle 1855 33–34, translated in Ospovat 1981 16).

16. For discussions of Lamarck's evolutionary views and the differences with Darwin's ideas, see (Hodge 1971, Burkhardt 1977, Ruse 1979, and Mayr 1982 343–359). Darwin's reactions to Lamarck are atypically harsh. See (F. Darwin 1888 II 23, 29, 39, 215).

17. Mayr (1982 382) points out that Chambers' writing outsold in the first decade the classic works of Lyell and of Darwin in their (respective) first decades.

18. For Baden Powell's judicious claims, many of which point towards the arguments of the *Origin,* see his (1855 360, 372–733, 376, 379, 387, 401, 413–427). These passages also make clear the influence of Owen's agnosticism about transmutation, which often leads Baden Powell to draw back from conclusions that Darwin would embrace.

Baden Powell's cautious defense of the possibility of transmutation made it apparent that Owen's ideas might be extended to an evolutionary account of the history of life. By 1859, Darwin had been at work on such an account for over twenty years.

3. Darwin's Innovations

Virtually everyone would agree that the *Origin* offers something new, a new theory—the theory of evolution by natural selection—or several new theories.[19] Attempts to specify what the novel theory is are inevitably influenced by general ideas about scientific theories. A battered and truncated version of the "received view" (scientific theories are axiomatic deductive systems) lingers on. Even those who are skeptical about the need for distinctive theoretical vocabulary, or correspondence rules, or axiomatizability, or any of the other ingredients of the logical empiricist picture,[20] are likely to suppose that any scientific theory worthy of the name must consist of a set of statements, among which are some general laws (laws that set out the most fundamental regularities in the domain of natural phenomena under investigation), and that such laws should be used to derive previously unaccepted statements whose truth values are subject to empirical determination.[21] When this residual thesis is applied to the case of Darwin, we are led to expect that the *Origin* advances some collection of new general principles about organisms. After all, what else could Darwin's theory be?[22]

19. Ernst Mayr (1982, 1991) distinguishes five theories that are advanced in the *Origin*. While it is important to separate Darwin's basic commitment to evolution (the thesis that species are connected by descent with modification) from his views about the mechanisms of evolutionary change (most prominently his claims on behalf of natural selection), it seems to me that Mayr multiplies theories beyond necessity.

20. For a sustained account of the "received view" and its problems, see Fred Suppe's comprehensive introduction to his (1977).

21. This minimal conception of scientific theories is often implicit in the historical literature, where it encourages the effort to find some short formulation of Darwin's new theory. See, for example (Mayr 1982, Ospovat 1981, Ghiselin 1969, Ruse 1979, and many of the essays in Kohn 1985). For many historical purposes little harm comes of this practice. But if one wants to be precise about the differences between Darwin and his predecessors and contemporaries and if one hopes to understand the reasoning of the *Origin*, then, as we shall see, the minimal conception poses severe difficulties.

The minimal conception can also be found in the practice of many philosophers who offer accounts of scientific change—it is, for example, espoused by Lakatos, Laudan, and Shapere. However, both Laudan and Shapere break with the popular supposition that the state of a science at a time can be encapsulated in the set of accepted theories. This is a salutary move and leads to positions with some kinship to that which I shall develop.

22. One answer, offered by Elisabeth Lloyd in her (1983), is that the theory is a collection of models (or, perhaps, model types). I had originally thought that Lloyd's advocacy of the "semantic conception" of scientific theories (a conception advanced and defended by Suppe (1972), Suppes (1965), Sneed (1971), and van Fraassen (1980) among others) was at odds with my own reconstruction. After numerous discussions with her, I am no longer sure about this. I have not found it helpful to use the standard terminology

The expectation is fostered when we turn to the opening chapters of the *Origin,* where we seem to discover exactly the kind of principles that were anticipated. Darwin's theory apparently rests on four fundamental claims:

(1)   At any stage in the history of a species, there will be variation among the members of the species; different organisms belonging to the species will have different properties (The *Principle of Variation; Origin* chapters 1–2, *passim*)

(2)   At any stage in the history of a species, more organisms are born than can survive to reproduce (The *Principle of the Struggle for Existence; Origin* chapter 3).

(3)   At any stage in the history of a species, some of the variation among members of the species is variation with respect to properties that affect the ability to survive and reproduce; some organisms have characteristics that better dispose them to survive and reproduce (The *Principle of Variation in Fitness; Origin* 80).

(4)   Heritability is the norm; most properties of an organism are inherited by its descendants (The *Strong Principle of Inheritance; Origin* 5, 13).

From these principles—more exactly from (2), (3), and (4)—one can obtain by a plausible argument

(5)   Typically, the history of a species will show the modification of that species in the direction of those characteristics which better dispose their bearers to survive and reproduce; properties which dispose their bearers to survive and reproduce are likely to become more prevalent in successive generations of the species (The *Principle of Natural Selection; Origin* chapter 4).

The justification for reconstructing Darwin's theory in this way is relatively straightforward. Darwin assembles the first four principles at the beginning of the *Origin,* and the main theoretical work then appears to be the derivation of the principle of natural selection from them.

Expositors of Darwin from T. H. Huxley to Richard Lewontin have reconstructed the "heart" of his theory in the way that I have done (Huxley 1896 287, Lewontin 1970 1). Nor will my own account entirely forsake this great tradition. But it should trouble us that the suggested reconstruction is at odds with an assumption that historians and philosophers of science often tacitly and legitimately make. We expect that the fundamental principles of a new scientific theory should be those statements introduced by the theory

---

employed by proponents of the semantic conception in stating my views, but it is possible that everything I say about Darwin could in fact be translated into their idioms. For the purposes of this chapter (and this book) I shall not explore the issue.

that most stand in need of defense and confirmation, and that the arguments assembled by the innovative theorist should be directed at the fundamental principles of the novel theory.

Virtually all of Darwin's opponents would have accepted (1)–(4). None of the great naturalists of the midnineteenth century would have denied—and none should have denied—that species vary, that the increase of a species is checked, that some variation affects characters relevant to the ability to survive and reproduce, that many properties are heritable. Moreover, they saw the force of the argument for (5), assenting to the idea that natural selection has the power to adjust the distribution of traits within a species, eliminating variants whose characteristics render them less able to compete in a struggle for limited resources. What was in dispute was not so much the *truth* of (1)–(5) as their *significance*.[23] For the committed Darwinian, these principles were the key to understanding a vast range of biological phenomena, and the principal theoretical and argumentative work of the *Origin* consists in showing how the seemingly banal observations about variation, competition, and inheritance might answer questions that had previously seemed to be either beyond the scope of scientific treatment or else only addressed in a partial and limited way by Darwin's predecessors.

Acceptance of (1)–(5) is compatible with the doctrine of fixity of species (which states that species are closed under reproduction). However, Darwin did not simply accept (1)–(5) and add the historical claim that lineages (ancestor-descendant sequences of organisms) have been modified to the extent that they embrace members of different species. Owen seems to have defended something like that view.[24] Nor did Darwin simply conjoin (1)–(4) with the denial of the fixity of species and the vague declaration that natural selection has been the primary force of evolution. The *Origin* contains a novel and well-articulated theory precisely because it fuses (1)–(5) with the suggestion that species are mutable to fashion powerful techniques of biological explanation. If we are to make plain the nature of Darwin's innovation, we need a new way of thinking about the complex entity that is the state of a science at a time. Reconstructing the *Origin* will provide an exemplar.

I shall start from the idea that the *Origin* proposes an explanatory device, aimed at answering some general families of questions, questions which Darwin made central to biology, by presenting and applying what I shall call *Darwinian histories*. To fix ideas, I shall characterize a Darwinian history in a preliminary way as a narrative which traces the successive modifications of a lineage of organisms from generation to generation in terms of various

---

23. I am grateful to Malcolm Kottler for pointing out to me that there is a parallel with the debates about Lyell's geological views, in that what was primarily in dispute was the significance of Lyell's claims about the time scale. See (Rudwick 1970).

24. Owen's complicated views about evolution and natural selection can be reconstructed from his anonymous review of the *Origin* (which frequently compares Darwin unfavorably with "Professor Owen"). See (Hull 1973b 175–213, especially 196, 201). I shall try to make clear in the text following what the differences between Darwin and Owen were and why Owen was so bitterly critical of Darwin's proposals.

factors, most notably that of natural selection. The main claim of the *Origin* is that we can understand numerous biological phenomena in terms of the Darwinian histories of the organisms involved.

Consider first issues of biogeographical distribution. For any group *G* of organisms—perhaps a species, a genus, or some higher taxon—we can identify the range of that group. With respect to any such group we can envisage a complete description of its history. From the Darwinian perspective, this historical description will trace the modification of the current group from its ancestors, revealing how properties change along a (possibly branching) lineage, and how, as these changes occur, the area occupied by members of the group alters. Darwinian histories provide the bases for answers to biogeographical questions.

Darwinian histories provide the *bases* for acts of explanation, and, confronted with a practical question of biogeographical distribution, *incomplete* knowledge of a Darwinian history will enable us to offer an answer. When we ask why a group *G* occupies a range *R,* we typically have a more particular puzzle in mind. For example, someone who wonders why so many marsupial species are found in Australia is likely to be perplexed by the fact that so few are found elsewhere. Perplexity is relieved by outlining the Darwinian history of the marsupials—describing how they were able to reach Australia before the evolution of successful placental competitors, how the placentals were able to invade many marsupial strongholds, and how the placentals were prevented from reaching Australia. Here and in other instances,[25] a general, unfocused, explanation-seeking question is determined in context as a more precise request for information. The request can be honored by abstracting from the Darwinian history, so that the needed information can be given, despite considerable ignorance about the details.

It is helpful to contrast the biogeographical proposals of the *Origin* with the approaches reviewed in the last section. One (creationist) possibility would resist the appeal to history altogether in favor of teleological answers to questions about distribution: because each group of organisms was created to inhabit a particular region, and because it has always inhabited that region, our understanding of the phenomena of distribution is advanced by recognizing those features of the organisms in the group that fit them to live where they do. As I mentioned previously, by 1859 this approach had fallen into well-deserved disrepute. It was well known that organisms transported by humans could thrive in areas that they had previously been unable to reach, and naturalists knew of other cases in which organisms seem ill-suited to their natural habitat (a popular example was that of those woodpeckers who inhabit treeless terrain).

More promising was the historical approach developed in most detail by

25. For example, the question why the birds known as "Darwin's finches" are confined to the Galapagos, a question that has played an exemplary role in the history of Darwinian biogeography. Ironically, as Frank Sulloway's brilliant detective work has shown (Sulloway 1982), it now appears that Darwin left the Galapagos without any appreciation of the significance of the famous finches.

de Candolle and by Forbes. On this view, we answer questions about bio-geographical distribution by seeing the current range of an organismic group as the result of a process in which an unmodified (or relatively unmodified) sequence of organisms has dispersed from an original "center of creation." The trouble is that this approach only postpones perplexity, for we need some scheme for explaining the distribution of original centers of creation—or at least some way of making the ways of the Creator seem less capricious. Darwin makes the point forcefully:

> But if the same species can be produced at two separate points, why do we not find a single mammal common to Europe and Australia or South America? The conditions of life are nearly the same, so that a multitude of European plants and animals have become naturalised in America and Australia; and some of the aboriginal plants are identically the same at these distant points of the northern and southern hemispheres? (*Origin* 352–353; see also *Origin* 394)

Darwin's challenge is to provide a comprehensible distribution of centers of creation that will allow for the disconnected distribution of the plants common to the hemispheres, while explaining the failure of the mammals to radiate into regions for which they are well suited. The thrust is that creationists will ultimately be forced into conceding that the distribution of original centers of creation is inexplicable. By contrast, as Darwin emphasizes, the theory of evolution claims for scientific explanation questions which rival theories dismiss as unanswerable.

I have begun with the example of biogeography, because it is the case on which Darwin often lays the greatest stress, suggesting that it was reflection on biogeography that originally led him to the theory of evolution.[26] Biogeography is only part of the story. The rest is more of the same kind of thing. With respect to comparative anatomy, embryology, and adaptation, Darwin also provided strategies for answering major families of questions.

Consider comparative anatomy. Here one major task is to provide answers to questions of the following general form:

> Why do organisms belonging to the groups G, G' share the property P (where G and G' will typically be acknowledged taxa and P will be some anatomical feature—bone structure in a forelimb, for example)?

Darwin proposes two forms of answer to these questions. In cases where P is a homology, the presence of P in both G and G' will be ideally explained

26. (F. Darwin 1888 I 336, II 34; F. Darwin 1903 I 118–119). Darwin's elder brother, Erasmus, confessed, "To me the geographical distribution, I mean the relation of islands to continents is the most convincing of the proofs, and the relation of the oldest forms to the existing species" (F. Darwin 1888 II 233). Huxley describes the importance of biogeography in his own reception of the theory of evolution (1896 276). Darwin's public account of the role of biogeography in his own thinking occurs in the opening sentences of the *Origin*.

by relating the histories of descent from a common ancestor (in which P was also present). In cases where P is a mere analogy, its common presence will be understood by tracing the history of the emergence of P along the lineages leading to G and G', *perhaps* showing how this was the result of similar environmental pressures.[27] Darwin reviewed the classic cases. Similarities in the bone structure of the forelimbs in various mammalian groups—moles, seals, bats, ruminants—are to be understood in terms of descent from a common ancestor. By contrast, the existence of wings in birds, bats, flying reptiles, and insects, is understood by recognizing the paths which these groups have followed in evolving the ability to fly.[28]

Plainly there are important similarities between the programs for comparative anatomy envisaged by Darwin and by Owen. Both employ a distinction between homology and analogy and propose different patterns of explanation for homologous and analogous traits. But where Owen invokes the presence or absence of a shared archetype and remains carefully noncommittal on the historical processes that have given rise to the groups under comparison, Darwin appeals to a shared ancestor and outlines a historical explanation. As in the case of biogeography, while relating the complete Darwinian histories of the groups involved would provide an ideal answer to a question about the relationships among them, our practical questions about the similarities among organisms do not require such detail. Quite frequently, the question of why two groups of organisms agree in a morphological property stems from puzzlement that organisms so different in other respects should share the morphological trait in question. If the property is a homology, the puzzle is resolved by outlining enough of the Darwinian histories of the organisms to reveal the main lines of their modifications from a common ancestor. Similarly, in the case of analogy, we need to tell enough of the Darwinian histories to understand how a similar feature has been produced in unrelated lineages.

As with biogeography, Darwin could easily contrast his approach to comparative anatomy with the available rivals. By pointing out such differences as those between the blind cave insects of Europe and America (*Origin* 138), he could pose a deep challenge to the sophisticated program for comparative anatomy advanced by Owen. On Owen's conception, we would explain the differences between North American and European cave insects by suggesting that the former realize certain "north american" archetypes while the latter

27. Note that at this point the explanation goes beyond simple recognition of ancestor-descendant relationships. Darwinian histories that trace the emergence of analogous traits by appealing to common environmental pressures must venture claims about the causes of evolutionary changes—perhaps by appealing to natural selection, perhaps by invoking one of Darwin's other mechanisms (e.g., direct induction by the environment).

28. Darwin provides a very clear account of the homology/analogy distinction and is well aware that his theory enables him to refine the concept of homology; see *Origin* 427 and (F. Darwin 1903 I 306). Note that, even though Owen's version of the distinction was more sophisticated than those of his predecessors, the analysis he offered still suffered from vagueness.

embody distinct "european" archetypes. But now there is a puzzle about why different archetypes are realized in the caverns of North America and in the caverns of Europe. Owen is committed to the idea that there are "secondary causes" that govern the emergence of the distinct species on the two continents, but he remains agnostic about whether the process of emergence involves reproductive continuity or creation de novo. However, the two options envisaged are not on explanatory par. If the cave insect species have evolved from ancestral species of insects, then we can understand the structural relations. But if they stem from separate processes of special creation, then the persistence of distinct "north american" and "european" archetypes in the two regions is quite mysterious—there is nothing in the environmental conditions to call forth different archetypes, and we seem to be forced to some odd hypothesis of parallelism.

The third example I shall consider is historically crucial, in that it represents the most promising field for the tradition of natural theology. Darwin confesses that his theory could not be admitted as satisfactory "until it could be shown how the innumerable species inhabiting this world have been modified, so as to acquire that perfection of structure and coadaptation which most justly excites our admiration."[29] However, he proposes that questions of adaptation, like questions of biogeography and comparative morphology, can be answered by rehearsing the historical process through which the adaptation emerged. The general form of question to be addressed is

Why do organisms belonging to group G living in environment E have property P?

where the property P is a characteristic which appears to assist its bearers in environment E. A complete answer to this question would trace the Darwinian history of G from the time just prior to the first occurrence of P, showing how the variation producing P first arose, how it was advantageous to its bearer in the original environment, and how that advantage enabled P to become progressively more prevalent in subsequent generations of the lineage.[30]

Since the problem of adaptation is the stronghold of approaches which emphasize the design of nature, the *Origin* contains numerous passages in which Darwin contrasts the explanatory power of his own theory with the deficiencies of its main rival. In some places, he stresses the difficulty of finding any coherent account of the creative design which will do justice to the aspects of nature which are "abhorrent to our ideas of fitness":

29. *Origin* 3. Richard Lewontin (1978) perceptively discusses the way in which prior emphasis on the problem of design made the discussion of adaptation central to Darwin's evolutionary thinking, and how this fact, in turn, led Darwin to stress the selectionist commitments of his theory.

30. Here I deliberately overdraw the adaptationist commitments of Darwin's theory. I shall consider below whether Darwin allows a more pluralistic approach to the evolutionary explanation of apparently beneficial characteristics.

We need not marvel at the sting of the bee causing the bee's own death; at drones being produced in such large numbers for one single act and then being slaughtered by their sterile sisters; at the astonishing waste of pollen by our fir-trees; at the instinctive hatred of the queen bee for her own fertile daughters; at ichneumonidae feeding within the live bodies of caterpillars; ... (*Origin* 472)

Other passages descant on the "Panda's thumb theme,"[31] the existence of many cases in which it is evident that natural contrivances fall far short of the standards of good design we would expect from a competent engineer, and in which it is more plausible to suppose that the available materials dictated a clumsy solution to a design problem.

He who believes in the struggle for existence and in the principle of natural selection, will believe that every organic being is constantly endeavouring to increase in numbers; and that if any one being vary ever so little, either in habits or structure, and thus gain an advantage over some inhabitant of the country, it will seize on the place of that inhabitant, however different it may be from its own place. Hence it will cause him no surprise that there should be geese and frigate-birds with webbed feet, either living on the dry land or most rarely alighting on the water; that there should be long-toed corncrakes living in meadows instead of in swamps; that there should be woodpeckers where not a tree grows; that there should be diving thrushes, and petrels with the habits of auks. (*Origin* 186)

This theme receives its most detailed treatment in Darwin's book on orchids— characterized as "a 'flank movement' on the enemy."[32] Again, it points toward the same moral: questions that rival approaches must dismiss as unanswerable can be tackled by adopting the Darwinian perspective.

4. Darwin's Patterns

I have claimed that the *Origin* offers a class of problem-solving patterns, aimed at answering families of questions about organisms by describing the histories of those organisms. The complete histories will always take a particular form in that they will trace the modifications of lineages in response to various factors—"Natural Selection has been the main but not exclusive means of modification."[33] I now want to give a more precise account of the notion of a Darwinian history.

31. See the title essay in (Gould 1980). I think that Gould is correct to view Darwin's sounding of this theme as central to his case for evolution.

32. See (Darwin 1862). Ghiselin provides a penetrating analysis of the argumentative strategy I have attributed to Darwin and of its significance in Darwin's defense of evolution. See his (1969 137). Darwin's military characterization of the role of the orchid book is given in (F. Darwin 1903 I 202).

33. *Origin* 6. The difficulty of interpreting Darwin's sentence is obvious. Natural selection might be heralded as the force that produces most evolutionary changes, or, perhaps, as the force that produces the most important evolutionary changes. An alternative con-

Darwinian histories are structured texts. I shall try to exhibit the structures that Darwin's paradigm explanations exemplify. Let us start with a notion of Darwinian history that is minimal in the sense of embodying the fewest assumptions about the tempo and mode of evolutionary change. This conception can be characterized as follows:

A Darwinian history for a group G of organisms between t_1 and t_2 with respect to a family of properties F consists of a specification of the frequencies of the properties belonging to F in each generation between t_1 and t_2.

This minimal conception allows for evolutionary change, for the property that frequencies may vary from generation to generation—indeed, traits initially absent may ultimately be found in every member of the group—but it does not offer any account of why this change occurs.

Consider, for example, some questions about biogeographical distribution. To understand why the Galapagos contains endemic species of finches which are similar to mainland South American forms it is sufficient to know the history of descent with modification without speculating about the modifying factors.[34] The structure of this explanation is as follows: we build on a minimal Darwinian history by using claims about the dispersal powers of the organisms and their current ranges to derive descriptions of the ranges in the next generation (or, more exactly, to derive statements that give the probabilities of ranges). Behind the explanation is a schematic form, instantiated in many Darwinian answers to biogeographical questions, in which we trace sequentially the changing geographical distributions of properties (including those properties that are crucial to species differences) and derive, at each stage, the current range of organisms with particular complexes of traits from ascriptions of powers of dispersal and the ranges in the previous generation.

Similarly, some questions about the relationships among groups of organisms can be addressed by rehearsing histories of descent with modification that do not explore the causes of the changes that have occurred along the lineages involved. Darwinian explanations of the presence of homologies take a very simple form, which we can exhibit as follows:

---

ception would be to suppose that the modifications produced through other causal processes are somehow impermanent, so that the large-scale course of evolution follows the trajectory laid down by selection. Finally, one might concentrate on those evolutionary changes that produce new taxa (for example, speciation events), understanding these as being effected by natural selection.

34. I do not claim that this will be possible with respect to all questions about biogeographical distribution. There are obviously many instances in which understanding the range of a species will involve recognition of the competitive relations with other species, and in which resolution of the biogeographical questions will turn on issues of coadaptation. Nonetheless, it is sometimes possible to answer such questions without investigating issues of adaptation, and in such cases the history of descent with modification will suffice.

COMMON DESCENT

Question: Why do the members of G, G' share P?

Answer:

(1) G, G' are descended from a common ancestor G_0.

(2) G_0 members had P.

(3) P is heritable.

(4) No factors intervened to modify P along the G_0–G, G_0–G' sequences.

Therefore (5) Members of G and G' have P.

Plainly, this style of explanation could be deepened by offering a derivation of (4) from claims about the environmental conditions encountered by the organisms (among others), thus introducing claims about possible causes of modification of the lineage. However, in providing explanations that exemplify COMMON DESCENT (and there are many in the *Origin* as well as in the writings of Darwin's allies), there is no need to venture into this territory. Darwin was thus correct in insisting that the "great truth" of common descent would open "a wide field for further inquiry."[35]

Cautious Darwinians (including Darwin in his most circumspect moments) adopt a position we may call "minimal Darwinism." Minimal Darwinian histories (and the patterns of explanation built on them) are to be used to answer biological questions (such as those about distribution and about homology), but, while we remain agnostic about the causes of modification in any particular case, we do regard ourselves as understanding the general ways in which evolutionary change is to be explained. Natural selection is identified as a *possible* agent of evolutionary change, in conjunction with such other agents as use and disuse, correlation and balance, direct action of the environment, and stochastic factors. On this approach we would not pretend to explain the modifications that have taken place along a particular lineage, and, quite evidently, we would have to forego attempts to tackle issues of organic adaptation.[36]

Numerous passages in Darwin's writings indicate that he usually preferred to be more ambitious.[37] A stronger conception of Darwinian history involves not only a specification of the changes that take place from generation to generation along a lineage, but also a sequence of derivations that will infer the distribution of properties in descendant generations from those in ancestral generations. These derivations will exemplify certain patterns, patterns which

35. Matters are different when we turn our attention to analogical traits. Were we to proceed in a parallel fashion to the treatment of COMMON DESCENT we would offer "explanations" exemplifying a pattern that does little more than restate the facts of convergence. Here an account of the factors that produced the same feature in independent lineages is crucial.

36. A position akin to this seems to have been advanced in our day by Niles Eldredge and Joel Cracraft (1980).

37. See, for example, *Origin* 3, 84, 170.

reflect ideas about the agents of evolutionary change. Prominent among them in Darwin's writings is the appeal to natural selection.

The most elementary form of selectionist explanation in the *Origin* instantiates the following pattern:

SIMPLE INDIVIDUAL SELECTION

Question: Why do (virtually) all members of G have P?

Answer:

(1) Among the ancestors of G there was a group of contemporaneous organisms, G_0, such that: (i) a small number of members of G_0 had P; (ii) none of the members of the generation ancestral to G_0 had P; (iii) each of the other members of G_0 had one of the variant characteristics $P_1, \ldots, P_n$; (iv) no other variant of P is present in any generation of the G_0–G lineage.

(2) *Analysis of the ecological conditions and the physiological effects on their bearers of P, P_1, ... , P_n*

Showing

(3) Organisms with P had higher expected reproductive success than organisms with P_i $(1 \leq i \leq n)$.

(4) P, P_1, ... , P_n are heritable.

Therefore (5) P increased in frequency in each generation of the lineage leading from G_0 to G.

(6) There are sufficiently many generations between G_0 and G.

Therefore (7) (Virtually) all members of G now have P.[38]

I suggest that this elementary pattern underlies the simplest illustrations that Darwin provides of natural selection: the explanations of the swiftness and slimness of wolves, of the presence of two varieties of wolf in the Catskills, the excretion of "sweet juices" by some plants, and many others.[39]

The intuitive idea behind SIMPLE INDIVIDUAL SELECTION is that one explains the presence of some traits (adaptations) by beginning with their initial appearance in some ancestral group, identifying the factors that favored them, and tracing their increasing prevalence. In many instances, especially when he is concerned with complex characteristics, Darwin's explanations have a

38. Obviously the elementary pattern that I have suggested could be refined in various ways that would make it more adequate as a representation of the simplest accounts of selection that Darwin gives in the *Origin*. In particular, it is evident that the notion of "sufficiently many" generations is tied to the amount by which the expected reproductive success of P bearers exceeds that of the competition (the fitness difference, as we would put it). Although Darwin was evidently aware of this point, his analyses in the *Origin* are nonquantitative, so that the use of the vague term "sufficiently many" is, in some ways, closer to his actual explanations.

39. The cases I have mentioned are presented at *Origin* 90–92. There are numerous others throughout the *Origin*.

more intricate structure. Instead of simply showing how the final version of a trait—perhaps an anatomical structure like an eye or a wing—became prevalent, the task is to understand how ancestral forms were successively modified in the direction of the current characteristic. We can think of this as a pattern that stitches together a number of instantiations of SIMPLE INDIVIDUAL SELECTION

The selectionist patterns I have discussed take variation for granted. There is no systematic attempt to explain the presence of the variant traits in G_0 by deriving (1) from some more ultimate premise. This, of course, corresponds to what is often regarded as a central tenet of Darwinism, namely the insistence on the randomness (more properly, the nondirectedness) of variation. Darwin himself was not quite so pure. The *Origin* allows for effects of "use and disuse." Typically, this is supposed to work in tandem with natural selection: organisms that do not use certain structures are more likely to generate offspring that lack those structures (tuco-tucos, burrowing rodents, are, it is claimed, frequently blind; see *Origin* 137), and, since the deficient organisms do not squander resources on developing a useless characteristic, they are at an advantage in the struggle for existence.[40,41]

Commitment to a stronger conception of Darwinian history makes it possible to answer questions, such as those which involve "perfections of structure," which lie beyond the scope of minimal Darwinian histories. This commitment may be undertaken more or less pluralistically. That is, one may allow as equally appropriate a number of different patterns for deriving changes in trait frequencies, or one may insist that a particular style of explanation should predominate. So, for example, Darwin's suggestion that natural selection is the major agent of evolutionary change can be interpreted as a commitment to preferring to understand the distribution of characteristics in a group of organisms by invoking SIMPLE INDIVIDUAL SELECTION. Or, more moderately, one can allow that the presence of many traits is to be explained by employing CONSTRAINED INDIVIDUAL SELECTION Or one may decrease the scope of selectionist explanations, by suggesting that the prevalence of some

40. Here Darwin's thinking seems to owe a debt to Geoffroy, who had insisted on development as a process in which a limited stock of resources is expended on generating adult morphology. In the first edition of the *Origin*, Darwin is quite cautious about other systematic explanations of the presence of variations. While he is inclined to allow that alterations in the environment are likely to increase *variability* (see, for example, *Origin* 134) he does not give much weight to the idea that the environment causes preferred types of variation (see *Origin* 131–134 for a restrained discussion).

41. Darwin's claims for the dominance of natural selection as an agent of evolutionary change are tempered by his frequent references to the—unknown—laws of correlation and balance. It is not always required to explain the presence of a trait by identifying the (hypothetical) selective advantage that it conferred as it spread through ancestral populations. For the characteristic may be prevalent in a group because it is correlated with another property whose presence is independently explained—perhaps by invoking natural selection. Thus another Darwinian pattern—CONSTRAINED INDIDIVUAL SELECTION—effectively applies Darwin's selectionist patterns to *suites* of traits, allowing that such suites may contain characteristics that are individually opposed by selection.

(even many) traits is to be accounted for in some other way—perhaps in terms of environmental effects, or use and disuse, or chance factors.

The *Origin* not only allows for the use of more or less ambitious notions of Darwinian history, but also covers a range of positions on the priority of selectionist explanations. This conclusion underscores a claim made by Gould and Lewontin, who note that "the master's voice" is often more tolerant of alternatives than is usually thought.[42]

So far I have indicated a number of different proposals that might be reconstructed from the *Origin*. These differ first in whether they attempt to explain changes in property frequencies along a lineage, and second in the forms of explanation that they admit or to which they give emphasis.[43] Minimal Darwinism triumphed relatively swiftly in the 1860s. The next sections analyze the nature of that triumph and the reasons for it.

5. Darwinian Practices

After the success of the *Origin* there are many changes in the ways in which biology is done. Most obviously, there are modifications of the sets of statements to which those who are counted as experts on the nature of living things would give their assent. There are also changes in biological language: the concept of species and the distinction between homologies and analogies are altered by Darwin's work. As I have been emphasizing in the last two sections, the questions that are viewed as important for biology are changed. New problems come to occupy biological attention, old issues disappear. Furthermore, the changes in questions are accompanied by changes in the sets of schemata that specify the ideal forms of answers.

However, these aspects, the alterations of belief and language that are

42. (Gould and Lewontin 1979). While I think that Gould and Lewontin are correct in their claim that Darwin is less selectionist than he is often interpreted as being, Mayr is right in his (1983) to point out that many of the alternative agents of evolutionary change discussed by Darwin are now widely discredited. What does remain—and this is sufficient for Gould and Lewontin to make an important case—is the variety of factors that Darwin would have lumped under "correlation and balance."

It is common for commentators to remark, parenthetically, that Darwin's commitment to pluralism increases in successive editions of the *Origin*. I owe to Malcolm Kottler the point that this quick assessment is misleading. Line-by-line comparison of the editions shows Darwin inserting disjunctions of possible causes of evolutionary changes where he had previously appealed to natural selection alone, but there is no addition of new mechanisms in later editions.

43. This does not exhaust the variety of versions of Darwinism. Nothing I have said recognizes Darwin's commitment to evolutionary gradualism, nor have I allowed for a possible Darwinian flirtation with selection of groups rather than individuals. Both these further variants can be accommodated within the framework I have proposed, the one by subjecting all Darwin's patterns to a gradualist constraint, the other by allowing for analogues of his selectionist patterns that take groups and group properties as targets of selection.

relatively familiar and the modification of problems and of explanatory patterns (which I have seen as central to the *Origin*), do not exhaust the novelties of Darwin's work. The *Origin* advances new ideas in methodology, ideas about what kinds of characteristics are important in successful theories. It suggests that certain types of observations and experiments are relevant to biology, and it allows epistemic authority to people who have not hitherto been counted as part of the community of scientists.

It will be useful to have a term to cover the disparate elements that were altered by the *Origin*. Let us say that the *practice* of an individual scientist is a multidimensional entity whose components are the language used in the scientist's research, the statements about nature that the scientist accepts, the questions that are counted as important, the schemata that are accepted for answering questions (together with assessments of how they are applied to "paradigm" examples and of how frequently they are likely to be exemplified in coping with unsolved problems), the methodological views that are specific to the research of the field of science, the canons of good observation and experiment, and the standards for assessing the reliability of others (together with particular views about the authority and reliability of contemporary co-workers in the field). Before Darwin, those who dedicated themselves to the study of living things differed widely in their practices. Not only were the practices of individual researchers different from one another, but there were groups of biologists whose practices had little in common. Champions of teleology differed in numerous respects from the *Naturphilosophen* and from compromisers like Owen. Even after the impact of the *Origin,* individual biologists still differed in their practices. Yet, although the practices are diverse, almost all of them are Darwinian.

The ability to appreciate both the diversity of individual practices and the unity of ideas that emerges at the level of the community will be crucial to the argument of this book. Thus, besides the collection of individual practices espoused by the individual scientists of the second half of the nineteenth century, each affected in different ways by reading the *Origin,* there is a *consensus practice,* something that represents the common elements of the individual practices and that becomes part of the cross-generational system of transmission of scientific ideas. In the latter nineteenth century, aspiring naturalists acquire, as part of their training, a practice that involves a Darwinian core. To understand the change wrought by the *Origin* we need to recognize both the weaker commitments to Darwinian themes that become integrated into consensus practice and the more ambitious versions of Darwinism that are taken up by individuals.

Darwin modified the language of biology in two important ways. First, a conceptualization of species that had been commonplace among his predecessors was discarded: the language of late nineteenth century biology no longer allows one to pin down the reference of species by declaring that "a species comprehends all the individuals which descend from each other, or from a common parentage, and those which resemble them as much as they do each other."[44] But if he abandoned old ways of saying what species are, Darwin and his followers largely agreed with the divisions of organisms into species that had been favored by his predecessors. Thus, although there was a shift in the way in which the referent of the term *species* was fixed (or, more colloquially, how one explains what species are), the term retained its old referent (the set of species taxa after Darwin is the same).[45] Similarly, there is continuity of reference of the term 'homology' (and its correlative, 'analogy'): Darwin calls *homologies* just those things that his predecessors have picked out as homologies. Here, however, his writings implicitly suggest a new descriptive account, one that would fix the referent of homology by declaring that a homology is a characteristic found in two descendant species as a consequence of their descent from a common ancestor.[46]

Darwin did effect large changes in the set of statements belonging to consensus practice. His work introduced a variety of new claims about the histories of particular organisms. The *Origin* is short on controversial new generalizations, but it is a hodgepodge of specific original theses about barnacles, pigeons, South American mammals, social insects, arctic flowers, Scotch fir, and so forth.[47]

As I emphasized in Sections 3 and 4, Darwin's primary achievement lies in introducing schemata for answering certain families of biological questions and identifying the questions that biologists should address. The mass of details is a cornucopia of illustrations. Consensus practice after Darwin absorbs the most striking examples from the *Origin:* every serious naturalist now has to know about the South American mammals and the distribution of arctic flora. The main alteration in the accepted statements of consensus practice thus consists in the incorporation of those descriptions that best illustrate the Darwinian schemata: these become the training paradigms for subsequent generations. Of course, as further instances of Darwin's patterns are generated and accepted, new examples may come to fulfil this pedagogical function: melanism in moths comes to be *the* example of natural selection.

44. (Cuvier 1813 120, as quoted in Beatty 1985 267). The account I offer here is a simplified version of Beatty's informative analysis.

45. The account of conceptual shifts suggested here will be developed in far more detail in chapter 4. Beatty (1985) provides evidence for the continuity of reference of 'species.'

46. As Mayr notes (1982 465), this analysis is implicit in Darwin and allows a descriptive way of fixing the distinction that Owen had drawn. The changes just mentioned, which are the principal conceptual shifts attendant on the debate inspired by the *Origin,* are epistemologically unproblematic. They do not threaten to break down the communication between Darwinians and their opponents, or even to create a situation in which "communication . . . is inevitably partial" (Kuhn 1962/70 149). Partly because they were talking about the same groups as species and the same traits as homologous, partly because the loci of disagreement were so readily identifiable, Darwin and his critics had not the slightest difficulty in communicating their ideas to one another.

47. Owen complained (Hull 1973b 170) that Darwin had introduced few "new facts." In one sense, this is quite correct, for Darwin often takes the phenomena laboriously assembled by others and makes historical claims about the organisms involved. After the acceptance of the idea that all organisms are related by descent with modification, these historical claims become new phenomena for subsequent biological research.

Meanwhile, individual scientists, concerned with specific biological questions or particular groups, have a rich body of more esoteric lore, itself formulated in Darwin's historical terms.

The introduction of the new schemata sets new questions for biology, in that, after Darwin, naturalists are given the tasks of (i) finding instantiations of the Darwinian schemata (that is, developing Darwinian explanations of particular biological phenomena), (ii) finding ways of testing the hypotheses that are put forward in instantiating Darwinian schemata, (iii) developing theoretical accounts of the processes that are presupposed in Darwinian histories (specifically such processes as hereditary transmission, and the origination and maintenance of variation). These tasks arise in different ways. The first, (i), is simply the result of Darwin's claiming of the questions about distribution, adaptation, and relationships, as legitimate and central questions for biology, to be tackled by giving explanations that exemplify his patterns. In attempting to instantiate the Darwinian schemata, biologists are compelled to advance hypotheses about the historical development of life, and it is incumbent on them to specify ways of testing these hypotheses (and hence to undertake (ii)), if they are to avoid the charge that evolutionary biology is simply an exercise in fantasizing.[48] Finally, (iii) stems from recognition of the fact that Darwin's theory is not only open-ended in provoking many specific inquiries into the properties, relationships, and distributions of particular organisms, but also in raising very general questions about the historical processes through which organisms have become modified.[49]

Although they are less central to the changes that Darwin recommends, the *Origin* also offers new ideas about the kinds of information that are relevant to biology, the ways in which that information can be acquired, and those who can be trusted to supply such information. Darwin suggests experimental projects that will be important in formulating Darwinian histories. Biogeographical accounts, for example, depend on assumptions about the dispersal powers of organisms, and these can be tested in various ways. So we find Darwin reporting his findings about the ability of various kinds of seeds to survive immersion in salt water (*Origin* 358–361) and describing an experiment on the survival of "just hatched molluscs" on a duck's foot suspended in air (*Origin* 385). Furthermore, tests of the variation that is latent in natural populations and of the powers of selection can draw on the craft

48. This line of criticism has surrounded Darwin's program since the beginning. It was articulated by several of Darwin's most astute early critics—particularly by Jules Pictet, William Hopkins, and Fleeming Jenkin (see Hull 1973b). More or less sweeping versions have been offered by contemporary critics: see (Ho and Saunders 1984) for a general indictment of neo-Darwinism along these lines, and (Gould and Lewontin 1979) for an attack on adaptationism. I have tried to disentangle what is correct about the criticisms from various types of overstatements in several places (Kitcher 1985a chapters 2 and 7, 1985b section VII, 1988).

49. Darwin was very clear about the open-ended character of his theory, and about its potential to inspire new kinds of scientific investigation. Not only is the last chapter of the *Origin* self-consciously prophetic, but Darwin's letters also indicate his hopes for the future development of biology (see, for example, F. Darwin 1888 II 128).

knowledge of plant and animal breeders. In using the reports of pigeon fan-
ciers and beekeepers, Darwin was enlarging the circle of those considered
reliable informants on scientific matters. Aware of the problem of establishing
expertise, Darwin engaged in elaborate procedures to "calibrate" his potential
sources.[50]

Finally, Darwin's writings make a methodological claim that was not
widely appreciated in his own day and that continues to be questioned by his
critics today. As we shall see in the next section, the ideas of the *Origin* were
defended by stressing their ability to unify the phenomena. Time and again,
Darwin's detractors insisted that this defense was inadequate, that the kind
of argument it offered was at odds with the canons of good science. In reply,
Darwin insisted that many well-respected theories—he liked to cite Maxwell's
electromagnetic theory, and it is clear that Lyell's geological theory also
furnished another, perhaps more immediate, example—postulated entities
and processes that were not amenable to human observation (because they
were too small, were too remote, or took too long), and that these entities
and processes could, nonetheless, be known to exist because of the traces
they leave in the observable world of the present.

Darwin thus changed the consensus practice of biology along a number
of different dimensions. The central modification concerns the problems of
biology and the ideal forms of answers to them. Attendant on this are alterations of the set of accepted statements and of the language of biology, as
well as changes in views of observational and experimental evidence and the
appropriate sources of that evidence. Finally, implicit in the *Origin,* and
explicit in some of Darwin's responses to reviews, is a methodological thesis
about the adequacy of certain kinds of scientific theories. To explore Darwin's
commitments here is to trespass on the topic of his argument for his proposed
modifications of biology, and it is to this topic that we now turn.

6. Darwin's Argument

Darwin claimed that the *Origin* is "one long argument." We have explored
the changes in practice that that argument was intended to support. Was
Darwin's reasoning up to the job?

Up to what job? There is no single powerful individual whose attitudes
are automatically those of the nineteenth century biological community, no

50. As Darwin emphasizes in a letter to Huxley, "The difficulty is to know what to
trust" (F. Darwin 1888 II 281). He attempted to solve the problems by comparing the
reports he received with those of established authorities (reputable scientists) and by becoming enough of an insider in the breeding subcommunities to have an understanding of
the professionals' estimates of professional skill. In effect, the result was to build a bridge
between biology and plant and animal husbandry that could enable his successors to exploit
the practical abilities of agriculturalists (and hobbyists) and that would (eventually) allow
for agriculture to be influenced by theoretical biology. For an informative discussion of
how Darwin solved the problem he related to Huxley, see (Secord 1985).

scientific dictator whose persuasion was necessary and sufficient for the change to take place. We can envisage different scenarios for the shift in consensus practice. Perhaps the argument of the *Origin* gave all (or almost all) contemporary naturalists reason to modify their individual practices to accommodate the changes Darwin recommended, and the change in consensus practice came about in consequence of a large collection of independent individual decisions. Or perhaps the argument persuaded only a small élite to modify their practices and the community shift resulted from the power (or prestige) of this élite. I mention these scenarios to bring into the open the point—obvious once it is made—that the claim that the argument of the *Origin* was sufficient to support a change in consensus practice is highly ambiguous. The task of disambiguation will occupy us in later chapters. For the present, my concern is to reconstruct that argument and to show that individual scientists with a number of different antecedent practices, including some who were previously quite at odds with Darwinian ideas, had excellent reasons for modifying their commitments to incorporate Darwin's central themes.

The reasoning of the *Origin* divides into three main parts[51]:

(1) An attempt to show that it is possible to modify organisms extensively through a natural process (natural selection).
(2) An attempt to show that, given the possibility of hypothesizing that organisms now classed in separate species (or higher taxa) are related by descent from a common ancestor, the introduction of such hypotheses would enable us to answer many questions about these organisms.
(3) An attempt to respond to difficulties that threaten hypotheses about common descent.

The early chapters are directed at (1), and it is in these chapters that the celebrated argument by analogy with artificial selection plays its crucial role. Darwin adduces a number of examples, most prominently examples of different kinds of pigeons, to show that the conscious selection employed by plant and animal breeders has been able to produce striking modifications of organisms. Claiming that the struggle for existence imposes a selective pressure analogous to the deliberate selection of the breeder, he concludes that it is possible to suppose that large modifications can also be produced in nature. Hence it is unwarranted to maintain that hypotheses asserting the modification of an ancestral species to produce a quite different descendant are, in principle, inevitably false.

At this stage, the way is clear for the work of the latter chapters of the *Origin* in which the explanatory power of the modifications of practice that Darwin recommends is extensively elaborated. Darwin himself seems to have seen these chapters as bearing much of the argumentative burden: he begs

---

51. The reconstruction I give here is a philosophical elaboration of a scheme for interpreting the reasoning of the *Origin* presented by Huxley (1896 72) and articulated in an illuminating review article by Jonathan Hodge (1977).

Lyell to keep his mind open until reading the "latter chapters which are the most important of all on the favourable side" (F. Darwin 1888 II 166–167). His approach is to marshall an impressive array of puzzling cases of geographical distribution, affinity of organisms, adaptation, and so forth, aiming to convince his readers that there are numerous questions to which answers fitting his schemata would bring welcome relief. Consider, for example, Darwin's partial agenda for biogeography. After describing the "American type of structure" found in the birds and rodents of South America, Darwin suggests that biologists ought to ask what has produced this common structure (or, in the terms of Section 2, why neighboring centers of creation continually bring forth embodiments of the same archetype). The similarities are too numerous just to be dismissed as beyond the province of scientific explanation.

> We see in these facts some deep organic bond, prevailing throughout space and time, over the same areas of land and water, and independent of their physical conditions. The naturalist must feel very little curiosity, who is not led to inquire what this bond is.
>
> This bond, on my theory, is simply inheritance, that cause which alone, as far as we know, produces organisms quite like, or, as we see in the case of varieties, nearly like each other. The dissimilarity of the inhabitants of different regions may be attributed to modification through natural selection, and in a quite subordinate degree to the direct influence of physical conditions. (*Origin* 349–350)

This passage, brilliantly crafted to disarm objections in ways that I shall discuss later, expresses a clear message. There are many details about particular organisms that cry out for explanation, and if biologists commit themselves to instantiating Darwin's schemata, then they will account for these otherwise inexplicable details.[52]

However, consideration of cases raises skeptical doubts to the effect that there are limits to the applicability of Darwin's schemata, so that his explanatory proposals are, at best, incomplete, or worse, incorrect. To construct a Darwinian history will typically involve scientists in advancing hypotheses about the existence of certain ancestral forms with particular properties. In many cases, the fossil record will contain no remnants of such organisms. How is this embarrassing lack of evidence to be understood?[53] Moreover, there are some properties of organisms which, in their final form, obviously assist their bearers, but which would appear to be at best useless if they were present in an incomplete state. Furthermore, Darwin's branching conception of the history of life presupposes the possibility that the descendants of a common ancestor may divide into two (or many more) groups whose members are not interfertile. In all these cases, well-known findings in biology and

52. For other important passages that deliver the same message, see *Origin* 318–319, 339–341, 394, 440–444, 452–453, 471–480.

53. Various forms of the poverty of the fossil record are posed forcefully at *Origin* 280–281, 287–288, 292, 301–303. One of Darwin's critics, Thomas Vernon Wollaston, saw the state of the fossil record as being "the gravest of all objections" to Darwin's theory, but he noted Darwin's frankness in admitting the facts (Hull 1973b/136).

geology seem to show that there are important and pervasive features of life that both raise the kinds of questions he intends biology to address and fall outside the range of his schemata.

Aware of the threat of limitation of his enterprise, Darwin devotes the third part of his argument to an attempt to show that the difficulties are only apparent. He offers an account of fossilization that is designed to explain why many of the personae in the history of life leave no traces in the record. He tries to show how complex structures—such as eyes—may emerge under selection, either because their incipient forms possess rudimentary advantages (perhaps, though not necessarily, the same kinds of advantages as those enjoyed by the final versions) or because they are correlated with advantageous traits. He gives a careful survey of the phenomena surrounding sterility and sketches a selectionist account of the emergence of a sterility barrier. By arguing that apparently large problems dissolve under close scrutiny, Darwin is able to defend the broad claim that the entire families of questions that he proposes to make central to biology can be answered in the ways he suggests.

Let us now suppose that Darwin's contemporary naturalists appreciated this chain of reasoning, and that they amended their practices in the following way. First, they accepted Darwin's delineation of the important problems of biology, accepting that questions of biogeography and comparative morphology should be addressed by instantiating the patterns of Section 4.[54] In consequence of their initial commitments to Darwinian explanations in biogeography and comparative morphology, all of them abandoned their previous general belief in the fixity of species and all admitted that natural selection is a *possible* agent of evolutionary change. Again, because of their initial commitments, all espoused Darwin's reconceptualizations of species and of homologies. Finally, all accepted the importance of the experimental and observational projects that the *Origin* recommends, all recognized the importance of the information on natural variation that breeders can supply, and all hailed as a major new task for theoretical biology the project of obtaining deeper insights into the phenomena of variation and heredity. This modification of their individual practices is the acceptance of what I have called *minimal Darwinism*.

Did the *Origin* provide these naturalists with good reasons for accepting minimal Darwinism? Yes. The attitude concerning questions and schemata for addressing them that I have supposed to be common to all naturalists, together with the conception that natural selection is a possible agent of evolutionary change, is well supported by the argument of the *Origin*. Moreover, once that attitude is in present, the changes elaborated in the latter part of the last paragraph follow automatically: the conception of species must be revised, experiments to determine dispersal powers become significant, the-

54. I shall assume that *some,* but by no means all, of them also favored the idea of using SIMPLE INDIVIDUAL SELECTION (and other selectionist patterns) to address questions about adaptation (and the modification of characteristics generally). Those who did so went beyond minimal Darwinism to a more thorough involvement with Darwin's ideas.

oretical questions about heredity and variation become major issues, and so forth. My claim, then, is that naturalists living in Britain in 1859, starting from any of the biological practices then popular, *ought* to have modified their practice to accept minimal Darwinism, in light of the reasoning that I have reconstructed from the *Origin*.

The best way to defend this claim is to see why a few people did *not* make the shift to minimal Darwinism, and to show how Darwin and his supporters had good responses to their objections. How did the critics attack the "long argument"? What replies did Darwin, Huxley, Hooker, Gray, and others make?

One important protest was that Darwin had no evidence of large-scale modifications by natural or artificial selection, so that there was no good reason to think that species could evolve in the way that he suggested. Typical were the comments of the entomologist Thomas Vernon Wollaston:

> there is no reason why *varieties,* strictly so called, . . . and also geographical "sub-species," may not be brought about, even *as a general rule,* by this process of "natural selection": but this, unfortunately, expresses the limits between which we can imagine the law to operate, and which any evidence, fairly deduced from facts, would seem to justify it: it is Mr. Darwin's fault that he presses his theory too far.[55]

Because Darwin could only suggest the *possibility* of unlimited modification, he was roundly chided by his critics for deserting the true path of science. Drawing an invidious contrast, the applied mathematician William Hopkins lauded the accomplishments of the physicists:

> They are not content to say that it *may* be so, and thus to build up theories based on bare possibilities. They *prove,* on the contrary, by modes of investigation that cannot be wrong, that phenomena exactly such as are observed would *necessarily,* not by some vague possibility, result from the causes hypothetically assigned, thus demonstrating those causes to be the true causes. (Hull 1973b 239)

In a letter to Asa Gray, Darwin explained how Hopkins had failed to appreciate the force of his argument:

> I believe that Hopkins is so much opposed because his course of study has never led him to reflect much on such subjects as geographical distribution, homologies, &c., so that he does not feel it a relief to have some kind of explanation. (F. Darwin 1888 II 237)

Although Darwin took some trouble in the *Origin* (and in his later *Variation of Plants and Animals under Domestication*) to show that artificial selection is capable of producing quite dramatic modifications of organisms, his principal response to the charge that variation is only limited is to shift the

---

55. (Hull 1973b 131). For similar remarks by Wollaston, François Jules Pictet, Samuel Haughton, William Hopkins, and Fleeming Jenkin, see (Hull 1973b 135, 145, 224, 253, 304ff.).

burden of evidence. The analogy with artificial selection is not intended to demonstrate—nor does it need to demonstrate—that variation is unlimited. Unless some reason can be given for supposing that there are limits to variation, then the explanatory power of the hypotheses that attribute descent with modification justifies us in accepting them, even though modifications as extensive as those which have been hypothesized have not been directly observed. Only someone insensitive to the ability of the novel theory to provide an account of phenomena that had hitherto seemed inexplicable (even arbitrary)—a nonbiologist like Hopkins, for example—will fail to realize that there is evidence for supposing that selection has quite extensive powers whose action cannot be directly demonstrated. The opening chapters of the *Origin* thus clear some space within which Darwin can defend his schemata for tackling biological questions by appealing to their power to unify the phenomena.[56]

Very well. But isn't there still some justice in the charge that "Mr. Darwin presses his theory too far"? The charge is presented forcefully in Richard Owen's version:

> We have searched in vain, from Demaillet to Darwin, for the evidence or the proof, that it is only necessary for one individual to vary, be it ever so little, in order to [be led to] the conclusion that the variability is progressive and unlimited, so as, in the course of generations, to change the species, the genus, the order, or the class. We have no objection to this result of "natural selection" in the abstract; but we desire to have reason for our faith. What we do object to is, that science should be compromised through the assumption of its true character by mere hypotheses, the logical consequences of which are of such deep importance. (Hull 1973b 201)

Both Owen and Wollaston view Darwin as accepting the proposition that the variation along a lineage exceeds any given limit, provided that sufficiently many generations are considered, on the basis of the observation that a small amount of variation can be observed over a small number of generations. They see this evidence as compatible with the proposition that variation only occurs within limits (say those marked by species boundaries) or as compatible with the proposition that Darwin accepts. Darwin, then, is jumping to conclusions prematurely, recommending that "mere hypotheses" assume the "true character" of science.

Darwin has a three-point reply to this criticism, a reply that is compressed in the discussion of the South American fauna that I have quoted (*Origin* 349–50; see p. 36). The first point is to emphasize the length of time available, and to draw (explicitly or implicitly) an analogy with geology. Natural selec-

---

56. Darwin's main argument stresses the unifying power of his schemata, but he cannot resist giving subsidiary arguments. So, for example, the early chapters of the *Origin* campaign against the idea that there is a natural boundary around species. This subsidiary argument becomes very important to certain versions of Darwinism—for example, those which take a nominalistic approach to species or those that emphasize evolutionary gradualism. However, it is incidental to the more cautious version of Darwinism that was widely accepted by British naturalists, the version of Darwinism with which I am concerned here.

tion can be observed to produce small effects, just as the forces invoked by
Lyell can be seen to make minor changes in the features of our planet, and
we should no more expect to observe selection transcending a species boundary than we should anticipate being able to observe erosion leveling a mountain range. The second point is to emphasize the explanatory power of the
theory and the perplexity that we ought to feel when confronted with the
phenomena—naturalists who do not wonder about the relationships among
the South American animals must "feel very little curiosity." The third point,
directed at the prior practices of his contemporary British (and, to a lesser
extent, American and European) naturalists, is that, on *anyone's* account of
the phenomena there has been some process by which new species have been
called into being. Owen and Wollaston prefer not to speculate on what this
process is. It is not for them to judge whether the members of the new species
are the descendants of the old or whether there has been an entirely new
creation. But, Darwin argues, these two accounts are not equally well supported when we discover a "bond" among the species living in the same
region at different epochs. We know nothing of the process of special creation
and have no reasons for believing that the products of special creations should
be bound by the organic similarities observed (for example) in the American
fauna. We do know *one* way in which resemblance among organisms is maintained—"that cause which *alone,* as far as we know, produces organisms quite
like, or, as we see in the case of varieties, nearly like each other" (my italics)—
to wit, through inheritance. Owen and Wollaston are thus cast in the role of
seeing an account of species formation in terms of an unknown process subjected to arbitrary constraints as equally credible as the suggestion that a
known process has been at work and that its effects are magnified on a time
scale that greatly exceeds human observational powers.

But Darwin's argument can be attacked from another direction. Should
his account be praised for its explanatory promise, or does it actually deliver
explanations? Darwin was sometimes inclined to make the stronger claim:

> Thus, on the theory of descent with modification, the main facts with respect
> to the mutual affinities of the extinct forms of life to each other and to living
> forms, seem to me explained in a satisfactory manner. And they are wholly
> inexplicable on any other view. (*Origin* 333)

Some reviewers were unconvinced. Hopkins lectured:

> A phenomenon is properly said to be *explained,* more or less perfectly, when
> it can be proved to be the necessary consequent of preceding phenomena,
> or more especially, when it can be clearly referred to some recognised cause;
> and any theory which enables us to do this may be said in a precise and
> logical sense, to explain the phenomenon in question. But Mr. Darwin's
> theory can explain nothing in this sense, because it cannot possibly assign
> any necessary relation between the phenomena and the causes to which it
> refers them. (Hull 1973b 267)

Hopkins's remarks make it clear that he regards Darwin's "explanations"
as falling short in two main respects: the hypotheses about descent with

modification which are invoked in answering biological questions are not independently confirmed, nor are those hypotheses linked by a gapless sequence of inferences to a description of the phenomena to be explained. The first demand is easily resisted. Darwin was fond of remarking that his proposal was no different from that of the physicists who introduced "the undulatory theory of light," without any direct demonstration of the passage of waves through the luminiferous ether, on the basis of its ability to explain the phenomena of diffraction, interference, polarization, and so forth (F. Darwin 1888 II 286).

The second point is more tricky. Darwin appreciated the fact that claims that a theory explains the phenomena are ambiguous. Explanations are responses to questions, actual or anticipated, and what is enough to answer one question may not suffice to answer another, even a question posed in the same form of words. To ask why a group of organisms shares a common feature may simply be to wonder about the nature of the bond that unites them, or it may be already to presuppose the character of that bond and to inquire how the feature in question has been preserved through a course of modifications. The latter question will require a different—and more detailed—answer than the former. The right response to Hopkins is to maintain that his conditions on explanation are too restrictive, that the *Origin* already offers some explanations and that it indicates the lines along which further explanations are to be sought.

This brief review of early objections to the argumentation of the *Origin* shows, I believe, that Darwin's reasoning was sufficient to justify a modification of biological practice to involve a commitment to minimal Darwinism. A naturalist who began with any of the main practices favored in Britain in 1859 and who had a clear view of the argument I have presented ought to have modified practice in the way I indicated earlier in this section (see p. 37). But the actual British naturalists of 1859 began from psychological states that involved far more than mere adoption of a biological practice: each of them had all kinds of opinions about all kinds of matters, as well as hopes, aspirations, fears, likes, and dislikes. Moreover, the argument I have reconstructed had to be recovered from the *Origin* and from the critical debate that surrounded the *Origin*. Finally, as naturalists struggled to make up their minds, other forces were at work upon them: their decisions were made in a matrix of personal, professional, and intellectual allegiances.

Recognizing the force of Darwin's argument was a tremendous cognitive achievement. To have modified one's practice by undergoing this reasoning is cognitively superior to doing so by rehearsing a simpler version (say one that did not consider and reply to one of the important objections); in its turn, using the simpler version is cognitively superior to basing one's modification on something more crude; and so it goes, until we reach the cognitive ground state of a hypothetical person who makes the Darwinian modification blindly, without any reasoning process behind it at all.

As I shall argue in some detail in later chapters, an emphasis on the use of reasons in the growth of science should not be accompanied by claims that those who do not fully appreciate the reasons are *irrational*. Darwin, I contend, saw the issues more clearly than any of his opponents. But it is surely true that there were many converts to Darwinism whose cognitive achievements were inferior to those of intelligent critics, who would have been unable to reply to important objections that they did not perceive. Hopkins, Wollaston, Haughton, Owen, and, especially, Fleeming Jenkin engaged in reasoning of considerable sophistication, even though they did not appreciate the possibility of Darwinian answers to their challenges. Labeling them as "irrational" is somewhat akin to calling the Empire State Building "short," simply because it is no longer the tallest structure in town. The debate about the *Origin* brought into ever sharper focus the structure of its central argument: Darwin's critics (and his champions) played an important *constructive* role in making it possible to appreciate the reasoning I have outlined here.

7. The Synthesis

After sixty years of acceptance of minimal Darwinism and serious debate about the hegemony of natural selection, the modern synthesis of the 1930s brought Darwin's ambitious proposals into the consensus practice of biologists. To understand the synthesis and how it extended the Darwinian ideas I have analyzed, it is convenient to divide it into two stages. First comes the mathematical work, the definitive reconciliation of Darwin and Mendel and the elaboration of theoretical population genetics.[57]

Mathematical population genetics takes as its central question the general problem, Given a population P with characteristics W and subject to an initial genetic distribution D_0, what will be the (expected) genetic distribution after n generations? In formulating concrete instances of the problem, a number of different kinds of features of the population must be specified: we must state whether the population is (effectively) infinite, whether the organisms reproduce sexually, whether the organisms mate at random. We must also identify the "forces" acting on the population—mutation, migration, natural selection—and offer a genetic description that lists the loci involved, the number of alleles at each locus, the linkage relations (if any), any disruptions of normal meiosis, and so forth. The very simplest case for a sexually reproducing organism is to suppose that we have an infinite, panmictic (randomly mating) population, with no mutation and no selection, in which we are tracing the frequencies of genes and binary gene combinations at a single locus, a locus at which there are two alternative alleles. In this case, it is easy to prove (as Hardy and Weinberg independently did in 1908) that if the relative frequencies of the alleles A and a are p and q (where $p + q = 1$), then *whatever* the initial frequencies of the genotypes (here the binary gene combinations:

---

57. That work began well before 1930. Its roots are in the neglected writings of Yule and in the independent discoveries of Hardy and Weinberg, and its first flowering is in a pioneering paper by Fisher (1918).

AA, Aa, aa) the population will move in one generation to an equilibrium in which the distribution of genotypes is p^2 *AA*, $2pq$ *Aa*, q^2 *aa*. Population genetics tackles the general problem by employing more intricate (often far more intricate) versions of the same type of combinatorial reasoning that is used in this simple case.

Selection is represented by assigning to genetic combinations coefficients that summarize their probabilities of making contributions to the next generation. We can fix ideas by thinking only in terms of survival. Selection may act upon a population because organisms bearing some genetic combinations are more likely to survive to maturity than others are. Assessing the transmission of alleles into the next generation, we shall need to take the survival probabilities into account. So, for example, if we are investigating an infinite panmictic population with two alleles at a locus, our original distribution of genotypes may be

$$p^2 \, AA \qquad 2pq \, Aa \qquad q^2 \, aa.$$

Suppose that the bearers of the genetic combination *AA* have a probability of 1 of surviving to mate, that the bearers of *Aa* have a probability w_{Aa} of surviving to mate, and that the bearers of *aa* have a probability w_{aa} of surviving to mate. Then the expected distribution at the time of mating is

$$p^2 \, AA \qquad 2w_{Aa}pq \, Aa \qquad w_{aa}q^2 \, aa.$$

The straightforward combinatorial reasoning used in deriving the Hardy-Weinberg principle now yields the distribution in the next generation:

$$\begin{array}{ccc} p^3(p + w_{Aa}q)/w & w_{Aa}pq & w_{aa}q^3(w_{Aa}p + w_{aa}q)/w \\ AA & Aa & aa \end{array}$$
$$\text{where} \quad w = p^2 + 2w_{Aa}pq + w_{aa}q^2.$$

Repeating the calculation, it is possible to compute the frequency distribution in any arbitrary descendant generation and to address the inverse problem of discovering how many generations are needed before the *a* allele is reduced to some specified frequency.

The algebra of the last paragraph translates into modern notation one of the simplest treatments of natural selection provided by Fisher, Haldane, and Wright: the case of a locus with two alleles with one of the homozygotes enjoying superior chances of survival. To achieve the full generality of their approach we need first to drop the idea that differences in expected contribution of genes to descendant generations are simply due to differences in survival abilities, letting the *weights* w_{Aa} (and so forth) stand for the *fitnesses* of the genetic combinations (or of the bearers of those combinations). In addition, we need to allow for more complex genetic systems, finite populations, internal structure of the populations (such as assortative mating), and combinations of selection with other factors that affect gene frequencies. Following the approach I have adopted in reconstructing Darwin's ideas, I shall present the contribution of Fisher, Haldane and Wright as a general pattern of argument.

GENETIC TRAJECTORIES

Question: Given a population P with characteristics W and initial genetic distribution D_0, what is the genetic distribution in the nth generation, D_n?

Answer:

(1) *Specification of Population Traits*

   (a) P is [infinite, of size N][58]

   (b)   The organisms in P reproduce [sexually, asexually]

(2) *Specification of Initial Genetic Distribution*

   (a)   There are n loci involved in the process under study

   (b)   At the ith locus there are m_i alternative alleles {specification for $1 \leq i \leq n$}

   (c)   The initial frequency of the jth allele at the ith locus, a_{ij}, is p_{0ij}; the initial frequency of the allelic combination $a_{11}a_{11} \ldots a_{n1}a_{n1}$ is $F_{01 \ldots 1}$, the initial frequency of the allelic combination $a_{12}a_{11}a_{21}a_{21} \ldots a_{n1}a_{n1}$ is $F_{021 \ldots 1}, \ldots$ {continued through all possible allelic combinations}.[59]

   (d)   The linkage relations are given by the recombination probabilities f_{ij} {specification for all combinations of alternative alleles}

   (e)   The fitness of the allelic combination $a_{11}a_{11} \ldots a_{n1}a_{n1}$ is $w_{1 \ldots 1}$,[60] the fitness of the allelic combination $a_{12}a_{11}a_{21}a_{21} \ldots a_{n1}a_{n1}$ is $w_{21 \ldots 1}, \ldots$ {continued through all possible allelic combinations}.[61]

(3) *Specification of Further Conditions*

   (a)   The probability that an allele a_{ij} will mutate to an allele a_{ik} is μ_{ijk} {specification for all i, j, k}

   (b)   The probability that an organism bearing the allelic combination $a_{11}a_{11} \ldots a_{n1}a_{n1}$ will mate with an organism bearing the allelic com-

---

58. Here, and in what follows, square brackets enclose alternative completions of a sentence.

59. If the population starts from a Hardy-Weinberg equilibrium, then the Fs can be computed from the p's. However, the general treatment that descends from Fisher, Haldane, and Wright does not assume that this is necessarily so.

60. It is not required that the w's be constants. Indeed the values of the fitnesses might depend on the frequencies either of alleles or of allelic combinations within P. Thus the schema allows for frequency-dependent selection.

61. Here one could also insert a clause allowing for the possibility of meiotic drive:

(f) If an organism has the genetic combination $a_{ij}a_{ik}$ at the ith locus, then the probability that the chromosome containing the allele a_{ij} will be transmitted to a gamete is g_{ijk} {specification for all pairwise combinations at all loci}.

It would be anachronistic to include this as part of the schema devised by Fisher, Haldane, and Wright, but it should be incorporated in the present version of the pattern for solving trajectory problems in population genetics. For discussion of meiotic drive, see (Crow 1979).

bination $a_{12}a_{11}a_{21}a_{21}\ldots a_{n1}a_{n1}$ is $g_{1\ldots1,2\ldots1},\ldots$ {continued through all pairwise combinations of allelic combinations}.[62]

(c) The frequency with which organisms bearing the allelic combination $a_{11}a_{11}\ldots a_{n1}a_{n1}$ migrate into the population is $u_{11\ldots11}$; the frequency with which organisms bearing the allelic combination $a_{11}a_{11}\ldots a_{n1}a_{n1}$ migrate out of the population is $v_{11\ldots11}$; {continued through all allelic combinations}.

(4) *Use of principles of combinatorics and probability to derive*

(5) The expected distribution of allelic frequencies and frequencies of allelic combinations in the next generation {if P is infinite}.

Or

The probabilities that particular distributions of frequencies (of alleles and of allelic combinations) will be found in P in the next generation {if P is finite}.

(6) *Iterating (1)–(5), one obtains*

(7) The expected distribution of frequencies in the nth generation, D_n {infinite case}.

Or

The probabilities that particular distributions of frequencies will be found in the nth generation {finite case}.

Fisher, Haldane, and Wright showed how important instantiations of this schema could be given, instantiations that are both biologically relevant and mathematically tractable.[63] Given this conception of their theoretical contributions, two questions naturally arise. How does the introduction of GENETIC TRAJECTORIES relate to the enterprise begun by Darwin? How does it provide a new agenda which is taken up by Dobzhansky and other architects of the synthesis?

Mathematical population genetics deepens our understanding of natural

62. In the case of random mating the g's vary directly as the products of the relative frequencies of the allelic combinations.

63. Instead of treating Fisher, Haldane, and Wright as introducing a general schema into population genetics (or, more exactly, founding the subject by proposing a schema) one could view them as specifying a general model type. The subsequent work of mathematical population geneticists would then be regarded as the specification of some of the features that are left open in the general model type, so as to yield particular population genetics models. This approach to the subject would result in a picture akin to that elaborated by Richard Lewontin in chapter 1 of his (1974) and articulated by Elisabeth Lloyd (1984, 1988). For certain purposes, this suggestion would be equivalent to the approach I pursue here. However, I am interested not only in the structure of the models but also in the relation between the work of Fisher, Haldane, and Wright and the families of problems addressed within evolutionary biology, and, for this endeavor, the focus on explanation and strategies for responding to explanation-seeking questions seems more straightforward. I would not be perturbed if Lloyd and others who favor a semantic conception of theories find it possible to translate what I say into their own preferred idiom.

selection by providing a detailed account of a process—hereditary transmission—that Darwin had to take for granted. We can capture this idea within the framework that I have offered by recognizing that Darwin's selectionist patterns, such as SIMPLE INDIVIDUAL SELECTION can be embedded within a pattern that contains GENETIC TRAJECTORIES as a subpattern. Instead of resting content with the bare claim that a property is heritable (which occurs as (4) in SIMPLE INDIVIDUAL SELECTION), theses about the inheritance of traits can be derived from specifications of the underlying allelic combinations. But this is only part of the story. Instead of simply considering cases in which a single trait becomes prevalent in a population, it is now possible to explore the fixation of a distribution of characteristics. In principle, we start from the presence of underlying genetic variation in a population, provide an analysis of the factors that modify frequencies of genes and of allelic combinations, and use GENETIC TRAJECTORIES to derive conclusions about subsequent genetic variation (from which we can arrive at claims about the distributions of phenotypic properties). The structure of the new pattern is as follows:

NEO-DARWINIAN SELECTION

Question: Why is the distribution of properties $P_1, \ldots, P_k$ in relative frequencies $r_1, \ldots, r_k$ ($\Sigma r_i = 1$) found in group G?

Answer:

(1)   Among the ancestors of G there is a group of contemporaneous organisms, G_0, such that (i) there was variation at n loci among members of G_0; (ii) at the ith locus there were m_i alternative alleles present in the G_0–G sequence.

(2)   In the environment common to all organisms in the G_0–G sequence, organisms with the allelic combination $a_{11}a_{11} \ldots a_{n1}a_{n1}$ have probability $s_{1 \ldots 1, j}$ of having the trait $P_j, \ldots$ {specification of the gene-environment-phenotype relations for all the allelic combinations}.[64]

(3)   *Analysis of the ecological conditions and the effects on their bearers of $P_1, \ldots, P_k$.*

Showing

(4)   The fitness of the allelic combination $a_{11}a_{11} \ldots a_{n1}a_{n1}$ is $w_1 \ldots _1$, the

64. Here I specifically allow that the forms of phenotypes may depend on aspects of the environment that *vary* from organism to organism. In many instantiations of the selectionist schema this possibility is ignored, and it is assumed that there is a uniform association of underlying allelic combinations with phenotypic traits. My schema could be complicated further by allowing for the possibility of changing environments. But, while I believe that Fisher, Haldane, and Wright would all have recognized the possibility that differences in microenvironment may make a difference to the phenotypic expression of an underlying genotype, consideration of fluctuating common environments seems to lie beyond the purview of their analyses.

fitness of the allelic combination $a_{12}a_{11}a_{21}a_{21} \ldots a_{n1}a_{n1}$ is $w_{21 \ldots 1}, \ldots$ {continued through all allelic combinations}.[65]

(5) GENETIC TRAJECTORIES

   (a) In step (1), (the *Specification of Population Traits*), G_0 plays the role of *P*.

   (b) In step (2), (the *Specification of Initial Genetic Distribution*), clauses (a) and (b) must be compatible with the information given in the identification of genetic variation in G_0 at (1).[66] The ascriptions of fitness must be those derived at step (4).

(6) Along the sequence G_0–G, there are t discrete generations.

(7) The expected distribution of phenotypic traits in G is . . . {infinite case}.

Or

   The probability of the distribution of phenotypic traits with relative frequencies $p_1, \ldots, p_k$ is . . . {finite case}.

Where (7) is derived from (6), (5) and (2) by using the conclusion about distributions of allelic combinations in the tth generation (D_t) obtained from (5) and the principles of genotype-environment-phenotype connection expressed in (2).

NEO-DARWINIAN SELECTION preserves the main structure of Darwin's selectionist patterns (while enlarging the scope to treat a broader family of questions) and simultaneously derives statements that had previously been taken as premises by using GENETIC TRAJECTORIES This, I claim, is the precise sense in which Fisher, Haldane, and Wright *extended the explanations* that Darwin had offered.

   NEO-DARWINIAN SELECTION brings together into one pattern of explanation various factors that are relevant to evolutionary change and enables their effects to be compared. In different instances it is possible to emphasize certain factors and ignore others: thus one may treat a large, but finite, population as if it were infinite (thus neglecting "sampling errors" and the effects of "genetic drift"), suppose that rates of mutation and/or migration are negligible, conceive of loci as independent, and concentrate solely on the consequences of fitness differences; alternatively, one may regard a large population as partitioned into an ensemble of small populations in which random "sampling errors" play a significant role. The two kinds of accounts I have just outlined (which can be specified more precisely in terms of preferred ways

65. (4) is derived from (2) and (3) by considering the probability that a bearer of an allelic combination will manifest a phenotypic trait and using the ecological analysis to compute the expected reproductive success of bearers of the trait.

66. Since there are m_i alternative alleles at the ith locus *in the sequence G_0–G*, this requires that the number of alternative alleles at the ith locus in G_0, m_i^*, be less than or equal to m_i. It is *not* required that $m_i^* = m_i$, since we should allow that new alleles may appear along the lineage.

of instantiating NEO-DARWINIAN SELECTION) were emphasized by Fisher and Wright, respectively. Thus we can appreciate that, while it settled some old debates, the formulation of mathematical population genetics raised new controversies. Exactly which kinds of instantiations of NEO-DARWINIAN SELECTION should most frequently be employed in accounting for the presence of the major features of natural populations?

The work of Fisher, Haldane, and Wright made it possible to compare the genetic trajectories of populations under various abstractly characterized conditions, and thus demonstrate that various popular views about the driving forces of evolution were mistaken. Using GENETIC TRAJECTORIES one can prove that even slight fitness differences can have significant effects over many generations, one can specify just how high mutation rates would have to be to counter a particular selection pressure, and one can compute the probability that a newly arising mutant will simply be eliminated by chance. These findings suggest new questions for empirical investigation. How large are the differences in fitness found among variants in evolving populations? How much genetic variation is present in natural populations? How high are mutation rates in nature? How fast do populations actually evolve?[67]

New questions thus arise as the new schema makes it possible to inquire which kinds of instantiations are most important to the emergence of various major features of living things and to pose precisely problems about the strength of selection or the tempo of evolution. In their turn, these questions inspire new types of observational fieldwork and experimental activity: it becomes important to find ways of measuring fitness values, mutation rates, genetic variation, and so forth.[68]

---

67. It is no accident that the issues I pose here are among those addressed by the main architects of the synthesis, and that the final question is formulated by Simpson at the beginning of his (1953).

68. From this perspective, we can make sense of the common biological project of measuring fitness differences by looking at the actual reproductive successes of phenotypically (or genetically) different organisms. This has frequently inspired complaints to the effect that it reduces natural selection to triviality (by turning some alleged "principle of natural selection" into a "tautology"). But we should distinguish two quite different projects within evolutionary biology. One of those projects consists in explaining the prevalent traits among groups of organisms (or, more generally, accounting for distributions of traits). In completing this explanatory enterprise we do not appeal to any "principle of natural selection." Instead, we instantiate some selectionist pattern (NEO-DARWINIAN SELECTION or one of Darwin's own precursors) and a critical part of the explanation is an anlysis of the ecological conditions that ground differences in expected reproductive success. For this enterprise it would *not* do simply to record our measurements of the actual reproductive success of organisms of different types: that would amount to nothing more than a trivial, nonexplanatory recapitulation of what is to be explained. However, biologists who aim to explain the prevalence of particular traits do not just count descendants and conclude that the bearers of the prevalent traits left more offspring than their rivals. They supply analyses of the relationships between variant traits and the environment that show, at least in a qualitative way, why this was to be expected. Where the counting of offspring (or some other method of measuring reproductive success) does play an important role is in a very different endeavor, namely the attempt to discern how large are the fitness differences among certain variants. This enterprise already presupposes that there is some (known or

The second stage in the synthesis consisted in forging connections between the mathematical work of Fisher, Haldane, and Wright and the study of evolution in nature. In part, this involved constructing paradigm instantiations of NEO-DARWINIAN SELECTION. Yet the architects of the synthesis should not be thought of as drudges, diligently bringing to nature the insights of their mathematical superiors. Large theoretical questions—such as the relative importance of mass selection (favored by Fisher) and random drift in small populations (favored by Wright)—were left unresolved by those who had articulated a precise scheme of explanation that could be exemplified in many different ways. In addition, there were the major problems that Darwin had raised (such as the problem of measuring variation in natural populations), as well as the question of relating neo-Darwinian explanations to the history of life and the issue of showing how the kinds of processes described in those explanations could give rise to species diversity.

The principals in the articulation of the synthesis effectively partitioned this cluster of theoretical questions, each confronting the task of bringing the neo-Darwinian understanding of evolutionary change to bear on the problems associated with some special biological subdiscipline.[69] Dobzhansky's central endeavor—articulated in his (1937)—is to articulate Darwin's branching conception of life from the perspective of genetics, and to understand how the schema articulated by the mathematical population geneticists can be instantiated to show how continuous genetic variation has given rise to distinct local populations that differ in gene frequencies, and how such differences could have been amplified over long periods to give rise to new species, and ultimately to higher taxa. Undertaking this project requires Dobzhansky to translate into modern terms many of Darwin's unresolved questions and to engage in observational fieldwork and laboratory experiments to try to resolve them.

Darwin abandoned existing concepts of species, without providing any descriptive specification of species. The connected tasks of delineating a species concept and analyzing the process of speciation were pursued in a second great work of the synthesis (Mayr 1942). Introducing the "biological species concept," Mayr defined species as "groups of actually or potentially interbreeding natural populations which are reproductively isolated from other such groups" (1942 120). The relation of reproductive isolation, for Mayr, required that there would be little breeding between reproductively isolated populations even if they were present in the same place. Reproductive isolation was thus quite distinct from geographical isolation, and the central theoretical questions for understanding the discontinuous diversity of life thus

---

unknown) ecological analysis that would form part of an explanation of the actual distribution of traits, and it seeks to uncover the strength of selection or to calibrate a proposed means of measuring the intensity of selection.

69. This becomes especially obvious in retrospective analyses of the synthesis—such as those offered in (Mayr and Provine 1980)—where particular authors can readily be assigned responsibility for particular fields. However, there is some overlap, and, in particular, (Dobzhansky 1937) not only addresses the questions arising from genetics but also surveys a number of other biological topics in less detail.

became those of fathoming the genetic changes required for descendants from a common ancestor to become reproductively isolated and of identifying which exemplifications of NEO-DARWINIAN SELECTION would be most frequently employed in accounting for these changes. Mayr extended his analysis by proposing that the attainment of reproductive isolation required the interruption of gene flow (conceived of as a homogenizing agent), so that geographical isolation is a necessary prior condition of reproductive isolation.

Perhaps the most difficult task was to integrate the new ideas about genetics, populations, and the explanation of evolutionary change with the phenomena of paleontology. Post-Darwinian studies of the fossil record had frequently inspired doubts about the importance of natural selection.[70] Simpson's accomplishment in his (1944) was to demonstrate that there was no need to extend the resources of the explanatory schema articulated by Fisher, Wright, Haldane, and Dobzhansky. NEO-DARWINIAN SELECTION would suffice for the phenomena of macroevolution.

Part of what Simpson accomplished was a "consistency argument" (Gould 1980b 161). However, he also set a new agenda for paleontology by demonstrating possibilities of measuring and comparing evolutionary rates and using the results to arrive at conclusions about evolutionary modes. Paleontology, in its new guise, could examine the effects of the microevolutionary processes analyzed by Dobzhansky and his co-workers as they were repeated over vast stretches of time.

8. After the Synthesis: Consensus and Controversy

The synthesis built upon Darwin, extending his schemata, answering questions he had introduced, enriching the stock of evolutionary claims and explanations. Contemporary biology builds upon the synthesis. My aim in this section is to examine the current commonplaces and the current disagreements, thereby providing a picture of the progress of Darwinism.

8.1. Multiplying the Phenomena

Distributed among the practices of numerous workers in the various subfields of biology is a large class of statements reporting evolutionary phenomena. Some of these are important to those who study particular groups of plants or animals: an account of the phylogenetic relations among various species of grains is part of the lore of the botanist, a history of the radiation of the marsupials belongs to the practice of the mammalogist. Others are elevated to the status of paradigms for the theoretical study of evolution because they offer unusually compelling or detailed instances of basic evolutionary pro-

---

70. (Bowler 1983) provides an extremely lucid and useful review of the paleontological worries about natural selection, and the search for alternative mechanisms of evolutionary change.

cesses. So, for example, the cases of industrial melanism in moths and of cowbird-oropendula parasitism furnish the standard exemplars of the action of natural selection in the wild. Contemporary biology thus enlarges the store of claims about the particular features of particular types of organisms that Darwin advanced in the *Origin*. We have greatly multiplied the phenomena of biogeography, comparative anatomy, and adaptation. In addition, the advances of the synthesis have been extended in study after study of the genetic variation in natural populations and of mutation, recombination, and selection both in the laboratory and in the field. Moreover, as we shall see, the triumphs of molecular biology have made it possible to add evolutionary phenomena of a qualitatively different type: for example, we can discuss homologies not only in terms of anatomical structures but by reference to protein composition as well.

We cannot think of evolutionary biology, even at the level of phenomena, as making progress through accumulation. Although a statement becomes part of the practice of every member of a subcommunity of biologists, the subcommunity deemed by the full community to be especially authoritative with respect to statements of that type, it may nonetheless be abandoned at a later stage. Consensus ideas about the dispersal of animals, including some accounts of current distributions that had been universally accepted, had to be revised in light of plate tectonics and its concomitant views of the former relations of the continents. Similarly, new discoveries of fossil hominids have forced paleontologists, primatologists, and physical anthropologists to scrap previously well-entrenched conclusions about the divergence of the hominid line from the great apes. Nonetheless, there is something apparently right about the idea that we know far more about evolutionary relationships than Darwin did. To capture that idea, I suggest we recognize that, while *some* descriptions of phenomena are revised or discarded, *many* become stable parts of consensus practice. The seeming growth in our understanding of biogeography, paleontology, species diversity, and species relationships is, I think, partly captured by the presence in later practices of an increasing number of stable reports of phenomena, with the rate of increase greatly outstripping the rate of revision.

8.2. Deepening Neo-Darwinism

There is a relation of *explanatory extension* between the pattern NEO-DARWINIAN SELECTION and Darwin's own selectionist patterns. Developments between the 1940s and the present have refined NEO-DARWINIAN SELECTION in two different ways.

First are advances in ecological analyses that attempt to substitute for qualitative claims about apparent immediate advantages—"longer legs help the animal to run away from predators"—a precise and detailed way of counting the costs and benefits of various traits over the course of an organism's lifetime. From David Lack's discussions of the effects of different clutch sizes in birds, to the analyses of lifetime consequences of reproductive strategies offered by George Williams, to the budgetary calculations of Bernd Heinrich, Geoffrey Parker, and other behavioral ecologists, there is ever increasing sophistication in attending to the factors that affect the reproductive successes of organisms and in finding ways to measure the precise impact on reproductive success.[71]

The achievements just mentioned can be understood as embedding NEO-DARWINIAN SELECTION within a yet more encompassing schema, one that uses insights of contemporary ecology to replace step (3) (the analysis of the ecological conditions) with general patterns of explanation in terms of costs and benefits from which differential fitnesses of different variants may be derived.

The second type of deepening of our understanding of selection (and other causes of evolutionary change) is based on ongoing work in mathematical population genetics. This work continues to revise and refine GENETIC TRAJECTORIES in several different ways. First, Brian Charlesworth, and others following his lead, have shown how to drop the simplifying assumption of discrete generations. Second, William Hamilton has made precise the intuitively obvious idea that an organism may increase the representation of copies of its genes in the next generation, not through the direct route of reproduction, but by behaving in ways that enhance the reproductive success of its relatives (who are likely to share copies of the same alleles). Finally, there has been renewed interest in the Darwinian notion of sexual selection, with rigorous treatments of the possibilities that alleles may spread in one sex as the result of the presence in the other sex of alleles that incline their bearers to favor particular types of mates.[72] In all of these examples, the new work can be conceived as removing limitations on GENETIC TRAJECTORIES replacing a restricted schema with something that is correct for a broader class of cases.

8.3. *The Impact of Molecular Biology*

The striking successes of molecular biology have all modified neo-Darwinian practice. One obvious consequence of these achievements is the incorporation of new processes and properties for which evolutionary explanations are needed. So, for example, we can now pose more clearly than before questions about the origins of life and investigate the evolution of various types of molecular systems. At the same time, molecular studies disclose to us processes (transposition, gene conversion) that may themselves play a significant role in the evolutionary process, thereby inspiring new refinements of GENETIC TRAJECTORIES.

A second way in which molecular biology affects the evolutionary practices

---

71. See (Lack 1966), (Williams 1966), (Parker 1974, 1976, 1978), and (Heinrich 1981). I should stress that these are only some among many outstanding examples of a flourishing tradition within contemporary evolutionary ecology. For important theoretical discussions, see (Maynard Smith 1978, 1982).

72. See (Charlesworth 1980, Hamilton 1966, Landé 1981) for representatives of work on these topics.

of biologists is in providing direct insights into the hereditary similarities and differences among organisms. Instead of resting content with the idea that similar anatomical structures indicate common descent—thus signaling the presence of shared genes—molecular biology can bring us closer to the genes themselves. So we can compare the amino acid sequences of the proteins in organisms from different species (or higher taxa) and use these as the bases for judgments about phylogenetic relationships. Even more directly, we can compare the nucleic acid sequences.[73]

8.4. Debates about Genetic Variation

I have offered a whirlwind tour of some prominent ways in which evolutionary biology has been modified in the last few decades. It is now time to turn to outstanding controversies. One important debate concerns the maintenance of genetic variation. Some contemporary evolutionary theorists hold that natural selection is responsible for the presence of multiple alleles at a locus. Others hold that such genetic variation signals the presence of neutral (or nearly neutral) mutations.

These disagreements are founded upon a deep consensus. Everybody believes that both the favored patterns of explanation can be deployed to account for *some* instances of genetic variation. The famous case of sickle-cell anemia reveals that natural selection can maintain allelic variation; by the same token, nucleotide substitutions which, because of the redundancy of the genetic code, do not affect the amino-acid sequence of the associated protein (so-called silent mutations) show the possibility of neutralist explanation. Thus within the framework of a commitment to Darwinian schemata, there can be residual disagreements about the extent to which particular schemata should be applied.

8.5. Adaptationism and its Critics

One of the chief worries of late nineteenth century evolutionists focused on the propensity of some of Darwin's followers to invent stories about possible reproductive advantages in situations where the selective significance of a trait is hard to fathom.[74] In the early years of the synthesis, biologists were cautious in discerning the action of selection. But, in the 1950s and 1960s, the synthesis

---

73. I should note that there are levels of comparison intermediate between the gross anatomical features and the fine structures of molecules. Advances in cytology have made possible comparisons of chromosome number, and even, in some cases, of banding patterns. The case of humans and chimpanzees provides a particularly fine example.

74. As Bowler notes (1983 141–146), the Darwinian penchant for fanciful tales about benefits fueled the search for alternative agents of evolutionary change.

"hardened."[75] *Part* of this hardening became manifest in the attitude that it is appropriate for a biologist, concerned to understand the prevalence of a particular trait in a particular group, to start from the idea that that trait has been separately fashioned by selection. The explanatory strategy of first resort, then, is to instantiate NEO-DARWINIAN SELECTION in such a fashion that only the selective value of the focal trait is taken into account and factors besides selection are downplayed.

To a first approximation, the Darwinian community divides into adaptationists and antiadaptationists. But this approximation is very crude indeed. As with the debate about variation and its causes, the issue is ultimately about frequencies. Everyone will concede that there are some instances in which selection of traits cannot proceed independently: cases of pleiotropy, allometry, and close linkage are well known. But, for some Darwinians, selection is very powerful. We should expect, they claim, that modifier genes will evolve to suppress unwanted side effects, that linkages can be broken, developmental associations modified. In effect, we can start by taking the traits of an organism to evolve separately unless or until we are forced to complicate our analysis.[76] Others believe that developmental constraints are likely to be omnipresent, and that we have little hope of arriving at adequate explanations of the presence of traits unless we probe the mechanisms of ontogenesis. For my purposes, the important point to note is that this debate, like that about genetic variation, proceeds within the scope of commitment to those ideas I have identified as central to Darwinism and neo-Darwinism.

8.6. Debates about Macroevolution

Nineteenth-century paleontologists sometimes believed that the causes of the large changes that have occurred in the history of life cannot readily be investigated by inspection of populations of living organisms (whether in the laboratory or in the wild). Dobzhansky was careful in opposing this attitude: the tenor of his (1937) was to explore how far the patterns of explanation used in understanding microevolutionary change could be deployed to account

75. The term is Gould's. See his (1983) for a perceptive analysis of the later writings of the architects which shows how the subsequent editions of their major works give greater emphasis to selection with concomitant downplaying of other ideas about the causes of evolutionary change. The trend was at least partly motivated by the perception that selective advantages could be found for what initially appeared to be nonfunctional traits. Particularly pivotal was the study by Cain and Sheppard of band polymorphism on shells of the snail *Cepaea nemoralis*, where it was shown how apparently trivial variations were correlated with changes in the color of the substrate, and that coloration affected rates of predation. See (Beatty 1986) for an account of how the selectionist explanation of these apparently nonadaptive variations led to increasing confidence in the power of selection.

76. This attitude is most obvious in the writings of Richard Dawkins (1976, 1982, 1986). Dawkins can legitimately claim, I think, to be elaborating with considerable clarity a view of selection that stems from Fisher (1930). He also represents the thinking of a significant number of evolutionary biologists whose strategies for dealing with particular examples tacitly embody his ideas.

for the large-scale features of evolutionary history. Both he and Simpson were initially drawn to the conception of population genetics articulated by Wright, a conception in which population structure may play an important role and in which drift in small populations is an important factor both in the process of speciation and in the breakthrough into new adaptive zones. In the 1940s and 1950s, Wright's ideas faded into the background. As the synthesis hardened, paleontologists became more confident in the power of natural selection, they began to conceive macroevolutionary processes as extrapolations of microevolutionary processes, and they offered exemplifications of NEO-DARWINIAN SELECTION that were simply extensions of those offered in understanding changes in allele frequencies in contemporary populations.

This situation changed in the early 1970s with the publication of a seminal paper (Eldredge and Gould 1972) which sparked the most publicized evolutionary debate of recent times. According to Eldredge and Gould, there is rapid evolutionary change (rapid relative to the geological time scale) while speciation is occurring. Once a species has formed, however, its morphology remains fairly constant. If Eldredge and Gould are correct, then the absence of intermediates is to be expected (since the segments of the record in which they would be found are small in comparison with the predominant periods of stasis) and there is an interesting further question as to how rapid speciation and prolonged stasis are to be understood in terms of NEO-DARWINIAN SELECTION.

Although the theory of punctuated equilibria begins from a thesis about the geometry of phylogenetic trees—*typical* phylogenetic trees, for this dispute is, like our previous controversies, concerned with the *frequency* of various scenarios in the history of life—two larger points of controversy quickly emerge. First, do the kinds of processes that occur in microevolutionary change, processes that can be studied on living populations and that have been studied ever since Dobzhansky's pioneering work, underlie the rapid speciation events? Conceding that the changes that occur when evolutionary stasis is punctuated are consistent with theoretical population genetics, some defenders of punctuated equilibrium (notably Gould) believe that they are importantly different from the familiar examples of microevolutionary change disclosed in postsynthesis research. Gould's claim can be captured in our terms by suggesting that instantiations of NEO-DARWINIAN SELECTION quite different from those that figure in orthodox neo-Darwinian explanations will be needed to understand the rapid speciation events (and, perhaps, even to understand the periods of stasis in cases where there are environmental changes).

A second major development of the theory of punctuated equilibrium is the attempt to expand the class of neo-Darwinian explanations by invoking higher-level selection. Specifically, Stanley, Gould, Eldredge, and Elizabeth Vrba have all proposed that major features of the diversity of life should be explained in terms of species-selection. If a species has a species-level property (the tendency to have a particular population structure, for example) which increases its propensity for leaving descendant species, and if this property is retained among the descendant species, then we can expect species with the property to increase in frequency. We can regard this as a claim to the effect that there is need for an *expansion* of neo-Darwinism by introducing schemata analogous to Darwin's, that would be instantiated in accounting for macroevolutionary phenomena.

9. The Progress of Darwinism

The history of Darwinism from 1859 to the present reveals a sequence of sets of contemporaneous practices. At each stage of the history there is marked individual variation. But at each stage we can pick out a consensus practice, something that all (or almost all) biologists share and that is integrated into the training of the subsequent generation of biologists. The consensus practice common to all biologists is relatively spare and is elaborated in different ways within different subdisciplines. To understand the growth of Darwinism we thus need to explore the ways in which biology is organized into fields and subfields and then recognize a hierarchical structure of consensus evolutionary practices. At the bottom of the hierarchy are the variant practices of individuals, above them the consensus practices of small groups of biologists (marine invertebrate paleoecologists or *Drosophila* geneticists, for example), above them the consensus practices of more inclusive groups, until we reach the consensus practice of biology. The history of Darwinism shows modifications both of the structure and the consensus practices at different levels.

That history shows neither simple accumulation nor sudden shifts in worldview nor indecisive oscillations. The picture is complicated, almost as complicated as the history of life, but, even on the basis of my cursory review of some highlights (Darwin's original forging of a new consensus, the modern synthesis, the contemporary situation), we can discern some major trends. First, even though reports of phenomena are revisable (and some are revised), there is a growing corpus of stable phenomena. Contemporary paleontology may be wrong about the details of the relationships among a particular group of species, but as we proceed to a coarser and coarser grain of description we find statements that have been firmly accepted for longer and longer periods of time.

Second, the phenomena that are sought and elicited by the various specialized disciplines obtain their significance because they play a role in explanatory accounts that instantiate broadly applicable Darwinian patterns. In the history of Darwinism those patterns have been extended, revised, and refined. As we trace the way in which this has been done, we see how an abstract and schematic picture of how the phenomena of nature are ordered is successively filled in, how the basic structure of Darwin's explanations in the *Origin* is preserved and extended as the questions about which he confessed his ignorance are answered.

Third, while there are recurrent themes that underlie disputes about evolution and the causes of evolutionary change, those disputes become transformed through the elimination of erroneous presuppositions. Anti-adaptationists still question the extent to which traits can be independently fashioned by selection, but their critique is precise and lends itself to empirical investigation. Their predecessors of the late nineteenth century, unaware of the mechanism of heredity, could do little more than announce their dissatisfaction. Similarly, while some paleontologists continue to wonder whether all the important processes in the history of life can be witnessed in microevolutionary settings, they can specify what kinds of alternatives they envisage. Darwinism makes *erotetic progress* by asking better questions, and, if the questions recapitulate old themes, they do so in more adequate ways, discarding faulty presuppositions, identifying points on which all disputants concur, and circumscribing more precisely what remains unknown or undecided.

3

The Microstructure of Scientific Change

1. Outline of a Naturalistic Philosophy of Science

With the history of science, as with human history generally, it is easy to become fascinated by the great dates—1543, 1632, 1687, 1859, 1905, 1953 and other favorites. Yet the sciences change continuously, week by week, day by day, because of numerous small incidents and decisions involving small groups of people. The goal of this chapter is to present a general descriptive framework for understanding scientific change, one that will do justice to the obvious fact that the large modifications that initially attract our attention, as well as the smaller episodes that are less conspicuous, result in complex ways from the thoughts and actions of individuals.

It will help, both in posing problems and in finding solutions, to idealize a bit, to break the history of a science into discrete, but relatively short, periods.[1] At the beginning of each period, there is a community of scientists, viewed by other scientists and members of the broader public as authoritative on a particular range of issues. Virtually everyone is prepared to defer to the scientific group on the range of issues in question and to accept resolutions of the issues when the group has achieved consensus. This group consists of two types of individuals, *veterans* and *apprentices*. All the veterans endorse the consensus practice, in the sense that for each the consensus practice is part of their individual practice, but each also subscribes to claims and commitments that go beyond those that are universally shared. There are differences in social position among the veterans. Some have greater credibility or authority than others, in the sense that, *on controversial issues,* matters that are *not* part of consensus practice for the group, their opinions are more likely to be credited by other members of the group.[2]

1. For concreteness, one might suppose that the period of the cycle I envisage is a few months. I believe that this idealization will suffice to capture the major features of scientific change even in fast-moving disciplines.
2. It is useful to distinguish two different relations of authority and deference here. First, there is the trust that is placed in the experts by other scientists and members of the broader public, the trust exemplified in Dobzhansky's willingness to defer to Wright's mathematics (Provine 1986 345–347) or the layperson's readiness to credit the views of paleontologists about the character of Silurian life. Second, there are the *varied* propensities

The apprentices are trained by the veterans. I shall suppose that each of them forms an initial individual practice that contains the consensus practice as a part, but that each extends this in ways that depend partly on the particular elaboration of consensus practice that is presented and partly on the cognitive state that the apprentice brings to the training process. Because different teachers elaborate consensus practice in different ways and because apprentices vary in their early intellectual ontogenies, there is variation in the individual practices with which they leave the training period and begin their work. Moreover, apprentices join the scientific community with varying degrees of initial credibility, determined (in part) by the status of the veterans from whom training is received.

During the work phase, individual practices (those of apprentices and veterans alike) are modified through conversations with peers and through encounters with nature. As information from others is accepted, modified, extended, or rejected, so assignments of credibility change. Those whose credibility declines sufficiently far are effectively excluded from further conversation, and they may drop out of the community altogether. Occasionally *outsiders,* people who have not been trained by community veterans, enter the community, earning credibility by advancing ideas that members of the community endorse. At the end of the work phase, the modifications of individual practices induce, according to rules that form part of the social system of the community, a change in consensus practice. Those who endorse this change may remain in the community to serve as veterans in the next period (although some of them may retire). Those who do not are excluded from the community and play no further role. The cycle thus begins anew with a population of veterans and a new crop of apprentices.

The philosophical problems about scientific change can be regarded as questions about iterations of this cycle. *Descriptive* problems are concerned with clarifying the notions I have employed. *Prescriptive* problems focus on the conditions under which iterations of the cycle yield a process that attains— or is likely to attain—the goals that we ascribe to the enterprise.

To provide a general descriptive account of scientific change we need a clear view of the states and processes that occur as constituents of the cycle.

Cognitive States. Science is not done by logically omniscient lone knowers but by biological systems with certain kinds of capacities and limitations. At the most fine-grained level, scientific change involves modifications of the cognitive states of limited biological systems. What are the characteristics of these systems? What kinds of cognitive states can they be in? What are their limitations? What types of transitions among their states are possible? What types are debarred? What kinds of goals and interests do these systems have?

Practices. The cognitive histories of individuals are highly idiosyncratic.

---

of some scientists in one group to defer to others (either in the same group or in a different specialty) with respect to issues that are not yet regarded as settled (not yet incorporated into consensus practice). The sources and effects of this latter type of variation will be studied in Chapter 8.

Abstracting from some of the everyday flux, we can identify a cluster of scientific commitments that remains *relatively* stable during a single period of the cycle. These commitments, collectively the scientist's individual practice, are manifested to others in her behavior: in her assent to certain statements, her pursuit of certain questions, her use of certain instruments or techniques, her production of texts with certain structures, her disposition to rely on some informants and not on others. What are the components of an individual practice? How should each component be understood?

Individual practices are still highly diverse. It is a fiction to suppose (concentrating on the simplest and most frequently acknowledged component of practice) that there is some corpus of statements such that each individual scientist has all members of the corpus and just those statements about the relevant aspects of nature stored inside his head, ready for assent under querying. Most scientists depend on tools and external aids (including other scientists) and they have private stores of information (in lab notebooks and on computers), their own very particular instruments and samples. But beyond this individual variation, there is surely something that all members of the scientific community, the expert group, share: consensus practice.

Consensus practice changes in response to modifications of individual practices; individual practices alter as a result of changes in individuals' cognitive states. What drives these latter changes? I shall start with an admittedly rough distinction. Sometimes scientists modify their cognitive states as results of *asocial* interactions, sometimes they change their minds through *social* exchanges. The obvious exemplars for the former are the solitary experimentalist at work with apparatus and samples and the lone field observer attending to the organisms—although I shall also take encounters with nature to cover those occasions on which scientists reflect, constructing chains of reasoning that modify their commitments. Paradigm cases of conversations with peers are those episodes in which one scientist is told something by another (and believes it) or when a change in commitment is caused by the reading of a text. The point of the distinction is evidently to separate those episodes that (very roughly) consist in finding things out for oneself from those in which one relies on others, but two important caveats need to be noted. First, there are many episodes in which cognitive states change in response both to social exchanges and to asocial interactions: consider, for example, occasions on which several scientists are using an intricate piece of apparatus and relying on each other for confirmation of its proper functioning. Perhaps such episodes can be decomposed into smaller incidents, each of which can be happily classified in terms of my distinction. But, for my purposes, there is no need to assume that this is so. I am primarily interested in understanding the dynamics of cognitive change in response to social and asocial sources. If exemplary and unproblematic instances can be handled, the project of understanding how to compound the "forces" can be undertaken later and applied case by case without any prior categorization of complex episodes. Second, in differentiating those occasions on which scientists "encounter nature," I have no intention of suggesting that the impact of stimuli from the asocial world is independent of the subject's prior state, including, very probably, her prior history of social interactions. The solitary experimentalist may bring to the latest trial all kinds of commitments, many of which were formed by relying on others—indeed, trust in the apparatus may be founded in social interactions with people to whom the experimenter defers. The point of selecting experiments as paradigm encounters with nature is not that a stimulus from the asocial world is the unique determinant of cognitive change but that it is *one relevant causal factor.*

Despite the important advances in our recent understanding of perception and cognition, despite detailed studies in the history of science and thorough investigations of some aspects of the sociology of science, we do not yet know enough to give all the details about the facets of science to which I have pointed. Nor do we yet know how much variation there is likely to be from field to field and from time to time. Nevertheless, for some philosophical questions, including those that concern me in this book, what we need are not so much detailed answers as a sense of the range of possibilities. We have to appreciate the options.

Our primary prescriptive tasks are to give an account of the goals of science and to derive from it a theory of what constitutes progress in science, to understand how individuals ought to behave and how their social relations should be designed to facilitate attainment of the goals. Even without detailed knowledge of all elements of the life cycle, we can address these issues, provided only that we can replace Legend's austere sentential framework with something which (for all its residual sketchiness) is closer to—but still an idealization of—scientific practice.

2. Individual Cognition

Legend was often elaborated as though a simple view of mental activity was presupposed. Think of the mind as a box whose contents are declarative statements or propositions.[3] At any given stage in the history of a field of science, we can suppose that all (competent) workers in the field have exactly the same propositions stored in their boxes. In response to new information— personal observation of some aspect of nature or reading of a text, such as the *Origin*—the contents of the boxes are modified and the housekeeping apparatus, the embodiment of the scientists' logical and methodological pro-clivities, goes to work to effect further reorganization, sometimes leading to the acceptance of large theoretical claims. One task of philosophy of science, traditionally conceived, is to understand the functioning of this housekeeping machinery. This simple tale of mental life, which, I claim, is taken for granted

---

3. I shall use these terms interchangeably, without any great attention to philosophical niceties. Issues about meaning, intention, truth bearers, and so forth, will only be relevant to my concerns in this book when the languages of scientific practices are under discussion (see Section 7 and Chapters 4 and 5). There my views will be made explicit.

not only by the adherents of Legend but also by many of their philosophical successors,[4] is so far from any realistic psychological account of the thought and behavior of scientists—as for example, in the responses to the *Origin,* reviewed in the last chapter—that questions about the roles of reason, subjectivity, and social forces cannot be satisfactorily posed, let alone tackled, without replacing it.

Although minds are not simply receptacles for storing representations, one facet of mental activity, prominent in the psychological lives of scientists, is representation (and the processing of representations). Thus one question that needs to be addressed concerns the representation of representations: should we think in terms of propositions (statements), of images, or of something quite different? Another issue concerns the organization of representations: how are they stored, how activated? Lastly, we must decide what other capacities, dispositions, faculties, to attribute to the scientific mind, and how they interact with one another. By focusing on these problems, we obtain a substitute for the simple propositions-in-a-box picture of mental life, something still highly idealized but (we may hope) sufficiently realistic to be useful in understanding the microstructure of scientific change.

There is no current psychological consensus about the nature of representations. Following *one* major approach in cognitive psychology and artificial intelligence, I shall assume that mental activity often involves the storage and retrieval of propositions, and that we can think of such processes on the

---

4. Supposing that traditional philosophy of science presupposes *any* psychological picture may well provoke howls of rage. After all, logical empiricism emerged from logical positivism, and the positivists prided themsleves on having heeded Frege's great insight to abandon psychologism in favor of the reformulation of epistemological questions in strictly logical terms. So, they claimed, they had advanced beyond earlier empiricists—notably Hume—who had appealed to crude psychology where they should have used refined logic. All this, I believe, was self-deception. Frege's own epistemological picture was heavily dependent on a view of mental functioning that he absorbed from Kant (see Kitcher 1979), and, in the writings of the positivists and logical empiricists, psychology went underground. Moreover, the psychology that lay buried was extremely simple: observation statements were inserted into the mind by experience (hence the framing of the problem of confirmation in terms of understanding the degree to which a "hypothesis" is confirmed by an "evidence statement"), and, once in the mind, all were equally accessible (hence the requirement of total evidence). A few proponents of Legend, such as Popper, dissented from the *details* of the tacit psychological consensus, but continued to subscribe to the same types of psychological themes, manifested in their ways of translating epistemological questions into issue of "logic." Ironically, the celebrated advance from Hume, consisting in the replacement of associations of ideas by statements and inferential links among them, erased one important realistic feature of Hume's account of mental activity, his differentiation of the accessibility of mental items in terms of the vivacity of ideas. All this is not to belittle the genuine achievements of logical positivism and logical empiricism, movements that exploited their idealized picture of cognition with great ingenuity. It is merely to note that the alleged repudiation of psychology was a myth, and that the gap between many traditional philosophical discussions and the behavior of scientists (and others) can be understood by recognizing how an impoverished view of mental life was taken for granted and, like the atheist's St. Christopher medallion, never publicly discussed.

analogy of the inscription, reading, and rewriting of statements in boxes. However, I shall not suppose that the *only* stored representations are propositional. Some experimental work has suggested that subjects sometimes store and operate with mental images, and there is continuing controversy about the extent to which imagelike representations play a role in cognition.[5] Because *public* images play so prominent a role in the conveying of certain types of scientific information—phylogenetic trees, diagrams of geological sections, models of molecules—it is tempting to think that part of the apprenticeship of a scientist consists in developing the capacity to produce appropriate types of images and to deploy private representations in the solution of problems.[6]

However, while I shall continue to adopt *part* of the simple "propositions-in-a-box" picture of the scientific mind, the part that views (some) representations as propositional, I shall break with its conception of the *organization* of representations. Here are some simple observations about human cognition: (1) Perception is a process in which a stimulus causes a modification of the cognitive state, sometimes consisting in the acquisition or strengthening of a belief. The content of the belief that is acquired or strengthened may depend not only on the perceptual stimulus, but also on the prior cognitive state. (2) Reasoning and problem solving may involve attempts to recall propositions (to which the subject would assent under cuing). These attempts are sometimes, but by no means always, successful. (3) Problem solving and decision making may be directed by desire to achieve certain goals. However, the goals that the subject aims to realize on one occasion may be quite different, even incompatible, with those that direct problem solving or decision making on other occasions. (4) In some decision-making or problem-solving situations, subjects are able to employ a form of inference that leads to successful attainment of their goal. In other situations to which the same form of inference would be appropriate, they fail to use it.

Now I take it that these relatively banal claims identify prevalent features of human cognitive life, indeed prevalent features of scientific cognitive life. They are also refined and made more exact by findings in experimental psychology. The simple propositions-in-a-box picture of the mind is unable to

5. Pioneering work on the role of imagery in cognitive tasks has been done by Roger Shepard and his associates. See, in particular (Shepard and Chipman 1970, Shepard and Metzler 1971). Stephen Kosslyn's (1981) is an important summary of the case for images, and Ned Block's (1981) collects many of the important contributions to the debate. Chapter 12 of (Goldman 1986) provides a useful overview and defends what I regard as a sensible ecumenical position.

6. I resist a much-debated current challenge to the propositional character of human cognition (see, for example, Paul Churchland 1989). I do so because I do not yet see how the envisaged alternative—thinking of cognition in terms of "patterns of network activation"—can do justice to the articulation of propositions and reasons that is so prominent in the growth of scientific knowledge. The difficulties of Churchland's account are apparent from his conflation of very different scientific notions—explanations, theories, and perceptual classifications are all treated as if there were no differences among them.

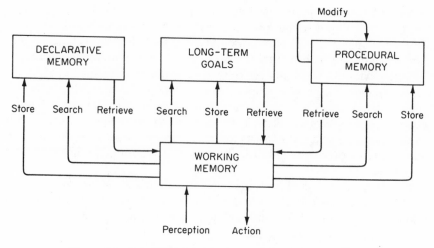

Figure 3.1. (Modified from Figure 1.2 of Anderson, 1983)

account for them, precisely because it treats the contents of the box as if they were independent atoms of equal status.[7]

We can do some justice to the simple observations by drawing inspiration from recent writings in artificial intelligence and cognitive psychology,[8] and by supposing that our minds are organized along the lines of the system depicted in Figure 3.1. This system is enclosed within a boundary across which information flows from the external environment and decisions are translated into action. The central division within the system is between those items that are activated, assumed to be transferred to *working memory,* and those that are stored in an unactivated condition. The current state of activation, the condition of working memory, may start a number of processes. Propositions (or images) already present in working memory may be transferred to *declarative memory.* Alternatively, the current state of working memory may initiate a *search* process for some desired representation from declarative memory. If the search is successful, the representation is transferred to working memory. In analogous fashion, currently pursued goals may be stored among the *long-term goals,* or the current state of working memory may promote a search for stored goals, which, if successful, issues in the retrieval of goals to working memory.

Figure 3.1 shows a third storage system. *Procedural memory* includes the

7. As we shall see (Chapter 7), Kuhn's critiques of the discovery/justification distinction and the theory neutrality of observation are more readily understandable in light of a picture of cognition that breaks with the "propositions-in-a-box" conception. Many of the alleged foibles of Legend and of logical empiricism can readily be traced to their common conception of mental activity.

8. I have been particularly influenced by the approach of John Anderson (1983). But I hope that those familiar with recent discussions in cognitive psychology (particularly accounts of memory) will recognize my approach as exposing common themes in varied, detailed, rival programs.

skills, know-how, and inferential propensities that are stored in the system. Many items in this subsystem are relevant in understanding scientific behavior. Consider for example the performances of trained experimentalists or observers. Think of the abilities to manipulate pieces of apparatus, to make the instruments work. There are also the inferential propensities of subjects. Under appropriate cuing a subject may form a representation of a particular problem—writing down the equation that governs a particular type of physical system or conceiving of an evolutionary situation by invoking a particular ecological schema—and then proceed to draw various conclusions. I shall understand these kinds of performances to be grounded in the person's inferential propensities.

This brief account of the subsystems and their contents does not specify the manner in which the representations or propensities are stored. With respect to the contents of working memory I shall suppose that representations may be *endorsed* or merely *entertained*. So, to continue an obvious (but fanciful) analogy, we might think of propositions as being entered in working memory either in ink (endorsement) or in pencil (entertainment). Operations on the contents of working memory can alter the mode of storage, promoting the entertained or relegating the formerly endorsed. For the contents of the long-term subsystems, the representations in declarative memory, the long-term goals, the stored propensities, I shall suppose that there are important differences in *strength* or *valence*. Valence varies directly with the probability of activation: items with high valence are relatively accessible, those with low valence inaccessible.

How seriously should we take this cognitive geography? It would be folly to claim that it is correct—or even approximately correct—for there is too little evidence and too many serious unanswered questions. But I do take it to have heuristic value, enabling us to pose clearly questions about the growth of scientific knowledge and to recognize distinctions and problems that have often been overlooked.[9]

3. Cognitive Constraints

Any conclusions about how cognitive systems ought to behave must be based upon an understanding of their limitations. The most obvious limitations on systems of the type I have described are those that result from considerations of size. Working memory is apparently subject to fairly severe constraints (Miller 1956), and although declarative memory, procedural memory, and the set of long-term goals may be very large, there is evidence to suggest that

9. As I maintained in footnote 4, and as I elaborate in (Kitcher 1992), every epistemology needs a psychology. The advantage of making one's psychological picture relatively explicit is that epistemological claims and concerns that are linked to some discredited element of the picture can readily be abandoned. And, of course, there is always the possibility of providing amusement for those who think that the picture is downright silly. Perhaps amusement will issue in attempts to achieve something better.

size limitations may affect how much of these subsystems can readily be searched (see Anderson 1983, Goldman 1986 chapter 10, Loftus 1979, Hyde and Jenkins 1973). Scientists facing complex decision problems—pondering, for example, whether the evidence is sufficiently strong to accept the main claims of punctuated equilibrium—must therefore find ways of organizing their consideration of bodies of evidence that are too large to be surveyed at once.

Size limitations can be relieved through the use of external aids, notebooks and computer memories for example, but the use of these aids only makes public the need to store information so that the scientist can gain access to it. It is, I think, quite likely that many of the sources of a scientist's belief are subsequently lost, in the sense that there is no possibility at a later time for the scientist to retrace his original route to the belief from the traces that endure in declarative memory, on paper, or inside his computer. If this is correct, then the philosophical ideal of the permanent possibility of rechecking may be unrealizable for an individual.[10]

More subtle constraints may result from the dynamics of the transactions among subsystems. The mere fact that an inferential propensity may be present in procedural memory does not entail that it will be activated under the appropriate circumstances (Nisbett and Ross 1980, Kahneman, Slovic, and Tversky 1982). Similarly, search operations through declarative memory may be unsuccessful because the state of working memory cues the search in the wrong way. The general problem here is that certain states of the system may be inaccessible from certain other states. I shall develop this point by considering how it might bear on some issues in the philosophy of science.

The general problem underlies *one* version of the notorious issue of the theory-ladenness of perception. We can formulate that issue as follows: External objects participate in causal processes which impinge on the cognitive subject: the subjectively proximal portion of such a process is the *stimulus,* and I shall suppose that stimuli can be individuated in some physicalistically unproblematic fashion. As a result of the presence of the stimulus and the prior state of the cognitive system, conceived as the totality of contents of the storage subsystems, the transactional processes among subsystems, *and* the state of activation (i.e., the condition of working memory), the system goes into a new state. If the new state of the system involves the insertion of a new believed proposition into working memory (possibly, but by no means necessarily, a proposition that was previously entertained in working memory), then I shall call that proposition a *perceptually induced* belief (or a PI belief).

The thesis that perception provides a theory-independent basis, intersubjectively available, for the checking and correction of theoretical claims can

---

10. Points akin to this have been made by Alvin Goldman (1986) and Gilbert Harman (1987), both of whom question the traditional supposition that it is epistemically required that people be able to trace the justifications for their beliefs by pointing out the difficulty or impossibility of doing so.

easily be articulated. Suppose that the mechanisms leading from the boundary of the system to working memory are functioning properly.[11] Then, whatever the cognitive state of the system, the presentation of the same stimulus will lead to the incorporation of the same PI beliefs. Moreover, provided that the system is in a properly functioning state (intuitively, that the subject is attentive, not hallucinating, and so forth) and that the external conditions are normal, then the PI beliefs will be true.

Numerous writers over the past two decades (Sellars 1956, Hanson 1958, Kuhn 1962/70, Feyerabend 1965, Hesse 1970, Churchland 1979, 1989) have denied this. The trained paleontologist, inspecting the strata exposed in a railway or road cutting, sees the advance and retreat of Paleozoic oceans; the novice sees only a rock face, traversed by slanting lines. Or, to revert to one of the most cited examples, Johann Kepler and Tycho Brahe, watching the dawn together, see, respectively, the sun coming into view as the earth turns and the sun beginning its daily journey around a static earth. The perceptual stimuli activate, either directly or through working memory, propensities that the paleontologist has acquired by dint of long training and that the novice lacks. The common stimulus that impinges on Kepler and Tycho activates different propensities in the two men because of the character of their background cognitive states, specifically because of their commitments to different systems of astronomy. It is possible that the paleontologist, Tycho, and Kepler should be unable to suspend the activation of the distinctive propensities and to inhibit the production of the PI beliefs that each of them forms.

The theory-ladenness of perception has seemed epistemologically threatening because it appears to block any chances of intersubjective agreement. One aspect of the problem is whether certain kinds of changes in cognitive state are impossible even for a normal scientific subject. The root worry is that there might be rival systems of cognitive commitment—involving special items in one or more of the long-term subsystems—such that both would accommodate the same perceptual stimuli in different ways so that no agreement would ever be reached. Instead of providing an independent check on cognitive commitments, perception would simply be a vehicle for persisting in the prior state, and the only ways of moving scientists to consensus would be through bribery, coercion, or some other form of conversion. Now the issues involved in this arena are surely partly empirical: findings from perceptual psychology can disclose whether people can come to adopt various kinds of PI beliefs under various kinds of stimuli if they start with certain kinds of prior commitments. However, as will become apparent in Chapter 7, it is important to have a general picture of the cognitive system, within which alternative hypotheses about theory-dependence can be formulated.

---

11. Of course, it is by no means an easy matter to say exactly what this means. Since my task here is to characterize the traditional notion of theory-independent observation from the perspective of my general approach to cognition, I shall be generous and assume that there is some way to make sense of the notion of proper functioning, perhaps in terms of the coherence of the deliverances obtainable from all (or an overwhelming majority of) the system's sensory receptors.

In particular, it is crucial to differentiate those learned propensities which distinguish experts from novices from beliefs that are stored in declarative memory (the kinds of beliefs that are supposed to make a crucial difference when Kepler and Tycho enjoy the sunrise together).

Perceptual inaccessibility is only one among a family of notions. Within our outline of a cognitive system we can pose the more general problem of the possibility or impossibility of various sequences of states. Thus we can inquire whether or not a certain initial cognitive state allows for the retrieval of some proposition from memory or whether it allows the activation of some goal or propensity. Just as certain kinds of perceptual beliefs may be precluded for those with particular theoretical commitments, so too a scientist's acceptance of the propositions, goals, and procedures associated with a particular doctrine may make her unable to engage in certain feats of memory or to be motivated by certain goals or to perform certain kinds of inference.[12]

4. Cognitive Variation

Are there particular forms that a cognitive system should take, particular propensities that should be present within it, or particular goals that it seeks to achieve? Legend conceives of scientific method as determinate. There are no options for the good scientist.[13] On one simple conception of scientific method, it would be quite impossible for it to be permissible to believe either of two incompatible propositions on the basis of some total evidence. If both members of a pair of contraries were equally reasonable, then the proper scientific stance would be agnosticism. Part of my rejection of traditional ideas about method will consist in celebrating human cognitive variation.

Scientists differ in their relations to the phenomena: what is known to one is not known to another, what is vivid for one is only a matter of "book knowledge" for another, and so forth. These differences in relationships to "the evidence" may prompt subjects to undergo quite different psychological processes when making up their minds on controversial issues. Thus, in the early twentieth century, naturalists who explored variation in wild populations of butterflies and the geneticists studying mutations in *Drosophila* drew quite different conclusions about the operation of selection, not simply because they assented to different "evidential reports"—indeed each group endorsed the findings cited by the other—but because their training disposed them to

12. As we shall see in Chapter 7, the problem of inaccessibility and the nature of the underlying dynamic constraints also bear on the question of the discovery/justification distinction.

13. Of course, options are allowed in subjectivist versions of Bayesian confirmation theory. If it is supposed that a necessary *and sufficient* condition for rationality is probabilistic coherence, then it is possible for rational scientists to be in extended disagreement because of differences in prior probabilities. However, it is precisely this feature of subjectivist Bayesian accounts that prompts the protest that they offer an inadequate representation of scientific method.

feel the force of certain kinds of considerations and not that of others.[14] At an even more fundamental level, very basic cognitive strategies may vary. Some people are cautious in drawing general conclusions, others far more daring; some try to integrate their conclusions, others are content to pursue independent lines of reasoning without thinking how they may be connected.

I want to contrast my ideas about cognitive variation with Legend's conception that there is a unique set of principles that ought to govern people's cognitive lives. This latter conception seems to me to generate the notion of a properly functioning cognitive system, such that, given the same inputs from experience, all properly functioning cognitive systems would undergo the same sequence of states. We can articulate this idea within my framework as follows: The cognitive system of a scientist—any scientist—contains the same set of goals and the same set of inferential propensities for working over input from experience. In principle, we can suppose that all scientists receive the same stimuli as inputs. Hence they should all produce the same output, storing the same propositions in declarative memory, acquiring the same skills in procedural memory, and so forth—unless, of course, they are not properly working scientific cognitive systems. There are two different ways in which they may fail to work properly: they may fail to receive certain stimuli or may receive misleading stimuli, so that they fail to acquire certain PI beliefs or even acquire wrong beliefs (in which cases the scientist is *ignorant* or *misinformed*), or their set of inferential propensities may lack correct propensities or contain wrong ones (in which case the scientist is *logically deficient* or *irrational*).[15] But, it is assumed, most scientific minds usually work well and the failures are typically failures of ignorance or misinformation. The philosophical task is to understand the proper working of the system of inferential propensities.[16]

Our (primitive, but heuristically useful) cognitive geography exposes *four* different types of cognitive variation. First, if scientists have access to different

---

14. In Giere's fruitful terminology, members of the two groups have different "cognitive resources" (Giere 1988 213–214). Mayr's discussion of post-Darwinian evolutionary theory in his (1982) is largely devoted to emphasizing the difference between the decision making of naturalists and experimental geneticists. Perhaps as a corrective to other historical accounts, he is concerned to praise the efforts of the naturalists. I would prefer to give a more symmetric treatment.

15. We might add a further type of lapse to cover formation of the wrong kinds of goals. Thus some scientists might put the interests of their careers or the advancement of some social cause before their dedication to the epistemic goals of science. For the moment, I shall ignore this sort of falling off since, on Legend's ideal of scientific behavior, those who devote themselves to the wrong goals may not count as scientists at all. But, in Chapter 8, I shall begin to explore the social consequences of such "lapses."

16. On subjectivist Bayesian attempts to articulate Legend's ideas about method, there is an important source of nonuniformity. Scientists differ in their prior probability distributions, so that there is one blind and arbitrary difference in cognitive state that would affect the ways in which inputs from experience are assimilated. It seems to me that this approach generates the wrong kind of cognitive variation, assuming that people are uniform in some relentless propensity for conditionalization but that there is enormous latitude in their initial judgments of probability.

bodies of information, then they may form quite different beliefs so that the contents of declarative memory (and possibly of the set of long-term goals and of procedural memory) will differ from scientist to scientist. Second, propositions are likely to be stored in different ways in the declarative memories of different scientists, so that what is easily accessible for one scientist is not readily produced in the problem solving/decision making of another. Third, depending on the *order* and *frequency* of exposure to different types of information, certain of the propensities may become more likely to be activated than others. Two scientists may share the same propensities, *in the sense that the same repertoire is present in both,* but different propensities may be facilitated in the two, as a result of differences in intellectual development. Finally, the scientists may differ in the contents of procedural memory, with one having inferential propensities that are not found in the other.

As I have already remarked, Legend sees exposure to a truncated or deceptive set of stimuli as producing ignorant or misinformed scientists. It is assumed that the limitations could easily be made up. However, our reflections on constraints should question this optimistic assessment. If no single scientific mind can store all the propositions (say) that are relevant to the further advancement of a field, then the differences among scientists are not accidental but essential to continued growth: the development of the field would be stunted if uniformity were imposed.[17] However, such differences may affect the ways in which further stimuli are accommodated, for, as we saw, there may be dynamic constraints on cognitive processing. But if Legend's treatment of the universal accessibility of information is too facile, its response to the fourth type of cognitive variation, the possibility of different propensities in the procedural memories of different subjects, seems to me to be too harsh. As I shall explain, it is highly likely that variation in these propensities plays a valuable role in the growth of science.

Before I take up this topic, let us look briefly at the types of cognitive variation that have traditionally been ignored. External aids, especially text-books, handbooks, and review articles, contribute to the homogenization of scientific experience, helping to foster the illusion that the same information is "in principle accessible" to all the workers in a scientific field. That this is an illusion becomes apparent when we reflect on situations of problem solving or decision making. For each scientist, there are particular items in declarative memory that are readily activated, others that are activated only with difficulty, particular pieces of information that can be assembled with ease from books and colleagues, others that are very hard for the scientist to track down. In thinking about evolutionary problems and scenarios, Darwin was surely guided by his direct experiences with South American mammals, Dobzhansky by his wild *Drosophila* populations, Gould by his fieldwork on the snail *Cerion*, E. O. Wilson by his deep knowledge of ants. Other connections are less direct: Darwin had secondary, but still privileged, access to information about pigeons and arctic flora, Gould to facts about trilobites, Wilson to

17. This theme has been taken up and elaborated by Miriam Solomon (1992).

knowledge about termites. At the other extreme, there are items that are now virtually inaccessible: collections of fossil shells that have remained for years in unopened drawers, beautifully prepared illustrations in nineteenth century works of natural history that no one has read in decades. In general, we can think of the declarative memories of scientists and their relations to others and to records (books, reviews, samples, etc.) as determining a system of nested spheres. The items in the interior spheres are easily recovered and play a prominent role in the scientist's reasoning. Those at the periphery are effectively beyond the scientist's ken. As I shall argue much later in this book, it is likely that efforts to impose a uniform ordering across a scientific field would severely hamper that discipline's further development.

Similar points can be made about differences in the activation of propensities. Co-workers in a scientific discipline often recognize one another as having distinctive cognitive styles. Faced with the same problem, different scientists will pose it and reason about it in different ways.

Once one has raised the possibility that there may be different propensities for activating modes of problem solving, representation, and inference that would be widely regarded as respectable, few people are likely to deny that such variation can be healthy in the growth of science. I now want to consider the far more radical claim that there may be significant differences in the fundamental inferential propensities stored in procedural memory, and that, contrary to the ideal of methodological uniformity assumed by Legend, this type of diversity can also be beneficial.

Any system that needs to make ampliative inferences (and human brains are among the many systems that are required to do so) has to strike a compromise between waiting too long and jumping to conclusions too quickly. It is quite possible that the tradeoff is made in a stable way for adult scientists, so that, for each scientist, there is a definite propensity (or spectrum of propensities) for that scientist to decide that she has enough instances, and that this propensity (or spectrum) is variable across the scientific population.[18] To put the point simply, some scientists may be jumpers and some inclined to hang back—Huxley and Owen serve us as possible exemplars.

There is no optimal height for a human male, one that will maximize reproductive success across all human environments. So too the variation in inferential success may vary, for jumpers and stick-in-the-muds, depending on the kinds of problem situations in which those propensities are called into play. If so, there will be no sense to the ideal of a single method, a unique description of the "properly functioning scientific mind." However, just as the recognition of biological variation does not prevent us from recognizing that certain alleles (and their bearers) are grossly nonfunctional, so too our

18. Howard Margolis (1987) offers some illuminating ideas about this topic, suggesting that there has to be a balance among four "subprocesses" he takes to be basic: jumping, checking, priming, and inhibiting. During ontogeny brains have to become configured in a dynamic system in which these four subprocesses interact. As I shall suggest later, there may be no solution that is optimal across all experiential contexts, and the compromises may be struck in quite different—adequate—ways in different instances.

appreciation of cognitive variation should not blind us to the fact that certain kinds of cognitive systems are (or would be) incompetent.[19]

I conclude that there are four important sources of cognitive variation and that in none of the cases should we take it for granted that there is a uniquely right way for the scientific cognitive system to be set up. Indeed, as I shall argue at some length in the final chapter, cognitive variation contributes in several important ways to the *community* project of investigating the world. Diversity of the kinds I have discussed serves the community as a hedge against becoming stuck—much as variation in natural populations can enable a lineage to survive events in which monomorphic populations would have gone extinct.

5. Goals

Legend takes it for granted that scientific behavior can and should be directed by the desire to attain cognitive (and only cognitive) goals. But there are important ambiguities here. Let us start by recognizing that scientists may have the goal of advancing the cognitive well-being of the large intellectual community to which they belong—the totality of scientists, past, present, and future. They may also have the goal of improving their own cognitive state. There is good behavioral evidence for attributing both goals. Biologists working on a particular group of organisms, *Drosophila* geneticists for example, sometimes conceive themselves as contributing to a community project that, as a whole, will promote our understanding of flies, inheritance, and evolution. But, on other occasions, they confess to more personal aims. They just want to understand (they say) how much polymorphism there is at such-and-such a locus in such-and-such a population.

This last declaration is itself ambiguous. Would the imagined geneticist be satisfied if someone else (perhaps someone drawing on the geneticist's own partial results) measured the extent of polymorphism and obligingly communicated the information? Probably not. A more precise formulation of the geneticist's aim is to say that he hopes to *find out* how much polymorphism there is at the locus. The geneticist's goal is not simply to reach a particular cognitive state but to arrive at it in a particular way—either as the result of his own efforts or through the work of a team to which he belongs.

Epistemic and nonepistemic considerations easily blur. What is it exactly that the geneticist wants besides knowledge that there are precisely so many alleles at the locus? The pure joy of discovery? The satisfaction of being the first to know? Recognition for being the first to know? The increased authority or increased resources that may accrue to the discoverer? All of these, perhaps. Part of the story is surely that answering outstanding scientific questions

19. I shall elaborate these points at much greater length in Chapter 6. Throughout this section, I have been much influenced by conversations with Stephen Stich and by his (1990). Stich would, I think, go much further in celebrating cognitive variation. Much of the difference can be traced to my attempt to defend the idea of broadly shared cognitive goals, and to use the attainment of those goals as a test of the adequacy of cognitive systems.

brings a number of types of satisfaction: intrinsic intellectual rewards, aware-ness of having completed a difficult task, public approbation of the effort.

Individual scientists have many long-term goals and some of their *non-epistemic* goals are relevant to their scientific work. They may desire ad-vancement, security, approbation by particular people, revenge on others (see Hull 1988 160), promotion of particular kinds of projects or causes. I shall suppose that the total set of long-term goals admits of fourfold division. One axis runs from the most evidently epistemic to the prototypically non-epistemic (without, I suggest in the spirit of the last paragraph, any sharp boundary). A second divides the personal from the impersonal. Thus a sci-entist may have the goal of contributing to the long-term community project of understanding some aspect of nature (an impersonal epistemic goal), the goal of advancing her own knowledge in a particular area (a personal epistemic goal), the goal of promoting a more egalitarian society (an impersonal non-epistemic goal), and the goal of attaining a position of eminence within her specialty (a personal nonepistemic goal).

Return now to our primitive cognitive geography. I shall suppose that a number of different goals may be present in the set of long-term goals, that these goals may conflict—in the sense that the subject would be required to pursue incompatible courses of action to attain them—and that the subject's actions are keyed to those goals that are currently active. More precisely, I imagine that goals may be retrieved into working memory, that conflicts among them may be resolved by the execution of various procedures, and that the actions that subjects perform are those that the subjects believe will lead to the attainment of those goals that are currently active.[20]

The traditional philosophical ideal of scientific detachment is reformulated as the demand that a properly functioning scientific cognitive system be one in which noncognitive goals are suppressed and only the epistemic goals become activated during the course of scientific inquiry. *Which* epistemic goals? I shall argue in Chapter 8 that the failure to see the possibility of conflict between per-sonal and impersonal epistemic goals may blind us to the function of various so-cial institutions within science. Such institutions may foster diversity, hampering some individual epistemic projects but promoting the community epistemic enterprise. Two serious questions arise. Can the activation of non-epistemic goals be suspended during the pursuit of long and complex inquiries (or, perhaps, during *any* inquiries)? (Is suppression possible?)[21] Must the acti-

20. My picture of goal activation can be extended to an account of decision making in several ways. Suppose, for example, that subjects have a number of different utility struc-tures, and that they behave as perceived-utility maximizers with respect to the currently active beliefs and the currently active utility structures. More precisely, I shall suppose that, when confronted with a situation calling for a decision, a subject activates a set of descriptions of potential strategies, a utility structure, and a set of probability judgments, and uses this representation to compute (almost always unconsciously) the expected utilities for each strategy. The strategy with the highest expected utility is then exemplified in behavior. Something akin to this has been proposed by Thomas Schelling (1984) and Howard Margolis (1982).

21. There is historical/psychological evidence bearing on this question. In an extremely

vation of nonepistemic goals always interfere with the attainment of epistemic ends, either personal or impersonal? (Is suppression desirable?)

6. Individual Practices

The most detailed picture of the life of a scientist would chart the moment-by-moment changes in her cognitive state, the ways in which items from long-term subsystems are activated and stored. At the opposite extreme, my idealized life cycle allows for a view of scientific change that abstracts from individual scientists altogether, concentrating on the sequence of consensus practices. But to understand why consensus practice changes in the way that it does we need to consider the actions and attitudes of members of the pertinent scientific community without taking into account all the fluctuations that occur. It is therefore useful to have an intermediate level of description, one that can be directly linked to the consensus practice of the community and employed to understand changes in consensus practice but one that can also be related to the psychological lives of individuals.

Generalizing from my account of the career of Darwinism, I shall take a scientist's practice to be a multidimensional entity whose components are the following:

1. The language that the scientist uses in his professional work.
2. The questions that he identifies as the significant problems of the field.
3. The statements (pictures, diagrams) he accepts about the subject matter of the field.
4. The set of patterns (or schemata) that underlie those texts that the scientist would count as explanatory.
5. The standard examples of credible informants plus the criteria of credibility that the scientist uses in appraising the contributions of potential sources of information relevant to the subject matter of the field.
6. The paradigms of experimentation and observation, together with the instruments and tools which the scientist takes to be reliable, as well as his criteria for experimentation, observation, and reliability of instruments.
7. Exemplars of good and faulty scientific reasoning, coupled with the criteria for assessing proposed statements (the scientist's "methodology").

In each of the last three components, there are two different levels to the scientist's practice—a commitment to cases (typically reflected in behavior) and an embryonic theory about why the behavior is correct. If pressed about his

provocative study (Sulloway forthcoming), Frank Sulloway reviews the attitudes of scientists at times of large ("revolutionary") change, arguing that scientists' responses are shaped by "psychologically formative influences operating within the family." I am extremely grateful to Sulloway for sharing his work with me.

commitment to certain sources of information, instruments, or instances of reasoning, the scientist may articulate the criteria which he takes to be the basis for his behavior, and these criteria may themselves become matters of debate.

7. Scientific Language

Scientists formulate their ideas in extensions of natural languages. Correlating the scientific discourses carried out in two different extended natural languages is frequently very easy: the technical expressions of one language (neologisms or old terms stretched to new uses) map smoothly onto the technical terms of the other. Yet, ever since Kuhn (1962/70), philosophers have been challenged by the suggestion that the languages employed by scientists who are separated by a major upheaval in their field are not intertranslatable. How should we think about scientific language so as to become clear about scientific communication and its limits and so as to recognize the dynamics of conceptual change?

The crucial issue concerns the semantics of the language. Syntax is relatively unproblematic, for we may regard the grammar of scientific language as that of the natural language in which it is embedded. Since Frege's seminal discussion (Frege 1892), it has been clear that linguistic expressions typically have two semantic functions. One of these is to refer (or to purport to refer) to entities: objects, sets of objects, relations, and so forth.[22] However, Frege argued that reference could not exhaust the semantic functioning of language, for, in many cases, replacing a constituent of a complex expression by a co-referential expression affects the "content" of the whole. Preservation of reference does not suffice for preservation of meaning. To use Frege's famous example, the expressions "the morning star" and "the evening star" both refer to the planet Venus, but, whereas the sentence "The morning star is the same as the evening star" is informative (and could be surprising to a hitherto uninformed astronomical observer), the sentence "The morning star is the morning star" is a trivial identity that will be accepted by the astronomical tyro.

Frege claimed that such examples revealed the existence of *intensional entities,* senses, that are expressed by linguistic expressions. The language user who understands an expression grasps its sense.[23] Sense also determines reference: two expressions with the same sense must share the same reference. Recent discussions in the philosophy of language have made it evident that

---

22. The standard semantics for the formal languages that logicians study, a semantics that descends from Frege but that modifies his views in important ways, takes names to refer to objects (people, organisms, physical things, and perhaps abstract entities: sets, numbers, concepts, ideas, and so forth). One-place predicates (e.g., "is wise," "is round," "has a mass of 30 grams") refer to sets of things—in each case the set of things that satisfy the predicate in question. Two-place predicates (such as "is larger than") refer to sets of ordered pairs of objects: an ordered pair $\langle x, y \rangle$ belongs to the referent of "is larger than" just in case x is larger than y. And so it goes.

23. The most extensive exposition of Frege's views about language in general and his ideas about sense in particular is provided by Michael Dummett in his (1973).

the conjunction of the two ideas is problematic (Kripke 1972, Donnellan 1974, Putnam 1973). Part of the problem stems from the fact that competent users of expressions are often in no position to identify criteria that the referents of the expressions must meet: people can talk about stoats and weasels, or elms and beeches (Putnam's example), without knowing what differentiates a stoat from a weasel or an elm from a beech. Apparently, it is enough for us to refer that there be experts, on whom we rely, who can make the distinctions. Another facet of the problem is that, even when a language user has a relatively rich set of beliefs about the referent, what she grasps ("what is in the speaker's head" in Putnam's phrase) may not suffice to isolate the referent under all possible circumstances. The slack may be taken up by the way the world is and by the relationship between the person and the world. Thus, to recall a thought experiment of Putnam's, a person on Earth and her molecule-for-molecule replica on twin Earth (her Doppelganger) use the term 'water' to refer to the colorless, odorless liquid that they drink and bathe in. Each of them has all the everyday beliefs about water that nonscientists standardly enjoy. However, while on Earth the liquid is H_2O, on twin Earth it is a different substance XYZ. Despite the identity of their conceptions, the earthling and her Doppelganger refer differently, the one to H_2O, the other to XYZ. "What is in the speaker's head" does not therefore determine reference. I shall articulate my approach to scientific language by building on the recent insights about reference.[24]

Imagine that a speaker produces a token of a term, "the morning star" say, and thereby refers to an object. Somehow a connection is made between the noises that the speaker produces and a part of nature. I shall call what makes it the case that that token refers to that object *the mode of reference of the token*.[25] For any case in which someone uses an expression successfully to refer, we can ask what makes it the case that the token refers to the entity it does—what is the mode of reference of the token?

The explanation may take the following form: the present speaker intends to continue her own prior references using tokens of the type; her own referential practice may have been begun by acquiring the term from others to whose usage she intended to conform; those others may have adopted the term

---

24. The recent literature on reference, sense, and the important problems raised by Kripke, Putnam, and Donnellan is enormous. Major contributions have been made by Gareth Evans (1975), Michael Devitt (1981), and others. I shall attempt to remain neutral on many of the controversial issues, since the problems that arise in dealing with conceptual change in science seem to be largely independent of questions that philosophers of language rightly take very seriously.

25. "Mode of reference" is a Fregean term, and Frege sometimes thinks of the sense of an expression as its mode of reference (1892). However, as several writers (most notably Hartry Field (1977) and Tyler Burge (1976)) have pointed out, the notion of sense is required to play a number of roles in Frege's system (the sense of an expression is to be that which is grasped by someone who understands the expression and also to be that which is the reference of tokens of the expression that occur within contexts of indirect discourse). Problems for Frege's approach (including Putnam's twin earth puzzle) arise because no single thing can play all the roles Frege attributes to senses.

from yet others; and so it goes until we reach a first user (or first users) who introduced the term by uttering it in the presence of an object with a present intention to give the name to that object. In this case, let us say that the *mode of reference* of the token that is presently produced by our imagined speaker is the complex causal chain that stands behind her current vocalization. That the token has the reference it does is *partly* determined by the speaker's intentions (if she did not intend to continue her own earlier usage and the usage of others, then the history of prior usage would be quite irrelevant to the reference of her token). It is also partly determined by events, states, and processes that are independent of her, to wit the transmission of the term and the connections between the first user(s) and the ostended object.[26]

Different tokens of the same type may be associated with different modes of reference. Imagine our speaker on a later occasion, scanning the early morning sky with a friend: "Look!," she exclaims, "the morning star is very low above the horizon." On this occasion, her dominant intention may not be to agree with her own prior usage or the usage of her community, but to single out a particular object in her line of vision. In effect, she is assuming the role of the first user(s) and *refixing* the reference of the term. She believes, of course, that the object singled out is the very same object that she has previously identified when taking over the usage of others, but, if she were to discover that she were wrong, her dominant intention is to talk about *that thing,* the heavenly body over there near the horizon. On yet other occasions she may intend to pick out an object by description, identifying the morning star as "the heavenly body that appears close to the sun in the morning" or "the second planet from the sun." Her intention may be revealed in her linguistic behavior, as when she reasons as if she were excluding, by linguistic fiat, the possibility that the morning star could fail to satisfy a certain description.[27]

Modes of reference thus fall into three types. A token's mode of reference is of the *descriptive type* when the speaker has a dominant present intention to pick out something that satisfies a particular description and the referent of the token is whatever satisfies the description. The *baptismal type* is exemplified when the speaker has a dominant present intention to pick out a particular present object (or a set of objects, one member of which is present). Finally, the *conformist type* covers those (many) instances in which the speaker intends that her usage be parasitic on those of her fellows (or her own earlier self), and, in this case, the reference of her token is determined through a

---

26. Those familiar with the seminal writings of Kripke and Putnam (especially Kripke 1972, Putnam 1973) will recognize that I am greatly compressing points that they elaborate in rich detail. Both writers see a role for ostension in the initial determination of reference, Kripke provides some excellent examples of how the reference of names can be fixed through causal chains extending back into history through the usages of earlier speakers, and Putnam emphasizes our dependence on experts ("the division of linguistic labor").

27. What is in the speaker's head thus determines *something.* As I shall suggest later, there are frequently several available ways of fixing the reference of a token of a given type, and the speaker's cognitive state—specifically an activated intention—picks out which of these fixes the reference of her current token.

long causal chain that leads back to an initial usage, a usage in which a token produced by a first user has its reference fixed either in the descriptive or in the baptismal mode.

Within a particular scientific community a particular term type is associated with a compendium of modes of reference that cover the variety of ways in which, for members of that community, the reference of tokens of that type may be determined. There may be a number of shared "first usages" that can serve as termini for conformist modes of reference. There may also be several descriptions (presumed to be equivalent) that members may employ to fix the reference of a token of the type, and, if inferences depend on the satisfaction of such a description, then, if this is made clear, the inference will not be challenged. Finally, although baptismal modes of reference may be idiosyncratic, in the sense that speakers may baptize the same objects on quite different occasions or may baptize different members of the same set or samples of the same stuff, there will be significant common properties of the baptismal events: "the morning star" will always be employed to dub an object visible in the sky, "water" will always be employed to dub a sample of a liquid, and so forth.

I shall call the compendium of modes of reference for a term (type) the *reference potential* of that term. An important part of my story about conceptual change in science (see Chapter 4) will be that reference potentials are typically heterogeneous: that the linguistic community to which a scientist belongs allows a number of distinct ways of fixing the reference of tokens of terms.

What, if anything, can be said about Frege's original problem? How do we explain the difference in cognitive content between "The morning star is the same as the evening star" and "The morning star is the morning star"? Instead of positing new objects, senses, and a mysterious process by which the speaker "grasps" them, let us develop our picture of cognition by introducing the notion of reference potential. For ease of exposition, I shall suppose that acquiring the reference potential of a term consists in incorporating a set of propensities into procedural memory. Imagine, then, that a speaker has acquired reference potentials for "the morning star" and "the evening star," and that these are not the same. If the speaker considers the sentence "The morning star is the same as the evening star," it is likely that the tokens of "the morning star" and "the evening star" will have their references fixed by differing modes of reference. If this is the case, then, I shall assume, the decision to determine whether or not the sentence is true activates a search of declarative memory. By contrast, it is probable that both tokens of "the morning star" in "The morning star is the same as the morning star" will have their reference fixed in the same way. In this case, consideration of the truth value of the sentence activates (at least in those speakers who have acquired the use of "is the same as" in English) a propensity to assent to it. Thus while the one sentence is assented to if it can be retrieved from declarative memory (and then on the basis of the process of retrieval) the other is judged correct by English speakers on the basis of a quite different process.[28]

28. I owe to the referee the interesting point that I can know the truth of "$a = a$"

My discussion of this example is both hedged and speculative. Qualifications arise because it is conceivable that presenting the sentences to subjects in different contexts could affect the psychological processes. *Typically* the modes of reference activated by "the morning star" and "the evening star," when they occur in these sentences, will be the descriptive modes (with the descriptions "the heavenly body that appears near the sun in the morning," "the heavenly body that appears near the sun in the evening"), so that there will be distinct referential modes activated in one sentence and the same referential mode in the other. But perhaps there are ways of subverting this. What is crucial to my account is the idea that *when* the expressions activate distinct modes of reference, determination of truth value (if it is to proceed without empirical investigation) must occur *via* search in declarative memory, whereas *when* they activate the same mode of reference, determination of truth value will proceed via a propensity based on internalization of lore about identity. This is admittedly speculative, but it seems to me to retain what is valuable in Fregean accounts while purging the invocations of mysterious intensional entities and connecting the explanation to a general account of cognition.[29]

A second concern about my account is the possibility of autonomous baptismal modes of reference. This arises most evidently in the case of general terms. We imagine a speaker producing a token of a term in the presence of an object, a sample of some stuff, or a member of some set. How does the reference of the general term become fixed so that it covers all the right stuff or so that it picks out the right set?[30] It is tempting to suggest, at this point, that the reference fixing must proceed by a tacit description. But, as several authors have cogently argued (Kripke 1972, Devitt and Sterelny 1987 51–53, 60), there are severe problems with the notion that all reference can be fixed through description. How is any link between language and nature ever established?

Consider an example of the problem. Imagine a brave soul ostending a tiger and courageously introducing "tiger." The object before him is a tiger—but it is also a carnivore, a mammal, a quadruped, a vertebrate, and a striped animal. What makes "tiger" refer to the set of tigers and not (say) to the set of quadrupeds? Without supposing that the linguistic innovator has stored and ready for production a description that would delimit the tigers, we can suppose that there are features of his cognitive state—propensities—that

---

even in instances in which I have no reference potential for '*a*' at all. I would suggest that this is because I can recognize that *whatever the modes of reference of the two tokens of '*a*' might be* they would fix reference to the same object. Again, there would be an important cognitive difference with statements of form "*a* = *b*."

29. My account shifts the focus from discussion of senses (abstract objects to which we stand in mysterious relations) to consideration of grasp of senses (conceived of as psychological processes that can be assigned places within our cognitive geography). Once we have linked these psychological processes to the ways in which reference is fixed we have a natural way of treating Frege's puzzle.

30. The problem is posed forcefully by John Dupre in his (1981), and I attempted to respond to it in my (1982), on which the discussion of the next paragraph is parasitic. Michael Devitt and Kim Sterelny have some illuminating things to say about this issue—which they call "the *qua* problem"—in their (1987).

would discriminate certain things as relevantly similar and others as not. There is no reason to think that these *discriminatory dispositions* as I shall call them are completely determinate. But, I shall imagine, if there is a unique natural kind that includes the object ostended and that conforms to the discriminatory dispositions, then the kind is the referent of "tiger."

Introduction of the notion of natural kind brings us to a third feature of my account that stands in need of explicit discussion. Some, but not all, of the predicates of the language are marked as referring (under a certain mode of reference) to natural kinds. Such predicates are associated with propensities for use in explanation and in inductive generalization. Learning a language, according to this conception, involves acquiring propensities for forming certain types of generalizations. Alternatively, the language embodies a view of where the divisions in nature are, and to learn the language is to acquire a propensity to see those divisions as natural. I shall have more to say about these issues later (see chapters 4, 5, and 7).

8. Significant Questions and Accepted Statements

Not all questions that can be posed about the subject matter of a science are significant. Not all significant questions are equally significant. Some questions are dismissed from the purview of a discipline because they are taken to have false presuppositions. Once the idea that all animals radiated to their present positions from Mt. Ararat has been rejected, biologists will abandon the question, How did the marsupials reach Australia from Mt. Ararat? Even questions with true presuppositions sometimes fail to count as scientifically significant. As far as I know nobody attaches any significance to the question, How many mammalian species have the same normal chromosome number as some *Drosophila* species?, although there is nothing obviously faulty about the presuppositions of this question. Of course, the significance of some questions can be a matter of debate, as, for example, in some contemporary discussions about the advisability of pursuing questions of adaptation.[31] Moreover, though a question may be universally recognized as significant, there may be strenuous debate about how significant it is. Some "whole organism" biologists wonder exactly how important it would be to answer the question, What is the nucleotide sequence of an entire human genome?, and, of course, many molecular biologists are skeptical of the value of those questions addressed by "whole organism" biologists.

One component of the practice of an individual scientist is an assessment of the significance of questions about the subject matter of the field. In fact, it is convenient to think in terms of a double assessment, corresponding to the scientist's impersonal and personal goals. The impersonal assessment consists of a partially ordered set of questions. Questions in this set are counted as significant for the development of the field, and, insofar as the

31. See, for example, Gould and Lewontin (1979).

scientist honors the impersonal goal of devoting his labor to the advancement of the field, he commits himself to working on these questions and to assessing their importance in the manner delineated in the partial ordering. However, the scientist has concerns of his own, matters that particularly interest him or to which he believes that his talents are especially suited. Such concerns generate a personal assessment, which can also be regarded as a partial ordering of a set of questions, the questions that the scientist views as significant for him to try to answer. We should not expect that these assessments match either in the membership of the sets or in terms of the orderings.

The third component of the practice of a scientist at a time is the set of statements about the subject matter of the field to which the scientist would assent at that time. This component recalls the traditional philosophical representation of the state of a science at a time. But there are some significant differences. First, I shall not insist that this set contain any general (lawlike) statements. As our study of evolutionary biology revealed, there are large subfields within biology that appear to lack any distinctive general principles of their own.[32] The generality that they achieve seems to be attained through the repeated use of explanatory patterns. Second, I shall not assume that the set is deductively closed—indeed, it may well be finite. A straightforward way to resist the demand for deductive closure is to recognize that the set of statements that scientists accept about the subject matter of their field might be subtly inconsistent (certainly, if one includes in the set reflective judgments prompted by epistemic modesty, such as the judgment that one of one's beliefs about nature is false, there will be an inconsistency). If the set is inconsistent then every statement of the language belongs to its set of consequences, so that if it were deductively closed the scientist would be committed to accepting all statements of the language.[33]

Yet, even though the set of accepted statements does not have the structure of an axiomatic deductive system, it nevertheless has an important structure. The scientist is disposed to take some statements he accepts as providing his reasons for accepting others: the psychological processes through which he would currently sustain his beliefs in the latter statements involve the activation of inferential propensities that start with beliefs in the former statements. The justificatory structure latent in the scientist's psychological life at any time may be radically incomplete. With respect to many statements the scientist may be able to say little more than "I remember it but I don't know where I learned it" or "I think I got that from an article" or "I learned that in graduate school" or "That was demonstrated by Crumkopf and Blitherstein." Sometimes a statement may be defended with the claim that some entity, state, or process was observed. Frequently, however, only the summary of the data, the product of much analysis, will be retained. The scientist supports her statement by declaring, "I (we) did an experiment that showed

---

32. I have also argued this point for the case of genetics in my (1984) and (1989).

33. For elaboration of this argument and some morals about requirements on human rationality, see (Stich 1990).

that," but, even if the circumstances of the experiment can be identified and the lab notebook is extant, the scientist will still have to rely on the authority of his own earlier recording of measurements and his own prior analysis of the data (unless, of course, he chooses to revise the latter).[34]

9. Explanatory Schemata

A crucial part of a scientist's practice consists in her commitment to ways of explaining the phenomena. Faced with explanation-seeking questions, the scientist is disposed to produce texts instantiating particular patterns. These patterns, or schemata, although they are not likely to be formulated by scientists themselves (at least not in the style of my reconstruction of Darwinian patterns), are implicit in scientific practice, and I would expect that practitioners could recognize them as underlying their own explanations.

What is a pattern? Let me give a general account of the conception I introduced with examples in Chapter 2. A *schematic sentence* is an expression obtained by replacing some, but not necessarily all, the nonlogical expressions occurring in a sentence with dummy letters. Thus, starting with the sentence "Organisms homozygous for the sickling allele develop sickle-cell anemia," we can generate a number of schematic sentences: for example, "Organisms homozygous for A develop P" and "For all x, if x is O and A then x is P" (the latter being the kind of pattern of interest to logicians, in which *all* the nonlogical vocabulary gives way to dummy letters). A set of *filling instructions* for a schematic sentence is a set of directions for replacing the dummy letters of the schematic sentence, such that, for each dummy letter, there is a direction that tells us how it should be replaced. For the schematic sentence "Organisms homozygous for A develop P," the filling instructions might specify that A be replaced by the name of an allele and P by the name of a phenotypic trait. A *schematic argument* is a sequence of schematic sentences. A *classification* for a schematic argument is a set of statements describing the inferential characteristics of the schematic argument: it tells us which terms of the sequence are to be regarded as premises, which are inferred from which, what rules of inference are used, and so forth. Finally, a *general argument pattern* is a triple consisting of a schematic argument, a set of sets

34. Large parts of the set of accepted statements in a scientist's practice are there through doxastic inertia. If the scientist now became committed to some Cartesian enterprise of overhauling his system of beliefs and resolved to throw out everything he could not justify *as of now*, then he would be left with a severely mutilated corpus. Gilbert Harman (1987) argues persuasively that there is an important difference between realizing that your old reasons are poor and realizing that you can no longer formulate what strike you as convincing reasons: in the former, but not the latter, instance there are grounds for undertaking belief revision. This is a helpful contrast, but it is only the beginning of the investigation of the desirability of doxastic inertia. As one might expect, the presence of some self-critical people within a scientific community, people who are prepared to give more weight to the traditional ideal of abandoning what is currently unjustified, can sometimes prove valuable for the advancement of the community project.

of filling instructions, one for each term of the schematic argument, and a classification for the schematic argument.

A particular derivation, the sequence of sentences and formulae found in a scientific work for example, instantiates a general argument pattern just in case (i) the derivation has the same number of terms as the schematic argument of the general argument pattern, (ii) each sentence or formula in the derivation can be obtained from the corresponding schematic sentence in accordance with the filling instructions for that schematic sentence, (iii) the terms of the derivation have the properties assigned by the classification to corresponding members of the schematic argument. The Darwinian patterns of the last chapter are general argument patterns in the sense introduced here, and the explanations produced by Darwin, Huxley, Fisher, Wright, Dobzhansky, Simpson, and other luminaries in the Darwinian tradition instantiate these patterns in the fashion just defined. Consider one of the simplest of my examples:

COMMON DESCENT

Question: Why do the members of G, G' share P?

Answer:

(1) G, G' are descended from a common ancestor G_0.

(2) G_0 members had P.

(3) P is heritable.

(4)   No factors intervened to modify P along the G_0–G, G_0–G' sequences.

Therefore (5) Members of G and G' have P.

This consists of five schematic sentences. We could equip it with Filling Instructions by requiring that G, G', G_0 be replaced by names of groups of organisms, and that P be replaced by the name of a trait of organisms. Similarly, the classification would declare that (1)–(4) are premises and that (5) is deduced from them.

Derivations may be similar in terms of either their logical structure or the nonlogical vocabulary they employ at corresponding places. The notion of a general argument pattern allows us to express the idea that derivations similar in either of these ways share a common pattern. However, similarity is a matter of degree. At one extreme, a derivation is maximally similar to itself and to itself alone; at the other, any pair of arguments can be viewed as sharing a common pattern. To capture the notion that one pair of arguments is more similar than another pair, we need to recognize the fact that general argument patterns can demand more or less of their instantiations. If a pattern sets conditions on instantiations that are more difficult to satisfy than those set by another pattern, then I shall say that the former pattern is more *stringent* than the latter.

The stringency of an argument pattern is determined in part by the classification, which identifies a logical structure that instantiations must exhibit, and in part by the nature of the schematic sentences and the filling instructions, which jointly demand that instantiations should have common nonlogical vocabulary at certain places. If both requirements are relaxed completely, then the notion of pattern degenerates so as to admit of *any* argument. If both conditions are simultaneously made as strict as possible then we obtain another degenerate case, a "pattern" which is its own unique instantiation. Relaxing the demands on nonlogical vocabulary (the conditions set by the schematic sentences and the filling instructions) while also requiring that the classification determine the precise inferential status of each term in the schematic argument yields the logician's notion of pattern.

My claim that a scientist is committed to a set of explanatory patterns is a formal (perhaps tedious) way of capturing what I view as an important idea. The scientist studying a particular aspect of nature has a conception of how the phenomena depend on one another. In old-fashioned terminology, we might say that the scientist has a view of part of the "order of being."[35] The task of explanation involves showing how particular phenomena fit within the order of being. Views of the order of being can be identified with commitments to using particular explanatory schemata. We have already seen how decisive transformations in science (Darwin's innovation) stem from changes in commitment to explanatory schemata. Chapter 4 will make plain the ways in which relations among successive sets of consensus explanatory schemata are critical for understanding scientific progress.

10. Other Components

Human knowers are not lone knowers: we can and do assign authority to others and base our judgments on what the authorities say. Complex and important projects, not only those of "big science" but most significant endeavors that have been undertaken in the history of science, would be impossible unless the protagonists were prepared to take some matters on trust.[36]

---

35. The phrase is Aristotle's, from whose *Posterior Analytics* stem the two main conceptions of scientific theory: that of a deductive system and that of a presentation of the dependencies in nature. My own emphasis, quite evidently, is on the latter conception of theorizing. Similar views about scientific theorizing are to be found in Kant (see Kitcher 1983b, 1986), and in the contemporary writings of Thomas Kuhn (1962/70), Sylvain Bromberger (1963, 1966), and Stephen Toulmin (1961). I have presented my conception of explanation in greatest detail in Kitcher (1989).

36. As I remarked earlier, there are varieties of trust. Each of us begins, in childhood, by absorbing the lore of our culture, and we do so, initially, without question. Later, apprentice scientists may submit themselves voluntarily to training by people whom the community regards as authoritative on certain matters. Still later, individual scientists identify certain people within the community as authoritative on issues that are not agreed on throughout the community. In each of these contexts there is the obvious worry that trust may inculcate errors that cannot later be removed through critical evaluation.

Recognizing the omnipresence of deference to others seems to me an important break with the Cartesian approach to epistemology that emphasizes the individual's responsibility for taking charge of his beliefs. The contemporary philosopher who has broken most decisively with Cartesian individualism is W. V. Quine (see especially his 1960, 1970, and

Recognition of our assignments of authority should serve as the beginning of epistemological inquiries. I shall prepare the way for those inquiries (to be taken up in Chapter 8) by explicitly including scientists' attributions of authority as a component of their practices. The component consists of a commitment to regarding certain people or groups as authoritative with respect to certain matters and a set of criteria for assigning authority.

Just as there are exemplary authorities, so too there are paradigm experiments (and observations) and paradigm instruments. Part of the practice of a scientist is a commitment to regarding certain kinds of experiments (or experimental setups) as revealing, certain kinds of instruments as reliable. As with the particular judgments about authorities, here too the claims about individual cases will be backed by general criteria. An instrument may be defended by citing its previous track record, or by offering an explanation of how it works. An experiment or observation may be supported by appealing to ideas about research design. I shall suppose that both the particular judgments and the principles that would be invoked in justification form a component of the scientist's practice, the *experimental lore*.

The experimental lore of an individual scientist reflects her propensities for interacting with nature in various ways. Differences in this component will generate three kinds of differences in scientific behavior: (i) a disposition to set up certain kinds of experimental or observational situations and to avoid others; (ii) a disposition to act in particular ways, rather than in others, when those situations obtain, and (iii) a disposition to respond in specific ways to the stimuli that result in those situations.

The final component of practice contains the scientist's views about proper inferential procedure in science generally and in the particular area of study. According to the view of cognition I have sketched, inferential propensities are stored in procedural memory. But the scientist may also retain, in declarative memory, judgments about the merits of various types of inferences and arguments, including both general and particular judgments. Moreover, it is possible that reflection of an explicitly methodological sort—the retrieval of certain statements from declarative memory—may activate inferential propensities that would not otherwise have been employed. Methodological ruminations may affect the inferences that are actually carried out.

It is just as important to distinguish between scientists' views about method and the rules (if any) that describe the proper way for them to proceed (*correct* method) as it is to differentiate their ideas about nature from the truth. There is little doubt about the fact that, at different stages in the history of science, different individual scientists have expressed different views on matters methodological. Does it follow from this that there is no set of methodological rules that applies to all sciences at all times? No. If we think of a set of

---

1974). Interestingly, the central point was never expressed better than by the onetime logical positivist Otto Neurath in the passage that Quine uses at the beginning of his (1960): "We are like sailors who must always rebuild our ship on the open sea without ever anchoring in dry dock."

methodological rules as formulating the optimal ways for scientists to form conclusions in certain contexts, then the mere fact that people have sometimes disagreed on what is best does not show that there is no optimum. I shall formulate my own conception of the proper goal of methodology later (see Chapters 6 and 7), but I want to scotch at the outset the notion that universal methodology is doomed by variation in opinion.

The *methodological exemplars* are inferences that are positively or negatively appraised in statements to which the scientist would assent.[37] Thus, for Darwin, Lyell's inferences about the gradual formation of the World from the current geological data and Maxwell's inferences from the electromagnetic phenomena to the electromagnetic field equations serve as positive methodological exemplars. Of course, Darwin also subscribes to a general *methodological principle,* claiming that what makes these inferences good is the ability of the conclusion to unify the phenomena reported in the premises— and he then defends his own appeals to common descent and natural selection by maintaining that they also accord with the methodological principle.

Methodological exemplars exhibit instances of reasoning that the scientist takes to be worthy of emulation or ripe for avoidance. Her methodological principles attempt to specify what makes the exemplars so virtuous or so disreputable. We should *not* assume that the scientist's methodology is correct, or that the principles she espouses genuinely cover her exemplars, or that either the exemplars or the principles conform to the inferential propensities that are activated in her own reasoning. Methodology may be a matter of sabbath observance.

The discussion of methodology brings into the open a general point, aspects of which have been touched on in earlier sections. There may be tensions within components of practice or among components of practice. The criteria invoked to justify paradigm experiments, instruments, or authorities may not support the items singled out for approval. Moreover, as I have noted in passing, the set of accepted statements may be inconsistent and there can be important differences between the impersonal assessment of significant questions and the personal assessment.

Besides such discrepancies that arise within components, different parts of a practice may be at odds with one another. The openness of a practice is a spur to further research. Scientists try not only to tackle unanswered significant questions, but to resolve tensions that they perceive within their practices.[38]

---

37. The positive methodological exemplars, the pieces of scientific work/inference deemed worthy of imitation, are part of what Larry Laudan calls "the canon" (1984, 1988), that set of past achievements to which current thinking about methodology or axiology (study of aims) must do justice. Positive methodological exemplars are used by many philosophers who turn to history for a guide to methodology. See, for example (Lakatos 1971, Worrall 1976).

38. These kinds of tensions give rise to what Laudan (1977) calls "conceptual problems." I believe that recognition of the components of a practice and the potential tensions among them can easily supply a useful taxonomy of the types of "conceptual problems" that

11. Consensus Practice

Individual members of the same scientific community differ in their practices. When disputes are resolved, when all the variants but one in some part of some component of individual practice are effectively eliminated, there is a change in *consensus practice*. If we are to understand the progress of science, we need to be able to articulate the relations among successive consensus practices.

What is a consensus practice? First, it is a practice with components like those of individual practices. But there are differences. For example, the assessment of significance of questions is impersonal, an ordering of the questions that are significant for the field, for the obvious reason that the consensus practice gives no weight to the personal projects and interests of particular individuals (unless, as we shall see, those individuals are licensed to speak for the field, in which case their interests become those of the field). Such small differences aside, it is easy to specify the *kind* of thing a consensus practice is: it is constituted by a language; an (impersonal) assessment of significant questions; a set of accepted statements with a (partial) justificatory structure; a set of explanatory schemata; a set of paradigms of authority and criteria for identifying authorities; a set of exemplary experiments, observations, and instruments and justificatory criteria; and, finally, a set of methodological exemplars and methodological principles.[39]

There are many things of this kind around in the community, many individual practices. How does the consensus practice relate to them? An obvious suggestion is that the consensus practice of a community is what all members of the community share. This would allow us to endorse the point (plain from the history of evolutionary biology) that each of the practitioners goes beyond what belongs to consensus, so that in understanding the decision making of individual scientists it will *not* do to conceive them as revising or acting on a state of knowledge which is consensus and nothing but consensus. So far, so good.

One obvious feature of the history of evolutionary biology (and on this count we can be confident that the example is typical) is that at each stage the community of biologists has internal structure. All members of the community share certain claims and commitments, but there are subgroups with richer sets of shared claims and commitments, *subcommunities*. For certain kinds of issues, particular subcommunities are considered authoritative: thus, if every

---

scientists face. For some exploration of this in the particular case of mathematics, see chapters 8–10 of (Kitcher 1983a).

39. There are obvious links between my notion of consensus practice and the Kuhnian concept of paradigm. Like Kuhn I take what is shared among scientists to be far more than the acceptance of a set of statements. However, Kuhnian paradigms are intended *both* to offer a richer description of the state of a science at a time *and* to divide the history of a scientific discipline into segments, such that there are significant epistemological differences between the course of science within a segment and the intersegment transitions. My approach bears no commitment to this second function. See (Kitcher 1983a chapter 7).

member of a subcommunity agrees that a particular question has a particular answer, then members of the entire community will agree that that answer is correct, and they will agree in virtue of the concurrence of opinion in that subcommunity. Not every member of the community of biologists is aware of the exact relationships of the species involved in the reptile-mammal transition. Not every member of the community has the discriminatory skills required to identify populations of organisms in the wild and to make the observations necessary for measuring intrapopulation variation. Each, however, is prepared to accept that the subcommunity of vertebrate paleontologists is authoritative on the question of the reptile-mammal transition so that if there is a thesis about the relationships shared by all vertebrate paleontologists, then it is part of the *virtual* consensus of the entire community. Similarly, the experimental lore of ecological geneticists is recognized by other members of the community of evolutionary biologists as constituting a special ability.

Moreover, the virtual consensus may be enriched by *deferred authority*. Perhaps not all vertebrate paleontologists agree on the reptile-mammal transition. However, if there is a subset of the paleontological community that is recognized by all members of the wider biological community as equipped to pick out the experts on particular paleontological issues (including the reptile-mammal transition) and if all those dubbed as expert by any member of this subset would agree on some claim about the reptile-mammal transition, then, by deferred authority, that claim belongs to the virtual consensus.

The intuitive idea behind virtual consensus is simple. Scientists do not clutter their heads with the details of other people's subspecialties. When they need information, they go to people who are regarded as authoritative with respect to the issue involved. If everyone agrees, then they accept that judgment. Even when there is not complete agreement, outsiders may still accept a resolution of the issue if they are prepared to defer to those appointed as experts by the Solomons of the field. Virtual consensus contains anything that a scientist could reach by following the judgments of a chain of authorities. Of course, a subcommunity may be in agreement about something with respect to which it is not granted authority by members of the wider community. In this instance, the agreement does not form part of the communitywide virtual consensus.

The *consensus practice* of a community at a given time is thus represented by (i) the *core consensus,* the elements of individual practice common to the individual practices of all members of the community, (ii) the *acknowledgments of authority* (themselves parts of individual practice) shared by all members of the community (including, perhaps, criteria for granting deferred authority); (iii) an organization of the community into subcommunities, resulting from (ii), with particular subcommunities recognized as responsible for and authoritative over particular types of issues, (iv) a *virtual consensus,* generated from (i) by the incorporation of parts of the consensus practice of subcommunities in accordance with the relations delineated in (ii) and (iii). At the widest level of all, the community itself may be constituted by the recognition of it as authoritative with respect to a certain range of issues. As I argued in the last chapter, one of Darwin's accomplishments was to convince others that the community of naturalists should be authoritative on questions of the origins of living things.

12. Changing Perspective

Legend conceives the growth of knowledge in terms of theories. More recent studies of science, inspired by reflection on the history of science, discuss paradigms (Kuhn), research programmes (Lakatos), research traditions (Laudan), domains (Shapere), and so forth. Although I owe debts to the insights of these writers, I have tried to build up a different type of framework. This is not, I hope, simply caprice. It is born of conviction that even the usual refined substitutes for Legend's blanket notion of theory are at a great remove from scientific work, so that the game of finding paradigms, protective belts, or research traditions in the actual course of scientific events becomes highly arbitrary and often unprofitable. Moreover, traditional conceptions of the units of change seem to me to confuse considerations from the levels I have distinguished, running together things that belong to the daily psychological lives of individuals, things that count among their more stable commitments, and things that are the property of the community rather than of any single member.

Conflation is most evident in slides between the individual and the community level. Instead of a population of varying individuals, whose practices are modified through interactions with one another and with nature, philosophers have usually treated the community as if it were a single knower whose initial state is *something like* consensus practice (or, sometimes, a probability distribution over the statements in the language of consensus practice in which very high probabilities are assigned to statements occurring in some tidied version of the set of accepted statements). The task is to specify principles that will revise this state in the light of experience (conceived, perhaps, as the insertion of "observation sentences" with high probabilities) so as to promote the acceptance of true statements, the solution of puzzles, or the uncovering of new corroborated empirical content. Formulating the issue in this way, certain kinds of problems are automatically excluded. What differences does the cognitive variation among individual scientists make to the growth of science? Is it possible that the practice of relying on authority could make it impossible for certain kinds of erroneous beliefs to be corrected? What systems of social rules for training and for consensus formation would promote a progressive sequence of practices? Can constraints on the recognition of novel evidence be overcome by multiplying observers? And—most generally—do the actual institutions of science, involving relative uniformity of training, division of labor, limited replication of results, deferral to the authority of others, hierarchies of power, and so forth, actually work? Do they advance or retard our cognitive goals?

Before we can try to answer these questions, we need an account of what those goals might be—a view of what constitutes scientific progress.

4

Varieties of Progress

1. A Problem Reposed

In conceiving of science as progressive we envisage it as a sequence of consensus practices that get better and better with time. Improvement need not be constant. Like the fortunes of a firm, the qualities of consensus practices may fluctuate. But, if a scientific field is making progress, there should be a general upward trend, comparable to the generally rising graph of profits that the firm's president proudly displays at board meetings.

The challenge is, of course, to say what "better and better" means in the scientific case, what corresponds to the accumulating dollars, yen, marks, etc. It is important to be clear from the start that the *first* task is to pinpoint those relations that should obtain among the members of a sequence of consensus practices if that sequence is to be progressive. To ask *how we know* that a field of science is making progress is to pose a separate question. There is an important connection between the questions: a definition of progress would be of little value if there were good reasons for thinking that it is impossible for us ever to know that we are making progress in the specified sense. Despite the connection, we should not confuse the issues, for example by lambasting a proposal for defining the notion of progress with the complaint that it has not yet been explained how we know that we are making progress. In my treatment, the definition will come first (this chapter). A response to various kinds of worries about concepts that I use will be presented later (Chapter 5).

Because one species of concern is so prevalent, it is worth forestalling related criticisms before we begin. I shall sometimes make free use of the concept of truth, explicating certain types of scientific progress as involving the attainment of truth. Moreover, I shall make judgments about the record of various scientists, past and present, in attaining truth, judgments based on the understanding of nature furnished by our current science. If the uses I make of the concept of truth are incoherent, then my account of progress will be wrong. Similarly, if the judgments about past and present attainment of truth that I make are incorrect or unjustified, then my errors will infect either my analysis of scientific progress or my claims about the ways in which science has made progress. Thus there are important objections that I shall ultimately have to overcome. However, before those objections can be properly posed and properly addressed, the account of progress must first be presented, so that it can become clear just where and to what extent I rely on conceptions that others find dubious. In Chapter 5 I shall try to give critical voices the hearing they deserve. Meanwhile, I hope that my (wild? naive? optimistic? flagrant?) claims about truth in this chapter can be read with patience by those who are moved by the criticisms, as I formulate a position whose coherence and tenability can later be debated.

Let me turn now to a second qualification. Sometimes in the history of science, fields split, merge, or give birth to hybrid progeny. In such cases the community relations within science (as a whole) are altered, and there are new consensus practices that have some kinship with more than one prior consensus practice, or old consensus practices that bear important relations to more than one later consensus practice. For most of this chapter, I shall ignore the special complications of such cases. I shall consider progress within a field, supposing that there is a single sequence of communities, uniquely determined by the fact that some members of an earlier community are veterans in the immediate successor community, and that we examine the relations among the consensus practices of the communities in this sequence.

We judge the progressiveness of a sequence of practices by thinking about the pairwise relations between practices. Just as we evaluate the fortunes of a firm over a period of time by looking at the returns in successive time segments, so too with fields of science. Things are looking up if for *most* pairs of adjacent time segments the later returns are greater than the earlier returns and if any subsequence in which the returns successively diminish is followed by a subsequence in whose final segment the returns surpass any level previously attained. Similarly, we can count a sequence of practices as progressive if *most* members of the sequence are progressive with respect to their predecessor and if any nonprogressive subsequences are followed by a subsequence whose final member is progressive with respect to any earlier practice in the sequence. Obviously, more determinate conceptions can be introduced by eliminating the vague reference to "most" practices in the sequence. For example, we can define *strict* progress in a sequence of practices by demanding that each successor practice in the sequence be progressive with respect to its predecessor. A sequence may be said to be weakly progressive just in case its final member is progressive with respect to its first member.[1] The fundamental point, however, is that any of these ways of characterizing a notion of progressiveness for sequences is dependent on an antecedent notion of progressiveness as a binary relation among practices.

Because practices are multidimensional it is possible that a change from P_1 to P_2 should be progressive along some dimensions but not along others. I shall not suppose that there is some way in which the dimensions are weighted so as to yield an overall measure of progress. Instead, I shall distinguish varieties of progress. Combining this idea with the discussion of the last

1. I am grateful to Kim Sterelny for suggesting this formulation.

paragraph, let us say that the sequence of practices $P_1, \ldots, P_n$ is *broadly progressive* just in case for every pair of adjacent members there is a component of practice with respect to which the change from the earlier to the later is progressive *and* the change from P_1 to P_n is progressive with respect to every component of practice.

The fundamental project, then, is to define a family of relations that hold between two practices P_i and P_j when, for some component of practice, the change from P_i to P_j would be progressive. Our definitions should be subject to criteria of adequacy in that they need to make it clear how achieving a progressive sequence of practices would realize (or realize more closely) our goals. The account that follows will presuppose that there are goals for the project of inquiry that all people share—or ought to share. The varieties of progress I describe will be understood in terms of the greater achievement of these goals.

In the terms of the last chapter, the goals in question are impersonal. Some are epistemic, others nonepistemic. We need a specification of impersonal goals for science, goals that can ultimately be defended as worthy of universal endorsement. On the basis of this, we must identify binary relations of progressiveness among practices, which hold with respect to the various components of practice. Third, we should be able to show that when the later practices of fields of science stand in the relations of progressiveness to earlier practices, then the fields in question are realizing the shared impersonal goals (or realizing them more fully).

2. The Practical and the Cognitive

One theme recurs in the history of thinking about the goals of science: science ought to contribute to "the relief of man's estate," it should enable us to control nature—or perhaps, where we cannot control, to predict, and so adjust our behavior to an uncooperative world—it should supply the means for improving the quality and duration of human lives, and so forth. At first, this conception of science as promoting impersonal nonepistemic goals looks relatively easy to understand. Suppose that, at some particular time, we have a practical project of doing something, and that this project is stalled because we cannot answer a particular question or make a particular device. If the development of science enables us, at some later time, to answer the question or to make the device, then it appears that we have made *practical progress*.

Unfortunately, precisely because our ideas about what we want to achieve may be modified in the wake of a scientific accomplishment, the notion of practical progress proves far more difficult than we might have thought. I believe that we can only make sense of it in the light of a very general account of human flourishing, one that lies beyond the scope of this book. By contrast, the notion of cognitive progress, which initially appears much more ethereal and elusive, can be approached far more easily because of the relative narrowness of the set of impersonal epistemic goals.[2] Although the task of providing an account of practical progressiveness is an important one, it should follow the more tractable project of understanding cognitive progress.

3. Cognitive Progress: An Overview

The most obvious pure epistemic goal is truth. Indeed, talk of truth—or approximation to truth[3]—has dominated philosophical discussions of scientific

---

2. Thus we can conceive of ourselves as recognizing the value of a large number of fundamental impersonal goals, mostly nonepistemic. Different people may give very different weights to these goals. Moreover, there may be substantial disagreement with respect to the weights assigned to the set of epistemic goals vis-à-vis the nonepistemic complement. All this is compatible with a very small degree of variation concerning the *relative* weights of the epistemic goals. My picture of the situation is illustrated by the following (obviously artificial) suggestion. Assume that there are two fundamental epistemic goals E_1 and E_2, and a much larger set of fundamental (nonepistemic) practical goals $P_1, \ldots, P_n$. Each person's conception of value is given by an assignment of weights, $u_1, u_2, w_1, \ldots, w_n$. I imagine that

(i) The *absolute* values of all the weights are variable from person to person.
(ii) The *relative* values of the w_i vary from person to person.
(iii) The *relative* values of the u_i are constant: i.e., u_1/u_2 takes some value k which is the same for (almost) all people.

To sum up this idea in a simple way, we may suppose that people differ in their practical values, differ in their understanding of the value of the epistemic, but, insofar as they commit themselves to any epistemic projects, share the same fundamental value system. Here I generalize the Kuhnian idea (1962/70 42, 1977 322) that scientists are bound together by shared values, by proposing that, in our dedication to inquiry, we endorse a shared system of fundamental goals. (To repeat, defense of this apparently optimistic view will come in Chapter 5.) Note that my supposition would automatically be correct if, instead of two fundamental epistemic goals, E_1 and E_2, there were just one.

3. The core of the problem originally posed by David Miller (1974) is this. Consider two theories (axiomatic deductive systems) T_1 and T_2. Let C_{T1} and C_{T2} be the sets of true members of T_1 and T_2, F_{T1} and F_{T2} the sets of false members of T_1 and T_2. According to the approach taken by Popper, which has an obvious motivation, we can count T_1 as having greater verisimilitude than T_2 provided that C_{T1} includes C_{T2} and F_{T2} includes F_{T1} with at least one of the inclusions being proper (and similarly, with substitution of indices for T_2's having greater verisimilitude than T_1). Assume now that T_1 neither includes nor is included in T_2. Then there are a statement s that is in T_1 but not T_2 and a statement r that is in T_2 but not T_1. Consider cases:

(a) Both s and r are true: then s is in C_{T1} but not C_{T2} and r is in C_{T2} but not C_{T1}; so neither does C_{T1} include C_{T2} nor does C_{T2} include C_{T1}; hence, by Popper's criterion, the theories are incomparable.
(b) s is true and r is false: then s is in C_{T1} but not C_{T2}, so C_{T2} cannot include C_{T1}; let f be any statement in F_{T1} (there must be such statements unless we have the trivial case in which all of T_1 is true); then $(s \& f)$ is in F_{T1} but not in F_{T2}; hence F_{T2} cannot include F_{T1}; it follows that neither theory can have greater verisimilitude.
(c) s is false and r is true: r is in C_{T2} but not in C_{T1}; let f be any statement in F_{T2}; then $(r \& f)$ is in F_{T2} but not in F_{T1}; so it cannot be that the required inclusions hold in

progress. But, in my judgment, truth is not the important part of the story. Truth is very easy to get.[4] Careful observation and judicious reporting will enable you to expand the number of truths you believe. Once you have some truths, simple logical, mathematical, and statistical exercises will enable you to acquire lots more. Tacking truths together is something any hack can do (see Goldman 1986 123, Levi 1982 47, and, for the locus classicus of the point, Popper 1959 chapter 5). The trouble is that most of the truths that can be acquired in these ways are boring. Nobody is interested in the minutiae of the shapes and colors of the objects in your vicinity, the temperature fluctuations in your microenvironment, the infinite number of disjunctions you can generate with your favorite true statement as one disjunct, or the probabilities of the events in the many chance setups you can contrive with objects in your vicinity. What we want is *significant* truth. Perhaps, as I shall suggest later (Section 8), what we want is significance and *not* truth.

Scientists in the tradition that extends beyond the seventeenth century to the ancient Greeks have been moved by the impersonal epistemic aim of fathoming the structure of the world. In less aggressively realist language, what they have wanted to do (as a community) is to organize our experience of nature. Seventeenth century thinkers such as Boyle and Bacon attributed special significance to this project because, seeing nature as God's Creation, to expose the structure of nature was to recognize the divine intentions. That further link, useful, possibly even essential, though it may have been in showing how science contributes to our overall conception of what is valuable, can be discarded if we merely pursue the internal project of understanding our impersonal epistemic goals. Our task is to make more precise the murky notion of uncovering the structure of nature (or of organizing experience), of discovering, insofar as we are able, how the world works.[5]

Historians have been struck by the apparent variability of the goals professed by different scientists, working in different communities at different times. But I want to start from the opposite end. It is hardly surprising that

---

either direction, and, again, T_1 and T_2 are incomparable.

(d) Both s and r are false: then s is in F_{T1} but not F_{T2}, and r is in F_{T2} but not F_{T1}; neither theory can have greater verisimilitude.

As Miller saw, these considerations show that Popper's approach to verisimilitude cannot be employed to compare theories (axiomatic deductive systems) except under very special circumstances. Inclusion relations among sets of true and false consequences are not sufficiently fine-grained to do the work required of them. But, since the sets of consequences in question are typically infinite, it is not clear what other kinds of comparison are possible. Hence the technical difficulties.

4. Again, I ignore those skeptical complaints that insist that truth is impossible to get, postponing discussion of them until Chapter 5. However, even the skeptic should agree that *if* truth is obtainable at all, *boring* truth is relatively easily generated.

5. As I shall suggest in response to claims that the aims of science vary from context to context, the reference to our limitations is important. Different conceptions of what we can hope to achieve yield different *derivative* goals. See the discussion of (Laudan 1984) in Chapter 5.

we can find considerable community of goals among contemporary evolutionary theorists, or that we can take Dobzhansky and Darwin (to cite two major figures) to share a conception of the aims of biological science. When we expand to a triumvirate, including Owen with Dobzhansky and Darwin, some of the kinship is lost: there are formulations of the goals of biology endorsed by Darwin and Dobzhansky, to which Owen would not subscribe. Nevertheless, at a more fundamental level, Owen shares their common vision. Like them, he contends that biology should explain the diversity of living things and trace the patterns in that diversity. He differs in his understanding of how this goal is to be realized, dissenting from the Darwinian conception of history as the key to life's diversity.

We can thus expand the group of those who share goals for science, including not only those in the Darwinian tradition but some of their predecessors. How far can we go? Can we include Goethe and Linnaeus, Buffon and Ray, Theophrastus and Aristotle? Given our recognition of what Darwin and Owen share, I see no reason why not. Moreover, even if the attempt does break down at some point—say at the division between modern and premodern science—there is surely still value in articulating the goals of modern science, and understanding its conception of progress.

I shall start with two varieties of progress that have received little attention in the literature, but that will be fundamental to my account of significance. The next two sections will be devoted to articulating the notions of *conceptual progress* and *explanatory progress*. Before we take up the details, it is useful to have a map of the overall view.

As science proceeds we become better able to conceptualize and categorize our experience. We replace primitive conceptions of what belongs with what, what things are akin, by improved ideas about natural groupings. Our language develops so that we are able to refer to natural kinds and to specify our references descriptively. In addition, we are able to construct a hierarchy of nature, a picture of what depends on what. Against the background of our categories and our hierarchy, we are able to pose significant questions, questions that demand more detail about our picture of the world. Significant statements answer significant questions. Significant experiments help us to resolve significant questions. Significant instruments enable us to perform significant experiments. Methodological improvements help us to learn better how to learn, to improve our evaluation of significant statements. Thus, to sum up the general approach, significance is derivative from the background project of ordering nature, a project that is articulated in our attempts to conceptualize and to explain.

4. Conceptual Progress

Conceptual progress is made when we adjust the boundaries of our categories to conform to kinds and when we are able to provide more adequate specifications of our referents.[6] Striking examples come from the history of all sciences: 'planet,' 'electrical attraction,' 'molecule,' 'acid,' 'gene,' 'homology,' 'Down's syndrome' are all terms for which faulty modes of reference have been improved. Consider, for example, pre-Copernican usages of 'planet.' What set of objects was singled out? One answer is that, for the pre-Copernican, as for us, planets are Venus, Mars, and things like that—in which case the planets would include the Earth. Another is to declare that the planets are specified as those heavenly bodies that are sometimes observed to give rise to erratic motions against the sphere of fixed stars (the "wanderers")— in which case the planets would be Mercury, Venus, Mars, Jupiter, and Saturn. Or, allowing the specification to take account of possible perceptual limitations, we might declare that the planets comprise not only those observed to "wander" but those we would observe "wandering" if we had keener senses, that is, all the things we would count as planets of the solar system except the Earth. If we adopt the first approach to the reference of 'planet' then pre-Copernicans succeed in marking out a natural division, but they fail disastrously when they come to say what they are talking about. On either of the second approaches, they are fully aware of what they are referring to but the set of objects they single out is not a natural category. Either way, post-Copernicans do better. And so it goes, not only for this example but for the other terms on my brief list and for many more besides.

As I shall explain later, the answers of the last paragraph are not entirely adequate, but they enable us to appreciate the possibility of understanding conceptual change and conceptual progress in terms of shifts in mode of reference. The thesis that I shall defend in this section is that the conceptual shifts in science that have caused most attention (and that are supposed to be troublesome) can be understood, and understood as progressive, by recognizing them as involving improvements in the reference potentials of key terms. Recall from Section 7 of the last chapter that the reference potential of a term for a community is a compendium of the ways in which the references

6. Here I shall take over, uncritically for the time being, Plato's famous metaphor of "carving nature at the joints." So part of the task of fathoming the structure of nature will be seen as an exercise in anatomy, exposing the lineaments of the objective divisions of nature and showing which things belong together. A strong realism about the structuring of nature provides a simple way of developing the fundamental epistemic aim selected in the last section, but it brings well-known epistemological difficulties in its train. I shall suggest later (Chapter 5 Section 8) that there are other metaphysical options, which seek to explicate the notion of natural kind by focusing on the groups to which reference is made in laws or in explanatory schemata, which can be linked to the project of fathoming the structure of nature, and which can avoid some of the epistemological difficulties. As I have suggested elsewhere (Kitcher 1986, 1989) I believe that this latter approach to the metaphysical issues has important advantages. However, the treatment of many questions about progress can profitably be undertaken by leaving the notions of natural kind and objective dependency (see Section 5) temporarily unanalyzed, using them to explicate conceptions of progress, and only returning to metaphysics later. So this chapter will remain neutral between the options of strong realism and the type of weakened (Kantian) realism that I have previously espoused.

of tokens of that term can be fixed for members of the community. I shall illustrate how reference potentials may shift by analyzing an apparently problematic case, and then offer a general view about improvement of reference potentials.

Kuhn and Feyerabend achieved the important insight that different communities of scientists, working in the same field, may organize the aspects of nature that concern them in different ways (Kuhn 1962/70, 1983; Feyerabend 1963a, 1965, 1970). Different conceptualizations make differences to observation, inference, and explanation (see Chapter 7).[7] But, most famously, Kuhn and Feyerabend have suggested that the phenomenon of "conceptual incommensurability" that they identify engenders problems of communication and of reporting of evidence. I shall begin by reviewing one of Kuhn's favorite problematic cases, with a view both to showing that there is indeed an interesting conceptual shift and to revealing that it does not have the dire consequences for communication with which it is sometimes credited.

There is a very simple story about the chemical revolution of the eighteenth century. Priestley, so the story goes, employed a language containing terms— 'phlogiston,' 'principle'—that fail to refer. Lavoisier used a language containing expressions—'oxygen,' 'element'—that refer to kinds that Priestley could not have identified. So there is a conceptual advance between Priestley and Lavoisier, one that involves the replacement of expressions that fail to refer by genuinely referring expressions and the introduction of terms that single out kinds for the first time.

Kuhn saw that this story is extremely implausible. If Priestley's language was so inadequate, then it is hard to understand how his chemistry could ever have appeared successful and how he could have contributed to the development of Lavoisier's own ideas. Yet Priestley was a successful chemist, one who appeared to his contemporaries to be extending the range and power of the phlogiston theory, and he was responsible for the experiment in which he (and, following his lead, Lavoisier) isolated a new gas, the gas Lavoisier called "oxygen." At times Kuhn seems to want to solve the problem of Priestley's success by declaring that the terms we take to be nonreferring ('phlogiston' and so forth) really did refer, picking out constituents of the "different world" inhabited by Priestley and his fellow phlogistonians. Divided by the chemical revolution, Priestley and Lavoisier inhabit different worlds (or, to put it another way, their theories have different ontologies). Unfortunately, this strategy of accounting for Priestley's apparent success will only accommodate part of the historical evidence. As we insist on the "many worlds" approach, the revolutionary divide between Priestley and Lavoisier looks ever more difficult to bridge, and communication between the two men

---

7. Thus, in the writings of Kuhn (1962/1970), Feyerabend (1963a, 1965, 1970), and Hanson (1958) there are suggestions about how differences in conceptualization issue in differences in perception. For the moment I am solely concerned with the relations among scientific languages, and with the thesis of *conceptual* rather than *perceptual* incommensurability.

appears partial, at best.[8] How then do we explain the dialogue between them, the role that Priestley played in the genesis of Lavoisier's ideas, the assessments and evaluations of Lavoisier's proposals that occur in the writings of Priestley and his followers? Sensitive to the historical phenomena, Kuhn does not fully commit himself to the "many worlds" interpretation, offering instead a position whose ambiguities have excited much subsequent discussion.[9]

We have a puzzle. If the central terms of Priestley's theoretical language do not refer, then it seems impossible to ascribe to him the achievements warranted by the historical record.[10] If those terms do refer, then his "world" must contain entities that we no longer recognize. An adequate solution to the puzzle must (i) recognize Priestley's contributions to the development of chemistry, (ii) avoid populating nature with strange entities, (iii) specify the exact way in which Lavoisier made a conceptual advance.

Begin with a brief overview of features of the language of the phlogiston theory. This theory attempted to give an account of a number of chemical reactions and, in particular, it offered an explanation of processes of combustion. Substances which burn are rich in a "principle," *phlogiston,* which is imparted to the air in combustion. When we burn wood, for example, phlogiston is given to the air, leaving ash as a residue. Similarly, when a metal is heated, phlogiston is emitted, and we obtain the *calx* of the metal.

Champions of the phlogiston theory knew that, after a while, combustion in an enclosed space will cease. They explained this phenomenon by supposing that air has a limited capacity for absorbing phlogiston. By heating the red calx of mercury on its own, Priestley found that he could obtain the metal mercury, and a new kind of "air," which he called *dephlogisticated air.* (According to the phlogiston theory, the calx of mercury has been turned into the metal mercury by taking up phlogiston; since the phlogiston must have been taken from the air, the resultant air is dephlogisticated.) Dephlogisticated air supports combustion (and respiration) better than ordinary air—but this is only to be expected, since the removal of phlogiston from the air leaves the air with a greater capacity for absorbing phlogiston.

In the last decades of the eighteenth century, the phlogiston theorists became interested in the properties of a gas, which they obtained by pouring a strong acid (concentrated sulfuric acid, for example) over a metal or by passing steam over heated iron. They called the gas *inflammable air.*

8. See (Kuhn 1962/1970 149, Kuhn 1970 200, 201, 204). Perhaps the most radical statement of the "many worlds" position is at (Kuhn 1962/1970 102–103).

9. For Kuhn's own retrospective on his views, see his (1983). Discussions and critiques of his claims about conceptual incommensurability can be found in (Shapere 1964, 1965), (Scheffler 1967), (Kordig 1971), (Davidson 1973), (Field 1973), (Kitcher 1978), and (Devitt 1981). The present discussion closely follows the line of my earlier attempts in my (1978, 1982, 1983c).

10. Even if one takes Priestley's principal accomplishments to reside in his actions rather than in his words, there is still a problem in coordinating his dicta with his behavior. If we assume that his central terms do not refer, then he has to appear as serendipitous— producing a stream of babble while simultaneously doing things (isolating oxygen, synthesizing water) that somehow enable others to repeat his doings and discuss them.

If we compare the descriptions of these familiar reactions given by the phlogiston theory, and by modern elementary chemistry (which descends from the work of Lavoisier), some identifications are immediately suggested. We might naturally suppose that dephlogisticated air is oxygen and that inflammable air is hydrogen. I shall consider the merits of these identifications in a moment.

In presenting the puzzle of Priestley's language, I spoke vaguely of Priestley's "success" and his "contributions" to chemistry. Can we be more precise? One simple suggestion is that Priestley enunciated various *true statements* which had not previously been accepted.[11]

Priestley was not occupied with some work of unwitting fiction. Inside his misbegotten and inadequate language are some important new truths about chemical reactions, trying to get out. A false presupposition, the idea that something is *emitted* in combustion, infects most of the terminology. The natural approach to Priestley's language is to take the faulty central idea of the phlogiston theory, the idea that phlogiston is the substance emitted in combustion, as fixing the reference of 'phlogiston.' Assuming that some expressions used by proponents of the phlogiston theory (e.g., '*x* is emitted from *y*') are coreferential with their modern homonyms, and that compound expressions that embed nonreferential constituents are themselves nonreferring (e.g., that 'the substance obtained by removing *x* completely from *y*' refers only if *x* and *y* both refer), it follows that virtually none of Priestley's statements is true. For since the reference of 'phlogiston' is fixed through the description "the substance emitted in combustion," since 'emitted' means what we standardly take it to mean, and since nothing is *emitted* in combustion, 'phlogiston' fails to refer. Because the reference of other technical terms is fixed by descriptions that contain 'phlogiston'—dephlogisticated air is to be the substance obtained when phlogiston is removed from the air—those terms also fail to refer.[12] Hence the statements that talk about the properties of phlogiston, phlogisticated air, dephlogisticated air, and so forth, cannot be true. So it appears that we cannot honor the idea that Priestley and his followers advanced some new, true statements.

11. In the wake of the work of Kuhn and Feyerabend, such simple references to truth are likely to provoke mirth, scorn, or shock. It is well to remember that *part* of the reason for viewing these suggestions as simplistic depends on accepting Kuhnian or Feyerabendian theses about what occurs in episodes like the transition between Priestley and Lavoisier, already adopting one of their solutions to our puzzle. Thus a partial defense of our natural way of talking about past theorists as articulating truths is to show that we can accommodate the historical material while continuing to engage in that talk. Other parts of the defense will be supplied in Chapter 5.

12. This consequence strikes me as intuitive, but it does merit some discussion. We could defend it by supposing (following Frege) that an expression formed by extensionally embedding a nonreferring expression fails to refer or by adopting a principle of substitutability for nonreferring names: a complex expression formed by extensionally embedding a nonreferring expression preserves its reference under substitution of nonreferring expressions that belong to the same syntactic category. The general idea of substitutability principles for nonreferring expressions is due to Hartry Field.

Even a brief reading of the writings of the phlogistonians reinforces the idea of true doctrines trying to escape from flawed language. Consider Priestley's account of his first experience of breathing oxygen.

> My reader will not wonder, that, after having ascertained the superior good-ness of dephlogisticated air by mice living in it, and the other tests above mentioned, I should have had the curiosity to taste it myself. I have gratified that curiosity, by breathing it, . . . The feeling of it to my lungs was not sensibly different from that of common air; but I fancied that my breast felt peculiarly light and easy for some time afterwards. (Priestley 1775/1970 II 161–162).

Surely Priestley's token of 'dephlogisticated air' refers to the substance which he and the mice breathed—namely, oxygen.

Similarly, it seems that Cavendish's uses of the terms refer in the following passage:

> From the foregoing experiments it appears, that when a mixture of inflam-mable and dephlogisticated air is exploded in such proportion that the burnt air is not much phlogisticated, the condensed liquor contains a little acid which is always of the nitrous kind, whatever substance the dephlogisticated air is procured from; but if the proportion be such that the burnt air is almost entirely phlogisticated, the condensed liquor is not at all acid, but seems pure water, without any addition whatever. . . . (Cavendish 1783/1961 19)

We readily understand what prompted this description. Cavendish had per-formed a series of experiments in which samples of hydrogen and oxygen were "exploded" together; in some cases, the oxygen obtained was not en-tirely pure, and the small amount of nitrogen, mixed in it, participated in a reaction to form a small amount of nitric acid; when the sample of oxygen was pure there was no formation of nitric acid, and the only reaction was the combination of hydrogen and oxygen to form water. In reporting the exper-iments he had done, Cavendish used the terms of the phlogiston theory to refer to the substances on which he had performed the experiments: that is, he used 'inflammable air' to refer to hydrogen, 'dephlogisticated air' to refer to oxygen, and so forth.[13]

---

13. These claims and my kindred remarks about Priestley will surely jar some historical sensibilities. I am using the ideas of modern science to provide a picture of the relationships between Priestley and Cavendish and the world. Isn't this Whiggish, illegitimate, question begging? I reply that without some picture of the relationship between past figures and the world, history is impossible: we cannot frame any hypotheses about what their discourse means. Thus the strategy of agnosticism leaves the historian in the presence of a parade of uninterpretable symbols, a detached text without significance. So where should we turn for a view of nature that will enable us to understand the actions and reactions of the protagonists? Is there some *other* account that would serve us better than modern science? Why not the best—the account we take for granted in planning our own activities?

One modish answer is to insist that we adopt the "actor's categories." To this it might be pertinent to inquire why we should favor a discredited account of nature rather than the one which we typically adopt. But there is a deeper point. Without an account of the relationship between the actor and nature already in hand we cannot interpret the language of past science, cannot even formulate the actor's categories. I claim that the account I

Our problems in specifying what Priestley and Cavendish were talking about arise from the presupposition that there should be a uniform mode of reference for all tokens of a single type. Once we liberate ourselves from this presupposition, adopting the notion, introduced in the last chapter, that a scientific term (type) may have a heterogeneous reference potential, we can begin to make sense of the language of the phlogiston theory, recognizing the achievements of Priestley and Cavendish and pinpointing the inadequacies of their language.

Successful ascriptions of reference should accord with a principle that Richard Grandy dubs the "principle of humanity." The principle enjoins us to attribute to the speaker whom we are trying to understand a "pattern of relations among beliefs, desires and the world [which is] as similar to ours as possible" (Grandy 1973 443).[14] In deciding on the referent of a token, we must construct an explanation of its production. That explanation, and the hypothesis about reference we choose, should enable us to trace familiar connections among Priestley's beliefs and between his beliefs and entities in the world, ascribing dominant intentions that we would expect someone in his situation to have.

Let us expose the way in which this strategy (a strategy that I take to be tacitly employed by sensitive historians) works in the case at hand. How did Priestley come to use 'dephlogisticated air' to refer to oxygen? Priestley began his discussion by talking about various attempts he had made to remove phlogiston from the air. He then recorded the details of his experiments on the red calx of mercury from which he had liberated a "new air." After a number of mistaken attempts to identify the gas, he finally managed to describe it in the terminology of the phlogiston theory.

Being now fully satisfied with respect to the *nature* of this new species of air,

---

provide in this section, far from riding roughshod over the actor's categories, actually displays those categories and enables us to see what Priestley and Cavendish meant.

Obviously, the historical sensitivities are raised by genuine worries. We should not assume that homonymous expressions have the same reference (or reference potential). We should not assume that the account of nature that we use in interpreting the languages of the past is infallible—it is not, and the history that we develop on the basis of the picture offered by present science inherits the troubles of modern science (and possibly more besides). I hope that thoughtful historians will appreciate the ways in which my use of present science responds to the points of real concern, and will not be misled into over-reaction, based ultimately on zeal for an impossible project—that of some kind of empathic telekinesis into the minds of past scientists.

14. I should emphasize that Grandy's principle diverges from the principle of charity in *not* requiring that we maximize agreement with those whom we interpret. The underlying idea is that we attribute beliefs by supposing that our interlocutors have cognitive equipment that is similar to our own, and using what we know about the experiences they have had. Those who have received very different stimuli can be expected to have formed very different beliefs; (Grandy 1973) provides a lucid elaboration of these points. I should note that for some interpretative projects—but probably not for those which arise in doing the history of science—the principle of humanity will need refining in much the way that it corrected the principle of charity: on occasion, background evidence may suggest to us that we are not dealing with interlocutors who share cognitive equipment similar to our own.

viz. that, being capable of taking more phlogiston from nitrous air, it therefore originally contains less of this principle [i.e., phlogiston]; my next inquiry was, by what means it comes to be so pure, or philosophically speaking, to be so much dephlogisticated; ... (Priestley 1775/1970 II 120)

From our perspective Priestley has misdecribed the new gas. His remarks on this occasion identify the gas obtained by heating the red calx of mercury as dephlogisticated air, *and set up a mode of reference for tokens of 'dephlogisticated air' that fixes the reference by the description "the substance obtained when the substance emitted in combustion is removed from the air."* When tokens are produced, with this mode of reference, they fail to refer. On other occasions, Priestley and his friends (Cavendish, for example) produce tokens of 'dephlogisticated air' with a different mode of reference. Their dominant intention is to refer to the kind of stuff that was isolated in the experiments they are reporting—to wit, oxygen. Thus, in the passages quoted earlier, Priestley describes the sensation of breathing oxygen and Cavendish proposes an account of the composition of water. *Because the referents of the tokens of 'dephlogisticated air' are fixed differently on these occasions, Priestley and Cavendish enunciate important new truths.*[15]

Of course, they do so in a misguided language. But it is not hard to pinpoint the troubles of their idiom. Not only is the reference potential of terms such as 'dephlogisticated air' (and even of 'phlogiston'; see Kitcher 1978 534) heterogeneous, but the modes of reference are connected by a faulty theoretical hypothesis. Thinking of combustion as a process of emission, Priestley and his friends are led to take it for granted that the gas liberated from the red calx of mercury has been obtained by removing phlogiston from the air. Their connection of what *we* see as two distinct modes of reference, that fix reference to oxygen and to nothing, respectively, rests on their acceptance of the following hypothesis:

H. There is a substance which is emitted in combustion and which is normally present in the air. The result of removing this substance from the air is a gas which can also be produced by heating the red calx of mercury.

Phlogistonians all accept *H,* and, in consequence, they are all willing to use tokens of 'dephlogisticated air' whose references are fixed either by the description "the substance obtained by removing from the air the substance emitted in combustion" or by encounters with samples of the gas obtained by heating the red calx of mercury.

We can now give a precise sense to the claim that a term employed by a

15. Note that applying a pure causal theory of reference to Priestley and Cavendish misdescribes those occasions (the initial misidentification of oxygen, for example) on which they fail to refer. The pure descriptive theory misdescribes those occasions (when they are reporting on samples of the gas that they have prepared) on which they succeed in referring to oxygen. So, I claim, the appeal to heterogeneous reference potentials is needed to account for the mixed record of success and failure.

community is 'theory-laden.' Theory-laden terms have heterogeneous reference potentials; the theoretical hypotheses with which they are laden are claims that are, in conjunction, equivalent to the assertion that all the modes of reference fix reference to the same entity. Theory-ladenness does not simply stem from scientific irresponsibility. Scientists inevitably court ambiguity in using the same term from occasion to occasion (Hempel 1965 123–133, 1966 Chapter 7, Kitcher 1978 542–543).

Phlogistonians communicate easily for they share the same reference potentials. From the perspective of the new chemistry, however, uses of central parts of the language of the phlogiston theory are laden with false theory. As a result there are no expressions in Lavoisier's language (or in ours) whose reference potentials match those of 'phlogiston,' 'dephlogisticated air,' 'inflammable air,' and other expressions. Kuhn is quite correct (1983) to declare that a certain style of translation is impossible: there is no way to take Priestley's text and replace the expressions in it that do not belong to our language with expressions of our language that have the same reference potentials as those they replace. If translation has to preserve reference potential, we cannot translate Priestley—for there is no term of the language of post-Lavoisierian chemistry that has the same reference potential as 'dephlogisticated air.' Moreover, the mismatch among reference potentials captures the intuitive (but vague) idea that phlogistonians and modern chemists "carve up the world differently" (Kuhn 1983).

Even if we cannot translate Priestley's texts (according to the standards of translation of the last paragraph), there is no difficulty in recognizing how we (and Lavoisier) can understand Priestley's claims. For, as my brief analysis of his language shows, we are able to recognize the reference potentials of his technical terms and to specify the referents of the tokens he produces. We could even offer a *reference-preserving* translation of his texts—although, since this would replace tokens of the same type with tokens of different types, it would hide certain inferential connections unless we employed one of the standard devices of translators (parenthetical identification of the term replaced, footnotes, prefatory glosses, etc.). Not only can we comprehend what Priestley said, we can also see that some of the statements he advanced are true and we can explain how to improve his language. Lavoisier made a conceptual advance by revising reference potentials so as to avoid presupposing false hypotheses.

I claim that the foregoing account solves the puzzle about the languages of the protagonists in the chemical revolution. It also motivates a general thesis about conceptual change in science. Conceptual change is change in reference potential. Dramatic examples are those in which the community becomes disposed to use tokens of a term (possibly an old term, possibly a neologism) to pick out a new referent, and those in which its members acquire dispositions to fix the references of tokens of old terms through description, where no such referential specification had been possible before. The cases on my original list command our attention because, at least at first sight, they belong to one of these types.

To understand conceptual *progress* we should recall my account of the
different kinds of intention that may dominate in the production of a token.
Perhaps for most linguistic usages, the general intention to conform to the
usage of others is far more important than any intention we may have to refer
to whatever fits a description. When we examine scientific usage, the intention
to conform is by no means the only one that has to be taken into account.
Scientists usually also have the general intention of referring to natural kinds,
picking out the real similarities in nature, and, in recognizing this intention,
we sometimes construe the descriptions they offer as *mistaken* identifications
of the referent rather than as successful identifications of a different referent:
thus, when he talks of the antics of the mice in the vessel and when he describes
the "lightness in his breast," we understand Priestley as talking about oxygen,
even though he would describe the gas as "the substance obtained by removing
phlogiston from the air." But there are also occasions on which scientists
intend that the referent should be whatever satisfies a particular description.
The *ideal* situation for a scientist would thus be to obey three maxims:

Conformity: Refer to that to which others refer.

Naturalism: Refer to natural kinds.

Clarity: Refer to that which you can specify.

There are many situations in which these maxims conflict. When they do,
the scientist "chooses" among them—in the sense that there is a dominant
intention to obey one rather than the others. If the choices are made in
different ways on different occasions of utterance, different tokens of the
same type can refer differently.[16]

Conceptual progress should be assessed in terms of proximity to the ideal
state. One of the goals of science is the construction of a language in which
the expressions refer to the genuine kinds and in which descriptive specifications of the referents of tokens can be given. For such a language the three
maxims are in harmony.

We can now introduce a notion of *conceptual progressiveness:*

(CP)   A practice P_2 is conceptually progressive with respect to a practice
P_1 just in case there is a set C_2 of expressions in the language of P_2 and
a set C_1 of expressions in the language of P_1 such that

(a)   except for the expressions in these sets, all expressions that occur
in either language occur in both languages with a common reference
potential

16. This should be easily understandable in terms of the psychological picture developed
in Section 2 of the last chapter. We can envisage scientists as subscribing to three different
long-term goals. In situations of conflict, one of these goals is activated, and, in conjunction
with other features of working memory it dictates the mode of reference of the term token
produced.

(b)   for any expression e in C_1, if there is a kind to which some token of e refers, then there is an expression e^* in C_2 which has tokens referring to that kind

(c)   for any e, e^*, [as in (b)], the reference potential of e^* refines the reference potential of e, either by adding a description that picks out the pertinent kind or by abandoning a mode of reference determination belonging to the reference potential of e that failed to pick out the pertinent kind.

In this definition, clauses (a) and (b) are necessary to eliminate the possibility that refinements of reference potential might be achieved at the cost of expressive loss. We isolate those parts of the language that do not change [in (a)], and demand that any kinds that can be discussed in the old language should be specifiable in the new [in (b)]. Improvements come about by abandoning modes of reference that are not in accord with one of the maxims or by adding modes of reference that would be in accord with both *clarity* and *naturalism.*

Numerous examples from the history of science reveal conceptual progress in the sense captured in (CP). The shift from Priestley to Lavoisier shows retention of the ability to refer to oxygen, with replacement of the flawed reference potential of the term Priestley employed to refer to oxygen ('dephlogisticated air') with the refined reference potential of Lavoisier's expression 'oxygen.' Similarly, in the Copernican revolution, a term, 'planet,' some of whose tokens previously referred to a natural kind (the set of things we would count as planets of the solar system) and some of whose tokens referred to a set that is not a kind (the set whose members are Mercury, Venus, Mars, Jupiter, and Saturn), underwent refinement of its reference potential. After Copernicus, reference can be fixed descriptively to a kind (the planets of the solar system) and the flawed mode of reference to the partial subset is dropped. Later in the history of astronomy, 'planet' comes to have a reference potential that shows further refinement: a descriptive mode of reference is used to fix reference to the kind that includes the planets of all stars, and the subkind of planets of the solar system is picked out by explicit restriction. Similar accounts can be provided for the other examples on my original list.

5. Explanatory Progress

Explanatory progress consists in improving our view of the dependencies of phenomena. Scientists typically recognize some phenomena as prior, others as dependent. For example, ever since Dalton, chemists have regarded molecular arrangements and rearrangements as prior to the macroscopic phenomena of chemical reactions, and, since the 1960s, geologists have viewed interactions among plates as prior to facts about mountain building and earthquakes. General ideas about these dependencies can be shared while specific theses about the details of dependence are debated. Commitment to Daltonian atomism in the nineteenth century persists through any number of claims about the formulas of chemical compounds.

Judgments about dependency can be understood as concerned with the forms of ideal explanatory texts that the scientists in question envisage. One component of practice is a collection of patterns or explanatory schemata. Fields of science make explanatory progress when later practices introduce explanatory schemata that are better than those adopted by earlier practices.

What does 'better' mean here? One answer, the response of robust realism, is to declare that there is an objective order of dependency in nature.[17] The character of chemical reactions is objectively dependent on underlying molecular properties, facts about mountain building and earthquake zones are objectively dependent on the facts of plate tectonics, the characteristics of contemporary organisms are objectively dependent on the evolutionary histories of those organisms, and so forth. Recognizing these dependencies— and how to extend them further—is an important progressive step.

The robust realism outlined here is not the only way to say what is meant by improving the set of explanatory schemata. An alternative is to tie advances to enhanced ability to understand the phenomena, for example by proposing that a particular collection of schemata offers a more unified vision of the world. I shall postpone comparison of these two metaphysical options until the next chapter (see Section 8, where I also take up the question of how to explicate the discussion of kinds that I took for granted in the previous section). For now it will suffice to see the general shape of the view. Explanatory progress consists in improving our account of the structure of nature, an account embodied in the schemata of our practices. Improvement consists either in matching our schemata to the mind-independent ordering of phenomena (the robust realist version) or in producing schemata that are better able to meet some criterion of organization (for example, greater unification).

Let us consider some examples from the history of science. Almost everything that Dalton maintained about atoms was wrong. Nonetheless, we recognize his formulation of atomic chemistry as an important progressive step because it introduced a new *correct* explanatory schema. Dalton proposed to explain facts about the course of chemical reactions (and about the weights of reactants and products) by appealing to premises about atoms, premises that specify the "fixed proportions" in which atoms of different elements combine. Chemists ever since have endorsed the claim that this is a correct picture of the objective dependencies, and they have further articulated Dalton's schema.

17. This conception is as old as the Aristotelian idea of an order of being. For robust realists, the aim of science is taken to be that of delineating the fundamental mechanisms at work in nature. If the achieving of adequate concepts is understood in terms of an anatomical description, the display of objective dependencies corresponds to physiology— showing how things work. For a modern attempt to give literal significance to these metaphors, see Wesley Salmon's defense of the "ontic conception" of explanation in his (1984). I discuss some of the epistemological problems associated with Salmon's approach—as well as some of its considerable merits—in my (1989). See also Section 8 of the next chapter.

One leading question of post-Daltonian chemistry concerned the ratios of the weights of substances that form compounds together. Within nineteenth century chemistry, we can discern the following simple pattern.

DALTON

Question: Why does one of the compounds between X and Y always contain X and Y in the weight ratio $m : n$?

Answer:

(1)   There is a compound Z between X and Y that has the atomic formula $X_p Y_q$.

(2)   The atomic weight of X is x; the atomic weight of Y is y.

(3)   The weight ratio of X to Y in Z is $px : qy$ ($= m : n$).

Filling Instructions: X, Y, Z are replaced by names of chemical substances; p, q are replaced by natural numerals; x, y are replaced by names of real numbers.

Classification: (1) and (2) are premises; (3) is derived from (1) and (2).

DALTON is elementary—although it was, of course, instantiated in many different ways during the early years of the nineteenth century by chemists who had very different ideas about the formulas of common compounds! What makes it important for our purposes is the way in which DALTON was preserved and extended by subsequent work.

An important nineteenth-century step in the extension was the introduction of the concept of valence and rules for assigning valences that enabled chemists to derive conclusions about which formulas characterized possible compounds between substances.

MOLECULAR AFFINITIES

Question: Why does one of the compounds between X and Y always contain X and Y in the weight ratio $m : n$?

Answer:

(1)   Molecules of X contain j atoms, molecules of Y contain k atoms.

(2)   The valences ("affinities") of X and Y are . . . .

(3)   The possible equations governing chemical reactions between X and Y are . . . .

(4)   There is a compound Z between X and Y that has the atomic formula $X_p Y_q$.

(5)   The atomic weight of X is x; the atomic weight of Y is y.

(6)   The weight ratio of X to Y in Z is $px : qy$ ($= m : n$).

The filling instructions extend those of DALTON by requiring that j and k be natural numbers, that the valence relations should be specified in accordance with a particular scheme, and that (3) be completed by presenting chemical equations in a particular way. The classification demands that (3) be obtained from (1) and (2) in accordance with the principles of balancing equations, and that (4) be derived from (3) by showing that the compound X_pY_q be obtainable on the right-hand side of some equation in (3). As in the case of DALTON (6) is to be derived from (4) and (5).

At this first stage, the attributions of valence are unexplained and there is no understanding of why the constraints hold. However, the original explanations of weight relationships in compounds are deepened, by showing regularities in the formulas underlying compounds.

The next stage consists in the introduction of a shell model of the atom to explain the hitherto mysterious results about valences. From premises attributing shell structure to atoms, together with principles about ionic and covalent bonding it is now possible to provide derivations of instances of (2). These derivations provide a deeper understanding of the conclusions than was given by the simple invocation of the concept of valence because they show us *in a unified way* how the apparently arbitrary valence rules are generated. Moreover, the appeal to the model of the atom enables us to derive instances of (3) from premises that characterize the composition of atoms in terms of protons, neutrons, and electrons.

SHELLFILLING

Question: Why does one of the compounds between X and Y always contain X and Y in the weight ratio $m : n$?

Answer:

(1)   Molecules of X contain j atoms; molecules of Y contain k atoms.

(2)   (a)   An uncharged atom of X contains g electrons; an uncharged atom of Y contains h electrons.

(b)   The first incompletely filled shell level around an uncharged atom of X contains g_1 electrons and g_2 free places, the first incompletely filled shell-level around an uncharged atom of Y contains h_1 electrons and h_2 free places.

(c)   The combinations of atoms of X and Y that will achieve shell filling either through electron transfer (ionic bonding) or through electron sharing (covalent bonding) are . . . .

(d)   The valences ("affinities") of X and Y are . . . .

(3)–(6) As for MOLECULAR AFFINITIES.

The filling instructions extend those of molecular affinities in obvious ways: g, h, g_1, g_2, h_1, h_2 are required to be natural numbers and to satisfy relations imposed by the general theory of atomic structure. The classification demands that 2(b) be obtained from 2(a) in accordance with the principles of the shell model of atoms, and that 2(c) be derived from 2(b) using the principles of covalent and ionic bonding. 2(d) is simply a rewriting of 2(c) in what now is viewed as old-fashioned language. (3) can either be obtained from 2(d), as in MOLECULAR AFFINITIES or directly from 2(c).

Finally, the derivations from premises about shell filling can be embedded within quantum mechanical descriptions of atoms and the shell structures and possibilities of bond formation revealed as consequences of the stability of quantum-mechanical systems. Although this is only mathematically tractable in the simplest examples, it does reveal the ideal possibility of a further extension of our explanatory derivations.

What this very simple, preliminary account of the development of atomic chemistry shows is how we can formulate precisely the intuitive idea that Dalton outlined a picture that was filled in in much more detail by his successors. Of course, they improved on his chemistry in two distinct ways. One trend (with which I am not concerned at this stage) consisted in the replacement of his erroneous ideas about chemical formulae, and, in consequence, the production of true instantiations of his simple schema. The notion of explanatory progress involves the improvement of the schema itself. This line of development is seen in the move from DALTON to MOLECULAR AFFINITIES to SHELL FILLING and beyond to the account in terms of quantum chemistry (where I have only indicated the schema).

The example of Darwinian evolutionary biology, considered in Chapter 2, shows a similar progression. *Some* of Darwin's schemata were correct; others—such as the schema underlying those of his explanations in terms of disuse that require Lamarckian mechanisms of inheritance—have been discarded. Particularly interesting is the fate of the selection schema, SIMPLE INDIVIDUAL SELECTION As we saw in Section 7 of Chapter 2, this is absorbed into NEO-DARWINIAN SELECTION a schema that offers a more articulated version of Darwin's picture of the workings of selection. That schema has been further completed in some of the work mentioned in Section 8 of Chapter 2 (in Hamilton's introduction of inclusive fitness and Maynard Smith's development of evolutionary game theory, for example). As in the Daltonian example, tracing the details of the relations among the schemata employed at successive stages enables us to make precise the idea that there is a cumulative process of extending, correcting, and articulating a basic picture of the order of some natural phenomena.[18]

Four distinct kinds of processes are at work in these examples of explanatory progress. First, we have the introduction of correct schemata, illustrated by the work of Dalton in recognizing the dependence of facts about the course

18. In the example from the history of chemistry I have been concerned to draw the line of development without introducing the historical details. The Darwinian example of Chapter 2 is meant to balance this by providing much more of the scientific background and linking the schemata to specific points in the history of biology. With the discussion of this section in mind, it should be possible for the reader to return to the more discursive treatment of Chapter 2, identifying the same structural relations and progressive picture that I display here. Another example, treated in more detail in (Kitcher 1989), focuses on the development of explanations of hereditary distributions from Mendel to Watson and Crick.

of chemical reactions (specifically about the weights of reactants and products) on the facts about atomic combination and by Darwin's insight that distributions and relationships among contemporary organisms are dependent on the course of descent with modification. Second, we have the elimination of incorrect schemata, such as Darwin's appeals to the inheritance of acquired characteristics. Third, we find the generalization of schemata, rendering them able to deal correctly with a broader class of instances: evolutionary theorists who appeal to classical individual selection alone are correct in identifying a certain type of dependence, but their proposals are less general—and therefore less complete—than those which allow for drift, migration, meiotic drive, inclusive fitness effects, developmental constraints, and so forth. Finally, there is explanatory extension, when the picture of dependencies is embedded within some larger scheme. The incorporation of Darwin's selectionist patterns within NEO-DARWINIAN SELECTION and the embedding of atomic chemistry in quantum physics show this process at work.

At the end of Chapter 2 I suggested that the history of evolutionary biology does not show an indecisive alternation but *something like* cumulative progress. The goal of this section and its predecessor has been to make that judgment more precise. There do seem to be cumulative processes of specifiable kinds in my examples, as schemata are introduced, refined, generalized, and extended.

But how typical are these examples? Am I cheating by simply drawing on periods that Kuhn would classify as belonging to "normal science," and that other students of scientific change would rank in analogous categories?[19] My reply consists of a historical reminder and a promise. The historical reminder is to note that the periods over which I have traced the development of schemata not only are long (of the order of 100 to 200 years at the stage of most rapid change in the histories of the relevant sciences) but also span "normal" and "revolutionary" episodes. In each case, even if all schemata after the first can be assigned to a single "normal scientific" tradition, the introduction of the first marks a progressive step over the accounts previously available: the schemata introduced by Dalton, Mendel, and Darwin provide answers to the questions they address that identify correct dependencies unappreciated by their predecessors. Moreover, if one wants to extend the notion of "normal science" (or some equivalent) to cover the entire periods through which I trace the refinement, generalization, and extension of schemata, then it will be necessary to downplay the significance of "revolutions" in science. These periods are so large, and the processes I identify so prevalent, that "normal science" will be everywhere once a field attains maturity.

But I do not want to claim too much for these examples. They serve to remind us of what scientists are sometimes brave enough to say, that domination of philosophical discussion by the vicissitudes of a few concepts from

19. These questions were posed forcefully to me by Larry Laudan and Rachel Laudan. I am grateful to both of them for much helpful discussion of this point, although I suspect that neither will be convinced by my response.

theoretical physics needs to be balanced by consideration of cases in which progress seems much more assured.[20] Instead of dogmatically claiming that my view will generalize across all cases, let me offer a counterquestion: How frequently do periods that do *not* exemplify the patterns I have discussed occur in the history of the (modern) sciences? This question leads me to my promise. At the end of the next chapter I shall explicitly discuss what seem to be the hard cases, the examples that critics of cumulative progress have turned to in an endeavor to expose the idea of explanatory losses.

I conclude by offering an explicit account of explanatory progress. Continuing to waive the large metaphysical questions that I postponed earlier, let us suppose that a schema is correct if it identifies a class of dependent phenomena and specifies some of the entities and properties on which those phenomena depend. (For the robust realist it formulates part of the objective order of dependencies in nature; for someone who does not believe in a mind-independent order of nature, it presents part of an ideal system for organizing the phenomena.) One schema is more complete than another just in case the former identifies a more inclusive set of relevant entities and properties or the former is correct for a more inclusive class of dependent phenomena. One schema extends another if and only if a schematic premise of the latter is derived from the former. The extension is correct if the properties attributed to the entities in instances of the conclusion depend on the entities and properties referred to in the corresponding instances of the premises. The historical examples I have discussed involve introduction of correct schemata, elimination of incorrect schemata, replacement of less complete with more complete schemata, and correct explanatory extension.

The following definition brings together the facets of explanatory progress in a straightforward fashion:

(EP) P_2 is explanatorily progressive with respect to P_1 just in case the explanatory schemata of P_2 agree with the explanatory schemata of P_1 except in one or more cases of one or more of the following kinds.

(a) P_2 contains a correct schema that does not occur in P_1.

(b) P_1 contains an incorrect schema that does not occur in P_2.

(c) P_2 contains a more complete version of a schema that occurs in P_1.

(d) P_2 contains a schema that correctly extends a schema of P_1.

Making explanatory progress in this sense advances our goal of recognizing the structure of natural phenomena (or, if you like, the best way of organizing

20. Thus Mayr protests the influence of a few theories from physics on the development of an allegedly general picture of science and scientific progress (1982 857). In a valuable review of Mayr's book, John Maynard Smith is admirably forthright: "Unfashionable as it may be to say so, we really do have a better grasp of biology today than any generation before us, and if further progress is to be made it will have to start from where we now stand" (see Maynard Smith 1989 11). I hope that my account explains clearly why Mayr and Maynard Smith are right.

various areas of our experience).[21] A suggestive (but not entirely adequate analogy) is to think of the work of children engaged on a large and complex jigsaw puzzle. Subregions of the puzzle correspond to the structure of dependencies among a particular class of phenomena. Identifying correct schemata is analogous to fitting a few pieces together, the correction of schemata corresponds to scrapping faulty efforts at fitting pieces, the completion and extension of schemata consist in putting the pieces already fitted into larger chunks of the puzzle. The ultimate aim, of course, is to complete the picture. (Here, perhaps, the analogy breaks down, for there may be no complete— or completable—picture.)

6. Derivative Notions

Significant questions arise against the background of the ordering of the phenomena captured in our explanatory schemata. There are two main ways in which significance can accrue to a question, generating *application* questions and *presuppositional* questions, respectively. Application questions are generated from projects of finding particular instantiations of the available schemata. Presuppositional questions investigate the conditions that must obtain if available schemata are to be instantiated.

In the early stages after a new schema has been introduced into consensus practice, almost all instantiations of it are important. Thus, in the early nineteenth century, chemists committed to Dalton's atomic theory regarded any problem of understanding the weight relations among compounds as significant. Because solutions to these problems required them to make judgments about the chemical formulae of compounds, the question of how to assign chemical formulae and to discriminate among proposed formulae obtained derivative significance. Similarly, in the years after the publication of the *Origin,* naturalists set to work to find convincing instantiations of Darwin's

---

21. My view of explanatory progress plainly has some affinity with the ideas of the logical empiricists about the unity of science. However, there are important differences— stemming from my rejection of the demand that there be accumulation *at the level of details*. For more exact discussion of the relationship, see (Kitcher 1984, 1989), which explore the question in the context of the growth of genetics.

The writings of Lakatos, Laudan, and Toulmin, all of whom hope to specify broader units (research programs, research traditions) behind individual theories are sensitive to the problems that beset the demand for accumulation at the level of details. But the emphasis on empirical prediction and the shying away from analysis of the explanations offered within a field of science at a time seem to me to interfere with the appreciation of the possibility of accumulative explanatory structure. Toulmin's conception of "ideals of explanatory order" comes closest to recognizing what I regard as the salient point: *the need to break away from concentration on accepted statements* (a feature of logical empiricism that survives in Lakatos and Laudan) *and to focus on the ways in which statements are* used *in answering questions.*

schemata, choosing questions for their apparent tractability and illustrative power.

But once a field has established a set of paradigm answers to application questions, further instantiations of its schemata are no longer on a par. Many questions to which an available schema could be directed are not regarded as significant because the record of success in instantiating the schema gives everyone confidence that they could (with time and effort) be answered, and the task of grinding out the details looks like hack work. The credentials of the schema are well attested by the roster of paradigm answers, and the questions that now appear significant are those that seem to involve special difficulties of producing instantiations. These questions raise the hope that when they are answered the community will obtain corrected, completed, or extended schemata. They are intrinsically significant. Other projects of application may be inspired by the needs of other fields. Questions arising in one field may be instrumentally significant because answers to them are needed for addressing some question of intrinsic significance for another field.

Intrinsic significance, I have suggested, often accrues to those questions that seem hard to answer—questions that challenge the ingenuity of a scientist (compare Kuhn 1962/1970 55; there are numerous points of contact between Kuhn's account of normal science and the proposals of this section). So, for example, the problem of "altruism" has loomed large in discussions of the evolution of behavior precisely because there were, until recently, good reasons for wondering how behavioral traits that appear to detract from the reproductive success of their bearers could be maintained in an animal population. When one of the explanatory schemata of a practice is intended to apply to phenomena, including some for which there are good reasons for believing that no instance of some premise of the schema is true, then there is a significant scientific question of showing how to instantiate the schema, how to complete it, or how to differentiate the problematic instances. Application questions can be inspired by the promise of explanatory extension as well as by the prospect of producing a more complete schema.

Presuppositional questions arise when instantiations of some accepted schema presuppose the truth of some controversial claim.[22] Darwin's selectionist schemata presupposed (as some of his critics emphasized) that the variation in natural populations could be maintained in reproduction. This lent considerable significance to the question of offering an account of variation and hereditary transmission that would show if (and why) the presupposition is true. That question was finally answered in the 1930s through the efforts of Fisher, Wright, Dobzhansky, and others.

Presuppositional questions have also inspired important research traditions in other areas of science. Consider the schema that extends MOLECULAR AFFINITIES, SHELL FILLING. That schema presupposes that it is possible for stable

22. There are filiations here not only to Kuhnian ideas but also to Larry Laudan's discussion of conceptual problems in his (1977).

atoms to satisfy the conditions of the Bohr model and generates the question of understanding how atoms meeting these conditions can avoid the kinds of collapse that would be expected from the principles of classical electromagnetic theory.

In these examples, the presuppositions of the schemata are problematic in that there are apparently cogent arguments from plausible premises, available to proponents of the practice, that seem to show that the presuppositions are false. What are required are a demonstration of the possibility of the problematic presuppositions and a diagnosis of the flaw in the apparently persuasive reasoning. Combining this idea with the earlier discussion of application questions, we can say that questions are intrinsically significant when (a) answers to them would exhibit the possibility of instantiating an accepted schema, or (b) would exhibit the possibility of instantiating an accepted schema in apparently problematic instances, or (c) would show the possibility of some problematic presupposition of an accepted schema. Questions are instrumentally significant when answers to them would answer some intrinsically significant question of some other field or answer some instrumentally significant question of some other field. (The last clause allows for the possibility of a chain of fields each of which turns for help to its neighbor; but, at the end of the chain, there must be some field with an intrinsically significant question.)

The account of significance I have offered is best seen as outlining the idea of an *apparently* significant question. Relative to a set of schemata certain questions should be regarded as significant because they demand instantiations for those schemata, or instantiations in apparently difficult cases, or demonstrations of the possibility of problematic presuppositions. We can say that a consensus practice is erotetically well grounded if the questions to which it assigns significance are indeed those that are significant relative to its schemata. *Genuinely* significant questions are those that are significant (in the sense I have indicated) relative to correct schemata. We make erotetic progress when we have an erotetically well-grounded consensus practice in which we pose genuinely significant questions that were not previously asked.

Scientists often describe some fields—especially those that are regarded as most exciting—by suggesting that they now know how to pose the right questions. A concept of erotetic progress should capture such descriptions. Sometimes erotetic progress can be a by-product of conceptual progress: Priestley's questions about the role of dephlogisticated air in various reactions are better formulated as questions about oxygen. On other occasions, as I have suggested in the last paragraph, incorporation of new explanatory schemata generates new genuinely significant questions: a striking example here is Darwin's initial introduction of his evolutionary schemata, which gave rise to a host of new application and presuppositional questions. However, there is a further facet of erotetic progress that has not yet been made explicit. We make progress by posing more tractable questions.

Scientific fields typically begin with big, vague questions. The introduction of explanatory schemata structures our large, cloudy wonderings, by suggesting new, more precise inquiries for us to undertake. Little of significance is achieved unless we have realistic hopes of answering these new questions. Thus, while atomists before Dalton had offered a correct schema, their explanatory pattern was so imprecise that it failed to furnish tractable questions of application. (Indeed, the history of eighteenth century efforts to develop Newton's program of dynamic corpuscularianism—outlined in the celebrated *Query* 31 to the *Opticks*—shows clearly how an explanatory advance can be too inspecific to allow for erotetic progress.[23])

Sometimes, we can make erotetic progress not only by adding significant new questions but by decomposing some of the significant questions of prior practice. Consider, for example, how the enterprise of finding instantiations of NEO-DARWINIAN SELECTION generates subsidiary questions within contemporary evolutionary biology. To complete an account of the maintenance of a trait under natural selection, one needs to be able to identify population structure, to assign selection coefficients, and to measure genetic variation. Thus, for example, questions of the form, What is the fitness of T for members of G in environment E?, become derivatively significant. Particular instances of such questions generate further investigations. If it seems that measuring the fitnesses of organisms of a particular species is tractable then scientists will bestow significance on questions about identification of certain types, assignment of organisms as offspring of others, measurements of fecundity, and so forth. Significance can ultimately accrue to quite technical, practical, and limited questions—How do you form a reliable estimate of the number of offspring produced by an ant colony?—precisely because those questions stand at the terminus of a chain of inquiries, each of which derives significance from its predecessor. In general, we can think of the significant questions of a field of science as hierarchically organized and represented by trees. At the vertex of each tree is some central question of the field, addressed by one of the most general schemata. Along each path, successive questions are generated because the provision of answers to them would aid in the resolution of a predecessor.[24]

At this stage, it is possible to broach an issue that may, quite reasonably,

23. For an illuminating discussion of the programs that descend from Newton, see (Schofield 1969). I have sketched an account of these programs in the terms of the present chapter in (Kitcher 1981 section 3). See also (Boscovich 1763/1966 and Kitcher 1986). In contemporary developmental biology, there is similar uncertainty about how to focus the big, vague question, How do organisms develop?

24. For an example of this hierarchy, see (Culp and Kitcher 1989). Obviously, there is a sense in which the accepted schemata of a practice play a similar role to the "hard core" of Lakatos or the "core assumptions" of Laudan. However, as I have already noted, one missing feature of the approaches of Lakatos and Laudan is their concentration on accepted statements rather than the ways in which accepted statements are used. A consequence of this seems to me to be that their views of science do not relate to much of the work in which scientists actually engage. They do not provide accounts that enable us to see how the very local and specific projects that occupy almost all scientists at almost all times come to be valued. Here I take Kuhn's analysis of normal science to be suggestive, and I have tried to articulate the suggestions.

appear bothersome. Close attention to my account of correctness of explanatory schemata and to the implications of (EP) will reveal that it is possible to have practices P_1, P_2 such that neither the transition from P_1 to P_2 nor the transition from P_2 to P_1 would count as explanatorily progressive. For, while P_1 may lack a correct explanatory schema present in P_2, it may also contain a correct explanatory schema absent in P_2. Nor is this merely a logical possibility. Devotees of the alleged phenomenon of "Kuhn loss" will insist that major shifts in science often involve the abandonment of explanatory insights that later developments in the field will reestablish. While Newton introduced the idea of deriving conclusions about accelerations from premises specifying the forces acting, he gave up the Aristotelian grounding of forces in the geometrical structure of space, an explanatory proposal that was later recovered by Einstein. Lavoisier's new chemistry sacrificed the possibility of accounting for the combustibility of the metals, which had been successfully undertaken by proponents of the phlogiston theory and which would be recaptured in the twentieth century accounts of chemical bonding. I shall not pursue these examples in detail here (see §9 of Chapter 5 and §9 of Chapter 7), but simply outline a solution to the problem of understanding the shifts from Aristotle to Newton and from Priestley to Lavoisier as progressive. Suppose we grant that Aristotle had a correct schema that was abandoned in Newtonian practice, and that Newton introduced a new correct schema that Aristotle had lacked (and similarly for Lavoisier and Priestley). From the perspective of (EP), then, Aristotle and Newton (Priestley and Lavoisier) cannot be ranked for progressiveness.[25] Nonetheless we may still score a considerable advance in erotetic terms by recognizing that the significant questions that are abandoned in the shift from Aristotle to Newton are intractable while that transition introduces a host of new significant questions. Indeed, we might even maintain that the possibility of decomposing the imprecise (but significant) Aristotelian questions requires the prior posing of the tractable (and also significant) Newtonian problems.[26]

The view I have sketched can easily be motivated by using the homely analogy that I introduced in dealing with explanatory progress. The task of arriving at a correct account of the structure of nature may require us sometimes to abandon primitive explanatory insights (which will much later be reinstated in articulated form), just as, in solving certain kinds of puzzles (particularly those of building three-dimensional models) pieces that actually fit together, and will eventually be rejoined, may have to be taken apart to permit the assembly of large configurations. *If* the sympathetic accounts of

---

25. Since I take progressiveness to be a relation between practices, and since practices are multidimensional, it is possible that rival practices could be incomparable on some dimensions but that one was superior to its rival along others.

26. Thus, if one views—as Kuhn (1970 206–207) seems to—the Aristotelian approach as containing an embryonic schema for addressing questions about forces, one that traces forces to underlying spatial structure, then it is quite pertinent to suggest that development of this insight only became possible through the detailed mathematical investigations begun by Galileo and Newton.

the achievements of Aristotle and Priestley are correct, then this approach seems to me to offer a satisfying diagnosis of what was progressive in the transitions to Newton and Lavoisier, respectively. The losses (if any) were vague insights that could not be articulated at that stage in the development of science; the gains, in both instances, were correct explanatory schemata that generated significant, *tractable,* questions, and the process of addressing those questions ultimately led to a recapturing of what was lost. Even if these transitions do not exhibit explanatory progress, they show erotetic progress.[27]

Once we have the concept of a significant question at our disposal, it is relatively simple to understand progressiveness with respect to other components of practice. Instruments and experimental techniques are valued because they enable us to answer significant questions.[28] One instrument (or technique) may do everything another does and more besides. If so, then we make *instrumental* (or *experimental*) progress by adopting a practice in which the former instrument (technique) replaces the latter. Making this conception of progress precise requires us to look more carefully at progress in the set of accepted statements: if we know what counts as improving the set of accepted statements, then we can characterize instrumental and experimental progress by recognizing the increased power of instruments and techniques to deliver improved statements.

We make progress with respect to the set of accepted statements in a number of ways. Sometimes, scientists eliminate falsehood in favor of truth, abandon the insignificant, add significant truths, or reconceptualize already accepted truths. Moreover, as I shall suggest in the next section, they often replace statements that are further from the truth with those that are closer to the truth. I shall start with the apparently naive idea that part of scientific progress consists in accepting statements that are both significant and true.

27. It is worth explicitly forestalling a confusion here. I claimed earlier that explanatory progressiveness was fundamental and erotetic progressiveness derivative. How then can one have erotetic progress without explanatory progress? The answer requires a distinction of levels. To make sense of the notion of erotetic progress, we need the concept of a significant question. The ultimate bestowers of significance are the correct schemata of a practice. Thus the *general notion* of erotetic progress is conceptually dependent on that of explanatory progress (more precisely on notions that figure in the analysis of explanatory progress). However, it is quite possible that, in specific cases, loss of a correct schema that cannot yet be instantiated should go hand in hand with the acquisition of the ability to pose, for the first time, questions that are significant and whose significance accrues from correct schemata (either those that are retained or some that are introduced in the new practice). Moreover, these questions may even pave the way for future refinement and instantiation of the now-discarded correct schemata. In either case, the transition will show erotetic progress without explanatory progress.

28. I owe to Gareth Matthews the observation that, while some of my relations of progressiveness concern the attaining of certain epistemic desiderata, others involve achieving the means for going further. Thus, if we are making conceptual or explanatory progress we have already gathered some good things (like the firm that is already making a profit). If we are making instrumental or erotetic progress then we have prepared ourselves well for gathering good things in the future (like the firm that has invested wisely in new ventures).

Philosophical reticence about the attainment of significant truth is the result of a failure of nerve, induced by thinking about the problem in a faulty way. Beguiled by the notion that scientific significance accrues only to systems of generalizations—theories, conceived as axiomatic deductive systems—the problem is framed first in terms of the truth of very general axioms. A pessimistic induction on the history of science (the product of reflection on the famous grand generalizations of our predecessors—notably classical physicists) instills a conviction that these axioms cannot be strictly true.[29] So the best we can hope to achieve is, apparently, scientific systems that are "close to the truth." Popperian moves to find measures of truth content become inviting at this juncture, and we are plunged into a technical morass of comparing the "sizes" of infinite sets of true and false consequences.

Tradition takes a misguided view of significance and so brings truth into the picture in far too ambitious a way, framing the crucial issues in terms of truth *for whole theories*. My approach circumvents these difficulties by offering a quite different view of scientific significance. A significant statement is a potential answer to a significant question. What we strive for, when we can get them, are *true* significant statements, that is, true answers to significant questions. Sometimes, and with differing frequencies in different sciences, those statements are universal generalizations. However, there are important fields of science in which exceptionless generalizations are not sought.

Consider some examples. "DNA molecules consist of two helical strands, wound around one another in opposite directions"—is this statement true? is it a universal generalization? what makes it significant? Molecular biologists (and many other scientists) would surely count the statement as among the most significant truths enunciated in the past half century (to be conservative about the time period). But if the statement is construed as a strictly universal generalization it is immediately clear that it is false. Not all DNAs are found in double helical form: contemporary molecular biology makes considerable use of single-stranded DNA. The existence of such molecules is not at all at odds with the intended interpretation of the original statement, which offered a *restricted* generalization about the DNA molecules typically found in cells in nature throughout most of the phases of the life of the cell. The restricting conditions are not formulated precisely, and, I suspect, nobody knows how to formulate them precisely. Nevertheless, the statement is both significant and true: significant because it answers the significant question, What is the structure of the genetic material?, true because counterinstances such as the one I have mentioned fall outside its intended scope.

Sometimes significant statements are even more obviously particular. Consider (1) "Part of Western California is at the junction of two plates that slide past one another," (2) "Gangs of male Florida scrubjays are sometimes able

---

29. I shall discuss the "pessimistic induction" in Sections 3 and 4 of Chapter 5. It is worth noting that Ian Hacking insightfully points out that our views about fluctuations in science would be rather different if we focused on *instruments* rather than on *theories* (see his 1983 55–57). See also the remarks by Mayr and Maynard Smith cited in footnote 20.

to expand the territories in which they reside." Such statements obtain their significance from the role they play in answering significant questions. (1) is an important part of an answer to the question, Why is there an earthquake zone in western California? (a question of cognitive significance for some, of practical significance for others, and of both kinds of significance for some of us!). Similarly, (2) plays a role in answering the question, Why do male Florida scrubjays often return to the parental nest and assist in the feeding of siblings?—since, as detailed research on these birds makes clear, there are beneficial consequences for inclusive fitness not only in providing relief for overworked parents but also in increasing the possibility of gaining a territory by imperialist expansion (see Woolfenden and Fitzpatrick 1984).[30]

Is it reasonable to believe that all the particular statements that fill contemporary scientific journals will endure as parts of the consensus practices of future fields of science? No. But the main problem is not with truth but with significance. If you turn to the pages of *Nature* (or other scientific journals) fifty years ago, they will present conclusions that are sometimes, by our lights, oddly formulated but substantially correct—conclusions, however, that no longer seem to be significant. Many of them never became part of consensus practice (or at least part of the consensus practice of a significant subdisciplinary group). Others enjoyed a brief career in consensus before being discarded.

Why does this occur? Because the explanatory schemata of a practice generate significant *primary* questions, which spawn *derivative* questions: answering an appropriate sequence of the latter is seen as a way to address the former. Not all routes succeed. A community striving to answer Q formulates the strategy of doing so by tackling q_1, q_2, and q_3 (in that order). The initial answer to q_1 is hailed as significant, but then the endeavor becomes stuck. Ultimately another route to answering Q is found and successfully exploited, and the answer to q_1 disappears from consensus practice. Moreover, even attaining the answer to a primary significant question by solving some derivative problems may not endure as a permanent achievement. The same work may be done more decisively or more elegantly by later scientists. Consensus practice is economical. The exemplary studies of one generation can be displaced by better exemplars.[31]

---

30. Examples can be multiplied by looking at any issue of any scientific journal.

31. Thus, at a particular point in time, the hierarchy of significant question *forms* corresponds to numerous hierarchies of significant *particular* questions. Scientists often work on specific projects in the hope that the instances they address will prove the key to tackling the question forms. Although they may reap all kinds of successes, their own favored instances may vanish from the subsequent discussions of the field simply because some rival hierarchy of specific projects provides a clearer or more elegant treatment of the question form.

However, as Rob Cummins pointed out to me, it is possible that work done on one sequence of questions should fail to be adopted as the definitive treatment of the ultimate issue but should prove useful in some other area of research. So, for example, some research efforts channeled originally toward problems within physics are attributed enduring sig-

Part of the story of progress in the set of accepted statements is that it consists in eliminating falsehood in favor of truth, eliminating the merely apparently significant in terms of the genuinely significant, and using improved language to reformulate antecedently enunciated significant truths. Significant instruments and experimental techniques enable us to make progress in the statements we accept. We make progress in our methodological principles by formulating strategies that give us greater chances of making conceptual progress, explanatory progress, erotetic progress, or progress in the statements we accept.

However, this is only part of the story. It is time to fill an obvious gap.

7. Verisimilitude

The friends of verisimilitude rise up to protest: not all scientific statements are true. They are right. Moreover, we sometimes make progress by improving the statements we accept, even though we do not attain truth. For all my efforts to avoid it, are we not finally stuck with the old problem of verisimilitude?

That problem has two parts. The first part concerns the search for exceptionless generalizations. Some sciences do offer exceptionless generalizations, and truth for such statements is typically harder to come by. However, there are some circumstances under which we naturally rank later generalizations as closer to the truth than earlier ones. Consider, for example, the bare neo-Darwinian statement—popular in the early days of critical thinking about group selection—that alleles disposing their bearers to forms of behavior that would aid others at costs to themselves would be selected against. For present purposes we can focus on two complicating conditions: (i) if the allele is associated with a disposition to favor only (or primarily) kin, it may be maintained in the population through inclusive fitness effects; (ii) if there is a structured population in which groups found descendant groups at different rates, dependent on internal conditions that are fostered by the presence of altruists, then, provided that the group founding process goes forward quickly enough, the altruistic alleles can be maintained.[32] Now we can formulate three generalizations: "Altruistic alleles are always opposed by natural selection," "Altruistic alleles are always opposed by natural selection except when there is an offsetting inclusive fitness effect," "Altruistic alleles are always opposed by natural selection except when there is a sufficiently strong effect from population structure

nificance because of their import for resolving debates within biology and geology about the age of the Earth. Significance can sometimes be serendipitous.

32. The first type of exceptional case is, of course, treated in Hamilton's classic pair of papers on inclusive fitness (1964). The second emerges from Maynard Smith's criticisms of group selection (1964) and has been further developed by a number of writers (see Roughgarden 1979 for a survey, and D. S. Wilson 1980, 1983 for the most general treatment; Sober 1984 provides relatively nontechnical expositions of the main ideas).

and group founding." A natural evaluation of these generalizations is to declare that the first is relatively close to the truth, that the second and third are closer to the truth than the first, and that the second and third are incomparable. However, anyone who assessed both the second and third as being closer to the truth than the original generalization would be able to construct a fourth generalization—"Altruistic alleles are opposed by natural selection except when there is either an inclusive fitness effect or a population structure effect"—and this would be ranked as closer to the truth than either the second or the third.

One can generate these results by appealing to the classical (Popperian) notion of verisimilitude.[33] In general, if there is a class A whose members have property B except under rare conditions $C, D, \ldots$, then the generalization "All A's are B" will be relatively close to the truth (exceptions will be infrequent). We will move closer to the truth by restricting the generalization to exclude one or more of the exceptional cases (C, D, etc.) and adding a true generalization about the properties of those A's that are subject to the exceptional condition.[34] In cases where there are *partial* treatments of the exceptions (as with my second and third generalizations, which cope with inclusive fitness and group structure separately) it is possible to combine them to produce a more inclusive account of the exceptions and so to improve on either of the partially successful generalizations.

Some cases, then, are equally unproblematic, either on my account or on traditional approaches to verisimilitude. Other kinds of comparisons are also easily made: some generalizations are preferable to others because they embody conceptual advances or because they figure in improved explanatory schemata. The residual context in which talk of verisimilitude is attractive can be studied by confronting a problem that arises for *particular* statements.

As I have emphasized, success in achieving exceptionless generalizations is by no means a sine qua non for good science. There are successful sciences in which accepted explanatory schemata contain full sentences (degenerate cases of schematic sentences in which all the schematic letters are replaced). In the Newtonian schema for giving explanations of motions in terms of underlying forces, for example, Newton's second law occurs. There are also successful sciences in whose schemata there are no such full sentences. Darwinian evolutionary biology has served us as an example.

The difficulties of achieving truth do not only arise when we generalize, nor do they always beset our generalizations. Chemistry has arrived at enor-

---

33. The reason is quite straightforward. The conditions for Miller's argument are not met in this case because the later generalizations are restrictions of earlier ones that avoid some of the original false consequences.

34. The process of "mopping up the exceptions" in science is typically not just a matter of restricting the scope of the generalization. This is clear in my original example, in which one does not simply declare that cases in which there are inclusive fitness or group structure effects are off limits but presents an analysis of how these factors would affect the process of selection. Scope restriction is only tolerable when the abandoned instances are of no significance.

mous numbers of true generalizations about the molecular structures of particular chemical substances—starting with Cavendish's poorly formulated, but correct, account of the composition of water. Conversely, in our ascriptions of values to magnitudes, we know very well that we are likely to make mistakes. The concept of verisimilitude seems to find a natural application here.

Imagine that an early scientist refers to a physical object and assigns a real number r as the value of a magnitude of that object. Later, another scientist assigns the value r'. Let the actual value be r^*. Assume that r and r' are both different from r^*. What both scientists say is false. But we feel a strong temptation to say that the utterance of the later scientist is closer to the truth than that of the earlier scientist, a temptation grounded in the simple fact that

$$|r - r^*| > |r' - r^*|.$$

There is no need for any very complex notion of verisimilitude to do justice to cases like this. Even though we admit that both statements are false, there is an obvious respect in which the later statement is superior to the earlier one. Both statements assigned a value to a physical quantity, and one of the assigned values was closer to the actual value than was the other. Instead of trying to achieve a linguistic ordering of statements, we let the world do the work of ordering for us. The philosophical problems of understanding progress are resolved by appreciating the multidimensionality of scientific practice, and thus focusing on truth for individual significant statements. Once this is done, the artificial problems that have been at the focus of much logically ingenious work on verisimilitude can be bypassed.[35]

Philosophers concerned with the history of science may applaud my dismissal of artificial puzzles, but they are likely to offer a different type of protest. How does my account handle the really difficult cases in which rival theorists both seem to have different pieces of the truth? If I am allowed my approach to conceptual progress then the comparison of Newton and Einstein does indeed turn on a simple comparison of generalizations. But how do we compare rival approaches to the character of light? Was Fresnel's assertion that light is a wave closer to the truth than the corpuscularian claims of some of his contemporaries?[36]

My *total* account of scientific progress should be able to compare rival *practices* in optics. That does not entail that I am committed to the task of saying whether Fresnel's "Light is a wave" is closer to the truth than Brewster's "Light is a stream of particles." Recall the motivation for the project of this section. Practices have many dimensions. Along *one* dimension, we

35. I believe that the apparatus I have developed for understanding conceptual and explanatory progress enables us to tackle difficulties that emerge in nonmonotonic approaches to the truth, and to understand the approximate truth of universal generalizations. But I shall not explore the ramifications of these questions here.

36. Larry Laudan suggested to me that my approach will only handle the easy cases, and offered the example of the competing claims about light.

can ask whether the set of accepted statements of one practice (Fresnel's) is progressive with respect to the set of accepted statements of another practice (Brewster's): that inquiry might be undertaken by considering whether Fresnel replaced significant falsehoods or insignificant truths with significant truths; or, using the ideas presented in this section, we might consider whether Fresnel's magnitude ascriptions were closer to the actual values than Brewster's. The account of progress I am offering is thus keyed to investigating the *fine structure* of what Fresnel and Brewster wrote and said (and what they did—for we can compare the instruments and experimental techniques they employed) and to resisting the comparison of slogans. "Light is a wave" and "Light is a particle" are, at best, gross advertisements for the two practices, slogans to be traded in the opening moments of debate.

"Difficult cases" can easily be manufactured for an account of progress (such as mine) by pressing the concept of verisimilitude where it should not be applied, or, more generally, insisting that progressiveness be gauged along some particular dimension. My account of verimisilitude does not allow us to rank "Light is a wave" versus "Light is a particle." Nor should it, for there is no plausibility in the claim that one of these is closer to the truth than the other. But my *total* account of progress does allow us to assess the practices of Fresnel and Brewster, enabling us to see that Fresnel made conceptual and explanatory advances, that he was able to answer correctly significant questions that Brewster could not (questions about interference and diffraction, for example).[37] We can *loosely* sum up all these advances by suggesting that the wave theory propounded by Fresnel was closer to the truth than Brewster's corpuscular theory—but this is only shorthand for a complex of relations of the types I have been at pains to characterize throughout this chapter, not a claim about the restricted notion of verisimilitude that figures in *part* of the story.

In the light of this section we can refine the account of instrumental and experimental progress. Very frequently in the daily practice of a science, there is a significant question of the form, What is the value of x for S? (What is the age of that stratum? How many neurons project from this nucleus into the telencephalon?) There are instruments and techniques designed to answer classes of such questions. An instrument or technique improves on an earlier instrument or technique if the values that it generates are closer to the actual values than those given by the earlier experiment or technique. But is it not possible that an instrument or technique could yield better results in some instances but worse results in others? Of course. Geologists know very well that certain kinds of isotopes are better for dating some types of rocks, other isotopes more reliable in other cases. The experimental practice of radiometric dating uses a motley of techniques precisely because there is no all-purpose decay process that can be used on all rock samples. Geologists combine separate approaches to fashion a mixture of techniques that will apply across

37. For illuminating discussions of the work of Fresnel and the acceptance of his ideas about light, see (Worrall 1978, 1989).

the entire domain they hope to investigate. The claim that experimental practice is progressive rests on the idea that the mixture of instruments and techniques now employed yields in each case a value that is at least as close to the actual value as that previously generated—or, if there are exceptions, that these are either rare, insignificant, or both.

8. Two Refinements

The account offered so far is deficient in two main respects. First, as noted in Section 1, I have sidestepped problems posed by the splitting and merging of fields. Second, in the last two sections I have overemphasized the goal of attaining truth (or of improving false statements) and neglected the important role played by idealizing theories. The present section will attempt to set these defects right.

When fields of science split, merge, or hybridize, comparisons among practices cannot properly be made by considering only a single ancestor and a single descendant. Evidently, if we compare an earlier undifferentiated practice of chemistry with a later practice of inorganic chemistry, we shall expect there to be losses. The obvious remedy is to consider *total* practice at the earlier and the later times. We can construct a map of the state of science at a time, and, for each ancestral field, identify its successor fields at the later time. Then we assess progress within a field by seeing whether the combined practice of the successor fields is progressive with respect to the practice of the ancestral field. Alternatively, we can take an ancestor-descendant unit to be generated in the following way: Start with some ancestral field and identify all its successors. Now consider all those fields which are ancestral to some one of the successors already picked out. Next, take each of the ancestral fields already singled out in the unit, and include all successors. Iterate the process until no new fields are introduced. Progress is then assessed within ancestor-descendant units by comparing the combined practices of the successor fields in the unit with the combined practice of the ancestral fields.

But this does not yet address one of the most significant aspects of the splitting and merging of fields, to wit the fact that the large-scale organization of scientific activity itself reflects a conception of the order of nature and that changes in this conception may be progressive. Here (as so often) Kuhn is suggestive (1977 31–65), using the histories of the physical sciences to show how some changes are not so much modifications within existing fields as redrawings of the map of science.

We can accommodate the idea of *organizational* progress, conceived as improvement of the accepted relations among the sciences, by extending our view of consensus practice, of conceptual, explanatory, and erotetic progress. I shall suppose that the consensus practice of a field includes some claims about the relationship of the phenomena of the field to phenomena studied by other fields, and that, in the fashion discussed in Section 11 of the last chapter, we may combine these partial views to a *broad consensus* vision of the structure of nature. This broad consensus vision underlies the organization of scientific inquiry, recognizing certain problems and projects as connected with one another and granting authority to a subgroup of the community which tackles the problems and pursues the projects. Organizational changes may redraw boundaries, modify ideas about the dependencies of fields on others, or incorporate methods of one field into those of another. All these kinds of change can contribute to varieties of progress that have already been described. The redrawing of boundaries may constitute conceptual progress, revision of claims about dependency can be explanatorily progressive, and integration of concepts and methods from one discipline sometimes provides strikingly more tractable versions of old problems, thus counting as erotetic progress. Examples are ready to hand, in Maxwell's unification of the theories of electricity, magnetism, and light; in the incorporation of the theory of the chemical bond within atomic physics; and in the development of economic models in ecology.

Let us now turn to the second deficiency that I diagnosed earlier. One of the most obvious features of some sciences (notably parts of physics, but also subdisciplines of ecology and evolutionary biology) is their employment of idealizations. The generalizations of phenomenological thermodynamics, of the kinetic theory of gases, of statistical mechanics, are not, strictly speaking, true.[38] Nonetheless, it is entirely legitimate to hold that physics made progress by first achieving the phenomenological generalizations, then those of kinetic theory, then those of statistical mechanics. Using the ideas of the last section, we can recognize the progress of thermodynamics by suggesting that successive generalizations specify functions that are closer to those actually involved in the dependencies of various quantities (pressure, volume, and temperature, for example). There is nothing wrong with this—as far as it goes. However, it fails to recognize a certain aspect of the progress of thermodynamics: the goal of the enterprise is not to advance a set of generalizations that are *exactly* true.

Start with a distinction. The statements of phenomenological thermodynamics, of kinetic theory, and of statistical mechanics are rightly counted as false when we interpret the terms that occur in them as referring to actual magnitudes of actual gases or of actual molecules. However, if we conceive of the referents of those terms as fixed through stipulation—a molecule of an ideal gas is to be a Newtonian point particle that engages in perfectly elastic collisions, and so forth—then the theories in question can be seen as simply elaborating the consequences of these stipulations. To use an old (and disreputable) terminology, the statements are true by convention. As I have argued elsewhere, when Quinean morals are properly understood, there is room for truth by convention (see Quine 1966, Kitcher 1983 chapter 4). What

38. For a penetrating exploration of this claim, see (Cartwright 1983). My own discussion of idealization in what follows is brief and plainly needs supplementing by attending to the important distinctions that Cartwright makes. For present purposes, however, I am concerned to show how the elaboration of idealizations can be accommodated within my general approach to cognitive progress.

is important is that the conventions should be grounded in appreciation of aspects of reality.

In the present context, the grounding is readily specified. Thermodynamics made progress by recognizing first that there are important relationships among the temperature, volume, and pressure of actual samples of actual gases, that these relationships can be complicated in various ways, but that the behavior of actual gas samples can be predicted and explained by comparing them with entities in a story (the story of ideal gases). Later, physicists made progress by recognizing that the properties of actual gases are dependent upon the mechanical interactions among the molecules of which they are composed. They saw that the task of describing these molecular interactions would prove formidably complex if the sizes of the particles and the possibilities of inelastic collisions were taken into account. Once again, the actual behavior of gases could be understood by comparing actual gases with the characters in a story. The story preserves the new insight into explanatory dependence—the dependence of thermodynamic properties on mechanical properties of molecules—while providing a simple way of highlighting the most important features of the dependence. In the kinetic theory, we find an idealized version of part of the actual explanatory dependence; statistical mechanics provides a more complete story.

In general, I propose that we view idealizing theories as true in virtue of conventions, and that we regard the conventions (if they are valuable) as grounded in a double achievement. The first part of the achievement is to recognize a hitherto unappreciated explanatory dependence. The second part is to see that the form of explanatory dependence can be articulated in detail by forgetting about some entities or properties that complicate the actual situations. Fields in which we do not idealize are those in which we aim to develop the explanatory dependence by using statements that are strictly true. Idealization is an appropriate substitute when we appreciate that the search for exact truth would bury our insights about explanatory dependence in a mass of unmanageable complications.

5

Realism and Scientific Progress

1. Facing the Music

The account of scientific progress that I presented in the last chapter promises old-fashioned virtues. It offers a broadly realist view of science: scientists find out things about a world that is independent of human cognition; they advance true statements, use concepts that conform to natural divisions, develop schemata that capture objective dependencies. Realism is suspect in many quarters, not only among those who are skeptical about the progressiveness of science but also for champions of alternative accounts of scientific progress.[1]

I shall not attempt to survey the varieties of realism that have figured in the recent philosophical literature nor to relate my own account to the realist proposals of others. Suffice it to say that realism is a doctrine with many sects.[2] The aim of this chapter is to investigate lines of criticism that have been developed against other broadly realist proposals, and to show that we can enjoy the old-fashioned virtues of my account of progress without abandoning important recent insights about the growth of science.

It is useful to divide my project into three main tasks. First, it will prove necessary to respond to those objections that question the coherence of the notions I employ, in particular the coherence of the realist conception of truth. Second, I must answer arguments designed to show that there is no viable account of knowledge that will fit with the claims of the previous chapter, that my picture of contemporary science as advancing truths about the world cannot be defended (at least not when truth is understood in the realist's preferred way). Finally, even if the attempts at impossibility proofs

---

1. Thus many contemporary sociologists of science, Bloor, Barnes, Collins, Shapin, Latour, Pickering, and others, will surely find my account of progress incredible, and see this as just another inevitable failure in the sequence of attempts to understand progress in science. Others, such as Laudan, Kuhn, and van Fraassen, will object to my realism but will suppose that an account of scientific progress can be given if it is framed in different terms. In some cases, members of both groups will offer the same criticisms, but it is always good to recall that the conclusions that they draw from the criticisms are not the same.

2. For some of the varieties, see the excellent anthology (Leplin 1984). Jarrett Leplin's introduction provides extensive coverage of the similarities and differences among rival views.

can be blocked, it remains to show that an account of scientific knowledge can be developed.

As will become apparent, there are many presentations of critiques of realism in which different kinds of objections blend. A criticism that begins by questioning the coherence of realism can shift to a demand for understanding how we can know that what scientists say is true. A specific objection to the possibility of certain kinds of knowledge may conclude with a challenge to provide a full account of knowledge. This chapter will take up criticisms of the first two kinds, endeavoring to defend the coherence of the realist concepts I employ and to turn back particular attacks on the possibility of attaining truth. The more ambitious task of advancing an account of scientific reasoning and scientific knowledge will occupy the next three chapters. Because of the blurring of boundaries among the types of objection, the treatment of some topics—for example, worries about underdetermination and theory-ladenness—will be incomplete. I can only promise that more will come later.

The challenges I shall consider can be distilled into a sequence of questions. (1) Is it coherent to suppose that the sciences aim at and attain the truth? (2) How can we know that we are making progress in any of the senses that I have delineated? (3) Doesn't induction on the past history of the sciences reveal to us that our current beliefs are overwhelmingly likely to be wrong? (4) Doesn't the history of the sciences show that the aims that scientists set for their enterprise vary from period to period? (5) Can I avoid a naive realism that takes us to have direct access to the facts about nature, unmediated by background theory? (6) Isn't my account subverted by the fact that our system of beliefs (including beliefs about our access to nature) results from interactions among people who "negotiate" what is to be accepted? (7) Does that account lead to an untenable form of realism in which it is assumed that nature has a determinate categorial and causal structure, the kind of mind-independent structure that has been dubious ever since Hume? I shall take up these objections in turn.

2. Rehabilitating Truth

Truth can seem commonplace, or utterly mysterious, depending on perspective. One type of concern about the kind of account of scientific progress I outlined in Chapter 4 proceeds by making appeals to truth look grandiose, even mystical. It is well to start by reminding ourselves of humbler uses of the idea.

Semantic facts concern the relation between language users and nature. In virtue of the state of the language user and the state of the rest of the world, there is sometimes a relation—the relation of reference—between the words spoken or written and items in the world. In consequence, the statement represents the world as being some particular way. The statement is true just in case the way in which the world is represented is the way it really is.

Philosophical concerns about truth, about the coherence of the notion of truth and of supposing that the sciences aim at truth, arise from this picture. Kuhn's own, highly influential, suspicions about talk of truth touch on some venerable questions:

> One often hears that successive theories grow ever closer to, or approximate more and more closely to, the truth. Apparently generalizations like that refer not to the puzzle-solutions and the concrete predictions derived from a theory but rather to its ontology, to the match, that is, between the entities with which the theory populates nature and what is "really there."
> Perhaps there is some other way of salvaging the notion of 'truth' for application to whole theories, but this one will not do. There is, I think, no theory-independent way to reconstruct phrases like 'really there'; the notion of a match between the ontology of a theory and its "real" counterpart in nature now seems to me illusive in principle. Besides, as a historian, I am impressed with the implausibility of the view. I do not doubt, for example, that Newton's mechanics improves on Aristotle's and that Einstein's improves on Newton's as instruments for puzzle-solving. But I can see in their succession no coherent direction of ontological development. On the contrary, in some important respects, though by no means in all, Einstein's general theory of relativity is closer to Aristotle's than either of them is to Newton's. (1962/1970 206–207)

This passage is both highly suggestive, and, I shall argue, deeply puzzling in certain respects.

Kuhn contends that a certain philosophical project, the project of "salvaging the notion of 'truth' for application to whole theories," is doomed. That project is abandoned in the account of scientific progress that I have given. But I do demand things that Kuhn seems disinclined to concede, to wit the conceptions of successful reference, adequate reference potentials, correct explanatory schemata, true statements, and improved false statements. All of these rely on the notion of a match between the scientist's representation of the world and what is "really there," a match Kuhn takes to be "illusive." Thus the first big issue, the one that will occupy us for this section, is whether we can make sense of the idea that conceptual/linguistic items (words, statements, schemata) match elements of reality.[3]

Kuhn continues by offering a different type of objection, one that tries to undercut our practice of claiming a match between our representations of nature and nature itself by pointing to large transitions in the history of science. As I shall try to show, the alleged instability of our conceptions

---

3. This project is a preliminary to answering questions that have been forcefully posed by Kuhn and many of his predecessors and contemporaries: how do we justify the claims that certain representations match what is "really there"? Attention to the worries raised by Popper, Duhem, Quine, and others will come later. But, as Popper himself clearly saw, the first issue to settle is the sense of realist appeals to truth by correspondence. My approach to this question will follow Popper's own lead, building on the work of Alfred Tarski (1936, 1944). (See Popper 1959 274 fn. *1; my own discussion is considerably indebted to Field 1972.)

of nature, supported by the invocation of Aristotle, Newton, and Einstein, leads into questions about the proper degree of pessimism we should have concerning our apparent achievements (to be discussed in the next section) and about the problem of losses in the history of science (briefly touched on in the last chapter and revisited in the final section of this).

Yet the exact force of Kuhn's criticisms is rendered unclear by his apparent sensitivity to the facets of science that motivate my account. Kuhn and I concur in avoiding the extremely global idioms that are employed in most discussions of progress, and Kuhn even seems inclined to limit his thesis to castigating "truth for whole theories." Why does he separate "the puzzle-solutions and the concrete prediction"? Why does he contend that later theories in mechanics (both Newton's and Einstein's) improve on their predecessors as "instruments for puzzle-solving"? Could it be that, at the level of individual statements, Kuhn is prepared to permit talk of truth and of approach to the truth? If so, how does such discourse avoid presupposing a match that is "illusive in principle"? If not, what exactly *is* a puzzle solution (since it cannot be a *true* answer to a question) and what does it mean for puzzle solving to improve?[4] Kuhn's view of progress appears to be directed at articulating some of the modes of progress discussed in the previous chapter. Yet there is an obvious drawing back, which ultimately generates a conception seemingly at odds with itself. Why? What is so bad about the idea of correspondence to (or match with) reality?

The correspondence theory of truth is often held to involve extravagant metaphysics, but, I claim, its roots lie in our everyday practices.[5] We explain

4. There is a general moral here. While truth is frequently dismissed sternly at the front door, it often seems to be warmly received at the back. Consider, for example, Laudan's (1977) attempt to explicate progress in terms of puzzle solving. Despite the centrality of the notion of puzzle solution to his discussion, Laudan says virtually nothing about what counts as solving a puzzle. There seem to be several possibilities: (a) something is a puzzle solution if it "works," (b) something is a puzzle solution if sufficiently many members of the community count it as a puzzle solution, (c) something is a puzzle solution if it is warranted to count it as a puzzle solution. The challenge for (a) is to explain what it means for something to "work" without presupposing some truthlike concept: this is not an easy task when the puzzles are closely linked to practical concerns (for reasons including but not exhausted by those indicated in Section 2 of the last chapter) and it becomes very hard when we consider the kinds of "conceptual" puzzles that occur in the theoretical reaches of the sciences. Option (b) seems to collapse Laudan's position directly into the kind of relativism he intends to avoid, and (c) poses the challenge of how to understand the notion of *warrant* without taking the concept of truth for granted. (I would make the same complaint against Michael Dummett's influential account of the semantics for mathematics— see especially Dummett 1977—and have sketched the argument in my 1983 142–143.)

5. The most sophisticated attempts to make correspondence truth seem mysterious and metaphysical are made in a more general context than that of the philosophy of science (see Putnam 1981, 1985, Stich 1990, Horwich 1990). I shall not provide a detailed rejoinder to them here for two reasons: first, dissatisfaction with Legend and with scientific progress stems from simpler—though related—considerations, and exploration of more purely "philosophical" arguments would lead into issues remote from my central themes; second, if the general attacks launched by Putnam (or others) were successful, my account of progress

and predict the behavior of our fellows by attributing to them states with propositional content (for celebration of this theme, see Dennett 1987). We explain and predict the differential successes of our fellows in coping with the world by supposing that there are relations between the elements of their representations and independent objects. Those with correct beliefs about spatial relations can navigate their way more successfully than those who have faulty beliefs, and they can do so because their beliefs correspond to the ways in which the constituents of the local environment are arranged. Simple psychological ideas form part of the explanation of everyday behavior. Simple semantical ideas, added to those simple psychological ideas, explain the consequences of behavior.[6]

Correspondence truth, I suggest, begins at home. Few are born antirealists, and those who achieve antirealism typically do so because it is thrust upon them by arguments they feel unable to answer. In the present instance, the arguments derive from the denial of the theory-independent perspective.

Kuhn's sympathy for this line of criticism is suggested in his phraseology: "There is no theory-independent way to reconstruct phrases like 'really there'; the notion of a match between the ontology of a theory and its 'real' counterpart in nature now seems to me illusive in principle" (1962/70 206). Now it is, of course, right to insist that any *description* of what is "really there," however abstract we try to make our formulation, will presuppose some language, some conceptualization of nature. Equally, if one thinks of a match between cognitive/linguistic items and independent nature as requiring some possible process of *matching,* then, because there is no Archimedean point from which both sides can be viewed, the notion of match will come to seem "illusive." The common root of both ideas is that we have no access to nature which does not involve some elements of some scientific practice—a point which is uncontroversial once one has recognized that the categories and beliefs of common sense are themselves parts of primitive scientific practices.[7] But why should this doom the idea that there is something independent of us to which we have access through processes that are dependent on the state of current science/common sense?

---

could be recast in the terms that the critic favors for talking about truth and reality (the "internal realist" idiom in Putnam's case). But, I believe, such attacks can be turned back by taking seriously the idea that semantic relations can be identified with thoroughly natural (physicalist) relations—that is, working out the program advertised in (Field 1972). For the present, that must simply remain a promissory note.

6. It is not hard to see that this perspective on the role of semantics in everyday practice foreshadows the popular argument for scientific realism in terms of the successes begotten of our scientific practices. For more discussion of this see §§3–5 of this chapter.

7. This theme is even more apparent in the writings of Feyerabend than of Kuhn (see Feyerabend 1963a, 1963b). The emphasis on common sense as itself a primitive theory is sharply directed against the partisans of "ordinary-language" philosophy (Feyerabend 1963a fn. 99 expresses this clearly and wittily). For many philosophers of science, the point hardly came as a shock. Popper, for example, had already detected theoretical involvement in our most basic, commonsensical observational claims (1959 111).

The move from the theory dependence of our *perception* of nature to the theory-dependence of *nature* is apparent in Kuhn's discussion of Priestley and Lavoisier.

> At the very least, as a result of discovering oxygen, Lavoisier saw nature differently. And in the absence of some recourse to that hypothetical fixed nature that he "saw differently," the principle of economy will urge us to say that after discovering oxygen Lavoisier worked in a different world. (Kuhn 1962/70 118)

Realists have their preferred picture of what happened to Lavoisier. There was a constant fixed nature. As a result of a series of interactions with this nature, Lavoisier's cognitive system underwent some changes (the changes we describe by saying that he acquired the concept of oxygen, came to believe that combustion involved the absorption of oxygen, and so forth). In consequence, his subsequent interactions with nature activated new propensities and induced PI beliefs which would not have been induced in his former self (or in his friend Priestley). Realists add that Lavoisier's new beliefs are better than his old ones because, in some instances, they match the way nature is. Kuhn's challenge to this story seems to be that its postulation of a fixed theory-independent nature is otiose. Since we have no independent access to nature, we are just making up an unnecessary story about the episode. How could we ever know that this is the way to view what occurs? How could we ever tell that Lavoisier's new beliefs match nature more closely than his old ones (the beliefs Priestley stubbornly retains)?

Kuhn's suggested economy is false economy. One way to recognize the work that the idea of a fixed nature does is to see how we are impoverished by giving it up. One difficulty is that it becomes troublesome to understand the improvements in "concrete predictions" and "puzzle solving." We believe that such improvements occur, for example in the transition from young Lavoisier to mature Lavoisier, because we recognize that there is increased harmony within our cognitive system: PI beliefs that are now induced in scientists can be accommodated more easily within the total set of beliefs. Realism provides a picture of the genesis of perceptual belief that attributes a causal role to something beyond our cognitive systems. That picture has the advantages that it offers an explanation of the apparent fact that some of our beliefs come to us unbidden, it allows for the contents of our perceptual beliefs to be partly determined by our prior cognitive state, and it enables us to understand our seeming ability to achieve greater cognitive harmony in terms of increased match between our representations and an independent reality. To demonstrate that the "hypothetical fixed nature" is unnecessary, antirealists must provide a rival picture that has similar virtues.[8]

8. The realist response does not claim to issue any guarantees here. There is no completing the Cartesian project of suspending all beliefs except those that are absolutely certain and then demonstrating on this basis that the realist account of nature and our relation to it is correct. Instead, realists counter with a challenge: what rival account will explain how we are able to achieve greater cognitive harmony in coping with inputs that

Serious epistemological questions remain. How do we know that the statements of the mature Lavoisier match reality more closely than those defended by his younger self? Obviously not by engaging in some out-of-theory experience that reveals to us both our own representations and the aspects of nature to which they are supposed to correspond. Our belief that Lavoisier made progress rests partly on the evidence for Lavoisier's chemistry, partly on our scientific understanding of the relationships between human cognitive systems and the world. As we shall see in later sections, significant challenges to claims about scientific progress can be generated by attempting to use science against itself, contending that we have good scientific reasons for thinking that our abilities to represent nature are inadequate in certain specifiable respects. Because these attacks are *internal* challenges, conceding the notion of a fixed, independent reality and arguing about our relationship to it, I shall postpone them for the present.

Let me close with a very brief reply to the worry that the notion of correspondence truth only makes sense for the simplest kinds of sentences— those which ascribe a property to an object, for example. Since Tarski's pioneering work on the concept of truth, we have been able to lay this worry to rest. Tarski showed (1936, 1944) how to reduce the notion of truth to that of reference (see Field 1972). Because we can hope to understand reference naturalistically, as a relation between people and other entities, there is nothing mysterious, magical, or supernatural about the concept of truth. But, even in advance of providing a detailed naturalistic understanding of reference, we can know that truth is as coherent as reference, thus turning back a more limited worry.[9]

3. Pitfalls of Pessimism

Contemporary biologists agree that human beings and chimpanzees had a common ancestor no more than a few million years ago. I claim that their assertion is true. Critics object. They demand to know why I think that what the biologists say is true. How should they be answered?

I shall start by distinguishing two forms of antirealism. *Global* critiques of realist epistemology contend that we have no basis for claiming that *any* statement is true (when truth is understood in the realist's preferred way).

---

we are apparently unable to control? For a more detailed articulation of this general line, see (Quine 1960).

9. This is not so much an explicit thesis in the writings of historians and sociologists as a pervasive attitude—indicated in eschewing talk of truth or reality or in uses of scare quotes. I have often heard it said in discussions that Kuhn has "disposed of notions like truth" or has shown that "there is no theory-independent world." Because such claims are so popular, any defense of a realist picture of science has to begin by showing that the notion of truth makes sense and that it is a mistake to slide from the theory dependence of our representations of the world to the theory dependence of the world. This is as far as I have hoped to go in the present section.

So the global antirealist would be skeptical about the truth of the biologists' claim as a consequence of a sweeping generalization. By contrast, *local* critiques of realist epistemology allow that we are entitled to maintain that some statements are true (in the realist sense) but they complain that statements which realists take to be true, statements with some special feature *U*, outrun this entitlement. So, one kind of local antirealist would deny our justification for accepting as true statements about entities in the remote past and would base skepticism about the biologists' thesis on this denial. The most famous form of local antirealism—one that has flourished in our century—identifies the special feature *U* as that of making reference to unobservables.[10] I shall argue against this version of antirealism later, but I shall also suggest that there are more congenial versions, indeed versions that give my account of progress everything that it needs.[11]

Imagine, then, that we face an antirealist who doubts that what the biologists say about our ancestry is true. An obvious first response is to elaborate the evidence that has convinced people about the recency of human-chimp common ancestry. So one can explain the general idea of a molecular clock, show how molecular clocks can be calibrated and how they give an estimate of about five million years for separation of the human and the chimpanzee lineages, and, as the coup de grace, rehearse the details of the fossil record. This will not satisfy a global antirealist. For beyond the issue of demarcating statements like this, which are accepted, from others (such as, Human beings were directly created along with all other animal kinds about 10,000 years ago) which are rejected, global antirealism's prime concern is how *any* statement might earn the title of truth, when truth is conceived as in the realist picture.

But there is an answer to this question. We have scientific views about the relations between ourselves and the rest of nature. In the light of these scientific views we can evaluate the likelihood that we are right about various kinds of things. So we examine the procedures that are involved in accepting the hypothesis about human-chimp separation and show, by appealing to background ideas about reliable detection of aspects of nature, that such procedures would regularly deliver truth (understood in the realist way).

10. Classical instrumentalism does not quite take this form, since, for the instrumentalist, it is not merely that we cannot justifiably assign truth values to statements purporting to be about unobservables but this epistemic remoteness deprives the statements of meaningfulness. The local antirealism that I characterize in the text is closer to the views of Bas van Fraassen (1980), which are discussed later, and, perhaps, to those of the many nineteenth century scientists who worried about atoms, electrons, and the ether.

11. There are also global critiques of the type of realism I favor which *might* allow my account of progress everything it needs, for example Arthur Fine's "natural ontological attitude" (NOA) (Fine 1986). I must confess to finding NOA elusive: in his attacks on realism, Fine seems to become an antirealist, and in his rejection of antirealism, he appears to become a realist. Plainly this assessment is at odds with NOA's professed goal of slipping between two antagonistic positions. If it can indeed find its way, then, perhaps, it can accommodate large parts of my account of scientific progress (assuming, of course, that those survive the criticisms considered later in this chapter).

Other procedures, such as those underlying the creationist claims that are rejected, would not.

Thoughtful antirealists will not be entirely convinced. A first worry will focus on the appearance of circularity. The envisaged defense of the truth of the claim about human-chimp separation requires acceptance of certain parts of contemporary science. Someone suspicious about molecular clocks or about the fossil findings will not be persuaded. Nor, by the same token, can we expect to satisfy those who think that our current physical-physiological-psychological conceptions of the relationship between human cognitive systems and the environment are mistaken. Skeptics who insist that we begin from *no* assumptions are inviting us to play a mug's game. Descartes's lack of success in generating an account of nature that would survive all possible doubt was in no way the result of deficiencies of intellect or imagination.

The global antirealist aims to challenge the status that is conferred on accepted science as a whole. To respond it is necessary to show how our *scientific* conception of the physical-physiological-psychological relationships between human cognitive systems and nature supports theses to the effect that certain types of processes and procedures yield true beliefs—representations that match nature—and to demonstrate that it is processes and procedures of these privileged kinds that are involved in the generation of accepted science. By hypothesis the skeptic accepts our current science, and that will include the conception of the relations between our cognitive systems and nature.[12] Hence the strategy seems entirely appropriate, for it starts with premises that are accepted by those we intend to address. Moreover, even though science is being used to defend itself, there is no guarantee of success in our project. Application of our conception of the relation between human cognitive systems and nature might convince us that some procedures crucial to various parts of accepted science are unreliable: suppose, for example, that we were able to show that large parts of contemporary genetics rest upon experiments in which people are required to discriminate mutants in ways that can be shown to outrun the abilities manifested in the test situations of perceptual psychology. As we shall see, considerations of this type generate versions of antirealism.

---

12. Nothing very grand needs to be involved in the early stages of this venture. Our dialogue with the global antirealist might begin with our outlining some findings about the performance of subjects in standard physiological and optical conditions when they are confronted with objects of various shapes and colors. If the global antirealist endorses claims about the regular covariance of belief with the stimulus objects, we can use this as an entry to claims that we have ways of attaining truth, conceived realistically. If not, then the global antirealist must dissent from so much of our contemporary scientific picture of the world that we have an insufficient basis for responding to his skeptical challenge. Here all there is to be said is that if the set of premises that he will allow is so curtailed, we have no chance of generating the conclusion he demands: this failure should be no more surprising than Descartes's inability to issue a complete guarantee for the truth of all those claims that Meditation I calls into doubt. A priori guarantees are not to be had, and if the antirealist refuses to allow us to appeal to relatively uncontroversial a posteriori premises, then the task that he sets is impossible, and this should simply be admitted.

Clever antirealists see that the issue is whether all or some of our accepted beliefs have a particular status, and they use the fact of scientific change to play reason against itself. Their best strategy is the so-called pessimistic induction on the history of science.[13] Here one surveys the discarded theories of the past; points out that these were once accepted on the basis of the same kinds of evidence that we now employ to support our own accepted theories, notes that those theories are, nevertheless, now regarded as false; and concludes that our own accepted theories are very probably false.[14] It should be apparent from my continued emphasis on practices rather than on theories that I shall want to reformulate this pessimistic induction before replying to it. But, even before doing so, I want to suggest that the history of science provides grounds for optimism as well as pessimism.

There are a number of ways to present an optimistic view of the history of science. One (offered in Devitt 1984 146) is to suggest that we look at the track record of successful uses of theoretical terms. Perhaps most of the posits introduced by our remote predecessors are entities we no longer countenance, but, as time goes on, we find that more and more of the posits of theoretical science endure within contemporary science. We attribute this increased success to improved abilities of scientists to learn about investigating the world. Hence, as Michael Devitt points out, the pessimistic observation that a lot of past theoretical science has been discarded is not enough. The skeptic needs to show that "the history of unobservable posits has been thoroughly *erratic*" (Devitt 1984 147).

Although this line of argument is suggestive, and although it makes the important point that confidence in the posits of contemporary science could be justified, even if our first efforts were discredited, provided that our practice of positing improved with time, it is vulnerable to some important challenges. Antirealists will deny that they have to show that "the history of unobservable posits has been thoroughly erratic," noting that confidence would be undermined if there were a relatively constant tendency, at all epochs in the history of science, to introduce posits which, roughly as often as not, had to be subsequently abandoned. They will also point out that if endurance simply amounts to temporal persistence, then there were many entities introduced within ancient science—epicycles, humors, and so forth—that survived for a very long time before being discarded. To develop Devitt's argument, one needs to capture the idea that the stability of modern references to molecules, genes, and extinct organisms throughout periods of great theoretical proliferation somehow redounds more to their credit than the persistence of Aristotelian and Galenic conceptions through the Middle Ages. Finally, anti-

13. Sometimes known as "the disastrous meta-induction" (see Putnam 1978). I shall examine the most sophisticated version of the strategy—due to Larry Laudan—in the next section.

14. At this point there appear to be two options. One *local* antirealist conclusion is to claim that this shows only that our claims about unobservables are suspect. Global antirealists, perhaps impressed by the theory-ladenness of all our beliefs, draw the more ambitious conclusion that all our current claims about nature are dubious.

realists will, quite legitimately, demand careful specifications of the careers of various kinds of posits that have been introduced in the history of science.[15] Ironically, this point will be central to my own defense against antirealist objections later.

For the reasons rehearsed in the last paragraph, I do not think that Devitt's insight furnishes a complete reply to the antirealist. Another line of response is to focus on comparative judgments about the merits of past theories (or components of practice). Consider the following optimistic induction:

> Whenever in the history of science there has been within a field a sequence of theories $T_1, \ldots, T_n$ (or explanatory schemata, statements, terms, instruments, . . .) such that, for each i, T_{i+1} has been accepted as superior to T_i, then for every j greater than $i + 1$, T_{i+1} appears closer to the truth than T_i from the perspective of T_j (more correct or complete in the case of explanatory schemata, equipped with a more adequate reference potential in the case of terms, and so forth).
>
> So, we can expect that our theories will appear to our successors to be closer to the truth than those of our predecessors.

The intuitive idea behind the optimistic induction is a very straightforward one. We believe that Priestley was wrong, Lavoisier was wrong, Dalton was wrong, Avogadro was wrong, and so on. But we also think that Lavoisier improved on Priestley, Dalton on Lavoisier, Avogadro on Dalton. So while we do not endorse the claims of our predecessors we do support their sense of themselves as making progress. In consequence, we expect that our successors will support our sense that we have made further progress.

Antirealists might reply that they too can endorse the progressiveness of the sequence of chemists mentioned, although they would try to explain this in different terms. But this misses the point of the optimistic induction. That induction is intended as a counter to pessimism. The structure of the pessimist's argument is as follows: "Suppose that we assign to the theoretical claims of contemporary science the status that the realist suggests. Notice, then, that our predecessors might have done the same. But the realist doesn't want to confer the privileged status on *their* claims, most of which are now discredited. So how do we justify treating the theoretical claims of contemporary science differently?" To this, the optimistic induction replies as follows: "Sensible realists should abandon the idea that contemporary theoretical claims are literally true in favor of the notion that there has been convergence to the truth. So we accept the pessimist's point. Notice however that this leaves untouched the idea that the historical development can be seen in terms of convergence on the truth, understood in the realist's preferred way. Using the realist's conceptions of increasing truthlikeness of statements (improved false statements, more adequate reference potentials, more complete or more correct schemata) we can take the very same examples that the pessimist uses to undermine our confidence in the literal truth of contemporary theoretical

---

15. In formulating this part of the debate I have been helped by the comments of Rachel Laudan and Larry Laudan.

claims and show that they do not invalidate reformed realism's contentions that our theoretical statements are improving, our schemata becoming more complete, our concepts more adequate, when all these are understood in the realist's preferred way. So the pessimistic induction has no force against a properly formulated realism." The optimistic induction, then, is not an attempt to show that the history of science *must* be interpreted in terms of the realist's conception of progress (although it is worth asking the antirealist to explain a rival account of progress in detail and to show that it does not make surreptitious appeals to truth) but to blunt an apparently damaging challenge to explain the difference between the science of the past and the science of the present.

Suggestive as they are, these observations strike me as altogether too weak. The pessimistic induction relies critically on the gross analysis of science as a constellation of theories which are supposed to be true and which are, historically, failures. I oppose to it a simple point. The history of science does not reveal to us that we are fallible in some undifferentiated way. Some kinds of claims endure, other kinds are likely to be discarded as inaccurate. *Furthermore this is exactly what we would have expected given our conception of the relationship between human cognitive systems and nature.* According to that conception we are relatively good at finding out some things, and discover others only with relative difficulty. We would expect that, in the latter sorts of endeavors, history would be peppered with false starts and plausible, but misleading, conclusions. With respect to the former projects we would anticipate doing better.

Instead of a blanket pronouncement to the effect that our current theories are probably wrong, it would be far more instructive to investigate the stability of various components of practice in various fields. *Perhaps* this investigation would pose a challenge to parts of contemporary science, by revealing that processes that we take to be reliable and that are critically involved in supporting those parts of contemporary science have, in the past, given rise to large numbers of erroneous or inadequate ideas. But this needs to be shown. I shall argue later that the usual stock of examples which is supposed to embarrass realists only issues in a form of antirealism that leaves my account of progress untouched.[16]

Before I turn to this project, let me list some of the questions that would have to be addressed in a more sophisticated study of the bearing of the historical record on the status of current science. (1) How stable are the reference potentials of scientific concepts? To what extent are refinements of reference potential cumulative? (2) How frequently do scientists discard—

16. In effect, the investigation that I recommend here turns the antirealist use of the pessimistic induction on its head. Instead of starting with a preconceived thesis about the feature U that distinguishes those statements that outrun our entitlement to claim truth, we might use the history of science as a *partial* source for discovering our limits. Appeals to history of science would be used in conjunction with studies of the cognitive performance of subjects under natural and artitifical conditions to find out where we are most likely to make mistakes and where we are legitimately confident.

rather than extending, correcting, or completing—explanatory schemata that have once become accepted? (3) What kinds of statements, once accepted, are abandoned as false? Is the process of improving false ascriptions of magnitude cumulative? (4) When are instruments and experimental techniques abandoned? How often does it happen that an instrument or technique comes to be viewed as inferior to the instrument or technique it originally replaced?

Nobody has made the systematic survey of the history of science that would be needed to provide adequate answers to these questions. In my idiosyncratic review, the preliminary observations should incline us to optimism rather than pessimism. Many scientific concepts—those of planet, acid, and Down's syndrome furnish good examples—undergo a cumulative process of refinement of reference potential (cumulative in the sense that later users endorse the reforms made earlier) before achieving reference potentials that endure unmodified through many changes in the sciences in which they are employed. As the last chapter indicated, the histories of chemistry and evolutionary biology reveal the endurance of explanatory schemata and a cumulative process through which those schemata are improved.

But of course people make mistakes. Our observational and inferential procedures for generating belief are fallible. They are not *equally* fallible. Nor are our foibles beyond correction. Contemporary science provides a picture of the physiological-physical-psychological relationship between human cognitive systems and nature, and we are able to use this picture to appraise and improve our performances.[17]

In the rest of this section, I want to take up the most sweeping use of the pessimistic induction, the global antirealist contention that the past history of science shows that we should not count *any* statement as true (in the realist sense). To defend antirealism about observational claims—either those based on unaided observation or those made with the assistance of instruments—it would be necessary to show how history of science reveals the presence of unanticipated sources of pervasive error in arriving at particular kinds of beliefs in particular kinds of ways. Now there are instances in which projects of measurement encounter difficulty after difficulty, others in which we seem to go from success to success. Microscopes, telescopes, electron microscopes, spectrometers all seem to have delivered a body of claims with high and increasing rates of stability. Define the success rate of an instrument or technique at a time to be the percentage of statements generated by users of the instrument or the technique at the time (judged as competent by the standards of the time) that have endured into the present (possibly with the kind of

17. It is important to recognize that while part of this picture is extremely recent and, in some respects, controversial, other aspects of our views about cognition are well entrenched and have survived millennia. They are implicit in our everyday practices of adjusting our movements so as to enter into causal contact with things about which we want to know. When I want to find out whether the plants on the patio are too dry I go outside and look at them, sometimes pushing my fingers into the soil. Doubtless millions of other people have done very similar things for thousands of years, and their behavior signals a tacit conception of the relationship between their cognitive states and the world.

reconceptualization that occurs when reference potentials are refined). Optimism about our current instruments and techniques is fostered if we discover that they stand at the end of a sequence of such techniques, whose initial members had high rates of success and whose subsequent members have exhibited successively higher success. For that discovery fosters the idea that we have learned about nature and, as Devitt suggests, learned ever more about learning about nature.

Further historical research could cast a pall over our optimism. We might find out, for example, that long sequences of apparent successes were punctuated by embarrassing reversals, prompted by recognition that what we took to be observational/instrumental/technical improvements were based on quite faulty understanding of the relationship between ourselves and nature, so that an instrument or experimental technique had misled us across the board. If this were to occur frequently in the history of science it would challenge our complacency about current instruments and techniques by raising the probability that there are unanticipated sources of error even for long sequences of high and improving success rates. Yet, as I have indicated before, the pessimist has to do considerable historical work to show that this is the case.

Even more work is required to demonstrate that the use of unaided observation to arrive at claims about the macroscopic properties of medium-sized objects is unreliable. For our descriptive lore about substances and organisms seems remarkably stable, enduring from antiquity into the present with, occasionally, episodes of reconceptualization. Given the account of conceptual change developed in the past chapter, we can thus defend against the suggestion that history reveals that we should not claim truth for our ordinary judgments about ordinary things.

Quite plainly we are more likely to be wrong when we advance general conclusions, when we make claims about things that are causally remote from us, and when we attempt precise specifications of magnitude. Our scientific picture of nature provides us with ideas of the limits of our accuracy (at least in some cases). The world may be such that our ability to achieve true generalizations about it may vary significantly from field to field—and we may use what knowledge we do acquire to differentiate those instances in which pessimism is warranted from those in which we are, justifiably, optimistic.

4. History Revisited

Global antirealism is not usually adopted as a direct consequence of a pessimistic induction on the track record of scientific theorizing.[18] Instead of

18. Nonetheless, global antirealism is not without supporters. There are many passages in his (1962/70) that seem to commit Kuhn to global antirealism (see the passages cited previously, and almost all references to the idea that scientists separated by a revolution live in "different worlds"). I take these to stem in large part from the worries about the coherence of realist notions canvassed in Section 1. Many sociologists of science are plainly

contemplating the pessimistic conclusions that philosophers *might* be able to generate by reflecting on the history of science, let us now turn to considering the *actual* arguments that have been offered, arguments that are typically concerned to reach *local* antirealist conclusions. One of the most vigorous opponents of the claim that science makes progress by generating theories that are successively closer to the truth is Larry Laudan (see his 1981, 1984 chapter 5).[19] A major part of Laudan's strategy is to assemble a parade of examples designed to embarrass those who think that successful science always involves theories whose central terms refer.

Laudan contends, "What the history of science offers us is a plethora of theories that were both successful and (so far as we can judge) nonreferential with respect to many of their central explanatory concepts" (1984 121). He bases this judgment on a survey of ether and subtle fluid theories in the eighteenth and nineteenth centuries, and on a list that includes the "humoral theory of medicine," "catastrophist geology, with its commitment to a universal (Noachian) deluge," "the phlogiston theory of chemistry," and several more examples. Laudan concludes:

> This list, which could be extended ad nauseam, involves in every instance a theory that was once successful and well confirmed, but which contained central terms that (we now believe) were nonreferring. Anyone who imagines that the theories that have been successful in the history of science have also been, with respect to their central concepts, genuinely referring theories has studied only the more whiggish versions of the history of science (i.e. the ones that recount only those past theories that are referentially similar to currently prevailing ones). (1984 121)

---

convinced of the barrenness of talking about truth—see, for example, Latour and Woolgar (1979), Shapin and Schaffer (1985), Collins (1985). Their global antirealism may simply be based on Kuhn's influence, or it may result from the idea that a pessimistic induction shows *at least* some version of local antirealism and that the recognition of theory-ladenness entails that no distinctions can be drawn among statements.

19. Although it is clear that Laudan does not question the claim that science makes progress, I am not certain whether he should be counted as a local antirealist who takes realism to overstep its bounds in defending the truth of "high theory" (a local antirealist who takes *U* to be the property of being a "theoretical" statement). If Laudan were to resist realist claims about the truth of nontheoretical statements, then he would seem to assume the burdens of global antirealism mentioned in the last section. Indeed, his arguments from the history of science do not seem to be at all directed toward the sweeping conclusions that a global antirealist would have to establish. Furthermore, accepting the idea of realist truth for nontheoretical statements would enable him to understand the idea of a puzzle solution in an obvious way, namely as the production of true, nontheoretical, answers to questions. The cost, quite evidently, would be to embrace a distinction very like that once drawn between observational and theoretical statements, and, since Laudan accepts the Kuhn-Feyerabend-Hanson critique of theory-independent observation, this would create considerable tension within his position. That tension is muted in practice because of his reticence in specifying the conditions that are required of a puzzle solution. In what follows, I shall ignore difficulties of interpretation, focusing on whether Laudan's historical examples serve to show that the kinds of statements with which he is centrally concerned fall beyond the range of realist entitlements.

Not so, I claim. Laudan's argument depends on painting with a very broad brush. We are to think of "theories," "success," and "referential central terms."[20] This discourse yields at best only spurious correlations.

To diagnose part of the trouble, let us look at an extension of Laudan's argument, intended to block the charge that matters have altered in the twentieth century.

> Consider, for instance, virtually all those geological theories prior to the 1960's which denied any lateral motion to the continents. Such theories were, by any standard, highly successful (and apparently referential); but would anyone today be prepared to say that their constituent theoretical claims— committed as they were to laterally stable continents—are almost true? (1984 123)

There are two points that ought to be conceded to Laudan at the outset. First, there were indeed large parts of early twentieth century geology that offered successful explanations and predictions of various phenomena, and which survive, sometimes reconceptualized, in current geological practice. Second, although there was, from 1915 on, a minority tradition that asserted the lateral motion of continents, the individual practices of most influential geologists, especially in the United States between the wars, contained the statement that the continents do not move laterally. Is this enough to support Laudan's antirealist conclusion? I do not think so. For the point of the excursion through the history of science is to convince us that the apparent successes of contemporary science might be misleading, that the terms of the languages of currently successful practices do not refer. To produce conviction on this point Laudan must establish more than the points I have conceded. It will be necessary to show that the denial of the lateral motion of continents actually plays some role in the successes of the old practice. No sensible realist should ever want to assert that the *idle* parts of an individual practice, past or present, are justified by the success of the whole.[21]

20. In Laudan's defense it might be noted that this is the vocabulary used by the proponents of realism whose claims he is out to discredit (see, for example, Putnam 1975). My own realist conception of progress builds on the framework introduced in Chapter 3, and, I shall argue, provides the means to distinguish what is correct in Laudan's historical claims from his antirealist conclusions in ways that were unavailable to the realists whom he attacks.

21. This point is akin to one of Hempel's great insights about the logic of confirmation: one cannot generate an acceptable account of confirmation by supposing both that (a) if e is a consequence of h then e confirms h, and (b) if h' is a consequence of h then any statement that confirms h confirms h'. Joint acceptance of these conditions yields the conclusion that any e confirms any h. For (by (a)) e confirms e & h; h is a consequence of e & h; so (by (b)) e confirms h. (See Hempel 1965, chapter 1.) One obvious moral is that confirmation does not accrue to irrelevant bits of doctrine that are not put to work in delivering explanations or predictions. Since sensible realists ought to accept this point, the extra demand made in the text is well motivated. As Laudan recognizes, at least one realist (Glymour 1980), perhaps the realist most attuned to Hempel's insights about confirmation, rejects the hyperextended holism that spreads success over all accepted sentences without asking which have been put to work.

Laudan's broad-brush way of discussing whole theories is reminiscent of the following "confutation" of those who think that successful basketball teams are typically those with tall players. We trot out examples of successful teams on which there is one diminutive person. It is, of course, important not to disclose the fact that this person has little or nothing to do with the team's success. Similarly, it is not enough to conceive a theory as a set of statements and distribute the success of the whole uniformly over the parts. One has to see how the statements are *used.*[22]

On the account of practices that I offered in Chapter 3 and used in analyzing scientific progress in Chapter 4, the set of explanatory schemata encapsulates the ways in which accepted statements are deployed. So, if this example is to do what Laudan requires of it, we need to show that the denial that continents move laterally figures in some schema of pre–twentieth century geology that was successfully employed in explaining and/or predicting. At this point, matters become difficult. The refusal to allow lateral continental motion is not so much a constituent of successful explanations and predictions—something that figures in the explanatory schemata of the practice—but a constraint, a boundary condition, with which the explanations/predictions of early twentieth century geology have to be compatible. To the extent that early twentieth century geology was successful, its achievements were in areas in which this constraint was irrelevant, areas in which, for example, recognition of the physical properties of rock strata (relative densities that we continue to accept) could be used to explain and predict geophysical features.

I can now offer a general diagnosis of what goes wrong in the examples on Laudan's list. Either the analysis is not sufficiently fine-grained to see that the sources of error are not involved in the apparent successes of past science or there are flawed views about reference; in some instances both errors combine. Let me now support this diagnosis by considering one of his central examples, the commitment to an all-pervading ether in nineteenth century electromagnetic theory/optics.[23]

Laudan contends that ether theories in the 1830s and 1840s were strikingly successful. This contention is a useful, but rough, description of a complex situation. Within the consensus practice of physicists in those decades there

---

22. Laudan (1984 116–117) offers some remarks that suggest that he can accommodate this kind of argument. He contends that any "weakening" of realist claims to allow that some "components" of successful past theories were false/nonreferential will prove disastrous for realism. But this is incorrect if the realist can distinguish between those parts of theory that are genuinely used in the successes and those that are idle wheels. Without that kind of distinction *everyone's* account of confirmation is going to be in trouble for the generic reasons presented in footnote 21. Moreover, I suggest that the move from thinking of "theories" to considering practices, and, specifically, to focusing on the set of schemata, will enable us to see how to make this distinction.

23. The geological case just considered is atypical of Laudan's inventory of examples, in that here Laudan does not make any claim about failures of reference of central terms. I had originally overlooked this difference, and I am grateful to Gary Hardcastle for convincing me that I was wrong.

were a number of accepted statements, well deployed in accepted schemata, which seemed to presuppose the existence of the ether. A number of eminent individual scientists also advanced claims about the nature of the ether, some of which secured wide acceptance ("The ether is an elastic solid"), others of which were controversial (the mathematical descriptions offered by Augustin-Louis Cauchy, George Green, and James MacCullagh). Before we can assess Laudan's thesis it is important to separate different enterprises within the physics of the period.

Laudan's central concern seems to be with the most obviously successful of these ventures

> Within the theory of light, the optical aether functioned centrally in expla-
> nations of reflection, refraction, interference, double refraction, diffraction,
> and polarization. (Of more than passing interest, optical aether theories had
> also made some very startling predictions. E.g., Fresnel's prediction of a
> bright spot at the center of the shadow of a circular disc was a surprising
> prediction that, when tested, proved correct. If that does not count as em-
> pirical success, nothing does.) (1984 114)

We need to ask two questions about this situation: first, what statements of the theory were deployed in providing these explanations and predictions? second, do these statements contain nonreferring terms—in particular, do they involve putative references to ethers or subtle fluids?

Consider, in this context, Fresnel's prediction of the Poisson bright spot.[24] On a superficial reading, Fresnel's derivation of descriptions of the fringes produced in diffraction is reassuringly familiar. He appeals to Christian Huygens's conception of the wavefront as a center of secondary wave propagation, applies the principle of interference, and computes the resultant wave motions at the screen (or the eye) using the mathematics that physics students learn from textbooks on optics. First, Fresnel shows how to resolve a single wave into two component waves that differ in phase by $\lambda/2$. He then calculates the wave motion at a point on the screen produced by an infinitesimal arc segment of the wavefront. Integration yields the entire contribution of the wavefront, and the resultant formula enables Fresnel to plot the expected distribution of illuminated and unilluminated zones (and, more generally, the intermediate degrees of partial illumination).[25]

24. Strictly speaking this consequence was not derived by Fresnel. It is a consequence of the general treatment of diffraction proposed in Fresnel's prizewinning memoir of 1818, published in 1826. (See Fresnel 1856/1965 247–382.) In the first note to this memoir, Fresnel explains, "M. Poisson communicated to [him] the singular result that the center of the shadow of a circular disc should be illuminated as if the disc were not there" (Fresnel 1856/1965 365). Although Arago and Fresnel himself subsequently confirmed this striking consequence, Poisson was never convinced (see Worrall 1989).

25. Geoffrey Cantor (1983) gives an accurate digest of Fresnel's method in the case of occlusion by a semiinfinite body (152–155). For Fresnel's own discussion, see the 1818 memoir in (Fresnel 1865/1956), especially 286–293, 313–316. Compare the contemporary treatments offered, at a relatively elementary level but with extraordinary lucidity, by Richard Feynman et al. (1963 I 29.1–29.7, 30.1–30.3), and, with rigorous mathematical detail by Max Born and Emil Wolf (1980 370–375).

Fresnel's treatment of particular problems reveals his introduction of an explanatory schema, applicable to questions of the form, What is the intensity of light received at point *P*?, which, omitting quantitative detail, can be understood as involving the use of Huygens's construction (the conception of the wavefront as a source of secondary propagation), together with the principle of interference to compute the intensity at the target point by summing the contributions along all possible paths. Fresnel's view of the dependency involved here (the intensity is dependent on the propagations along all possible paths from the wavefront) is endorsed in contemporary physics, his mathematics for articulating the dependency is enshrined in elementary texts and is embedded in a richer mathematical framework in advanced discussions. So, by contemporary lights, it is hardly surprising that his discussions of interference and diffraction were so strikingly successful. He was right about so much!

But, as Laudan's use of this achievement indicates, we are missing part of the story. Suppose that we probe a little more deeply, asking what Fresnel was talking about when he discussed wave propagation and offered his mathematical calculations of resultant intensities. Contemporary physics texts are very clear in specifying what *they* use Fresnel's mathematics (or the more refined treatment stemming from Kirchhoff) to discuss the propagation of the electromagnetic field in cases where the frequencies are very high (Born and Wolf 109–132, 370). Modern treatments do not make any reference to an all-pervading ether in which the light waves are propagated. But, for Fresnel and many of those who followed him, the existence of such an ether was a presupposition of the successful schemata for treating interference, diffraction, and polarization, apparently forced upon wave theorists by their belief that any wave propagation requires a medium in which the wave propagates. All the successes of the schema can be preserved, even if the belief and the presupposition that it brings in its train are abandoned. That, to a first approximation, is what occurred in the subsequent history of wave optics.

Do the schemata successfully employed by Fresnel and other wave theorists of the early and middle nineteenth century contain inextricable commitments to the ether? Do the terms figuring in them fail to refer because of Fresnel's understandably mistaken belief about the necessity for a medium of propagation? First, let Fresnel speak for himself. Here is his introduction of the general approach to light which he champions in the prizewinning memoir:

> Descartes, Hooke, Huyghens, Euler, thought that light resulted from the vibrations of an extremely subtle universal fluid, set in motion (*agité*) by the rapid movements of the particles of luminous bodies, in the same way that the air is stirred (*ébranlé*) by the vibrations of sounding bodies... (1818, in 1865/1965 248)[26]

26. Here and elsewhere, translations from Fresnel are my own. In the paragraphs that follow, I shall simply give page references to Fresnel's memoir as reprinted in (1856/1965 I).

Moreover, the ether makes a second entrance when Fresnel discusses the interference of light waves. First, his statement of the problem to be solved:

> Given the intensities and the relative positions of any number of systems of light waves of the same wavelength, which propagate in the same direction, to determine the intensity of the vibrations resulting from the joint action (*concours*) of these different systems, that is to say the velocity of the oscillations of the molecules of the ether (*la vitesse oscillatoire des molecules étherées*). (286)

Similar vocabulary occurs in his solution, as in the formulation of the equation of propagation.

Thus, representing by u the velocity of the molecules of ether, we have

$$u = a \sin[2 \, (t - x/\lambda)] \quad (288)$$

However, by the time Huygens's principle is applied to calculate the intensities of light received at a screen, partially occluded by a semiinfinite obstruction (313–316), his discussion is carried through in terms of light waves (*l'onde lumineuse*) and intensity (*l'intensité des vibrations lumineuses*) rather than by explicit references to the vibrations of the molecules of the ether.

Historians have rightly insisted that the ether played an important role in Fresnel's thinking.[27] Even my brief survey of some key points in the prize-winning memoir should enable us to see why. But we should not jump too quickly to the conclusion that the terms employed by Fresnel in his successful schemata fail to refer. Consider two hypotheses about the mode of reference of Fresnel's term 'light wave,' or, more exactly, about the tokens of that term that figure in his solutions to problems of interference and diffraction.

> *HR.* Fresnel's dominant intention is to talk about light, and the wavelike propagation of light, however that is constituted. He has, of course, a false belief about the medium of propagation. But, since his primary aim is to discuss light and its wavelike qualities, his tokens of 'light wave' in the solutions of the diffraction and intereference problems genuinely refer to electromagnetic waves of high frequency.

> *HN.* Fresnel's references are explicitly fixed through the descriptions he gives at the beginning of the memoir in characterizing the wave theory of light and in introducing the wave equation. Tokens of 'light wave' that occur later in the memoir—for example in the solutions to the problems of diffraction—thus have their references fixed by the description "the oscillations of the molecules of the ether." Since there is no ether, they fail to refer.

27. See the illuminating discussion by Cantor in his (1983), especially 158, and also Jed Buchwald's (1981) and (1989, especially 306–310). Buchwald is primarily concerned with Fresnel's treatment of polarization rather than with his discussions of interference and diffraction.

To generate Laudan's preferred conclusion—that some central terms of successful optical ether theories fail to refer—one must accept *HN* and reject *HR*. It is not enough to claim that the term 'ether' fails to refer,[28] for this term does not figure directly in the schemata that Fresnel employs so successfully. Laudan must contend that the vocabulary that does occur in the striking problem solutions is so thoroughly permeated by faulty references to the ether that that vocabulary is nonreferential.

There is a lot to be said for *HR*. The sole function of the ether in the prizewinning memoir—and throughout *most* of Fresnel's writings—is to answer to the felt need for a medium in which light waves propagate. Fresnel typically makes no detailed claims about the nature of this medium.[29] Why, then, should we give priority to a description making reference to an entity about which Fresnel would surely have admitted his almost total ignorance, rather than seeing his dominant intention as that of talking about the wavelike features of light, *however they happen to be realized*? Fresnel's situation is, after all, not so uncommon. People sometimes find themselves in the position to understand the internal structure of a system without knowing its constitution or its origin. Geneticists did not fail to refer to DNA even when they believed that the genetic material was composed of proteins, and physiologists were able to identify functions while thinking in terms of divine design and not the action of natural selection. Fresnel's success is based on his insights into the propagation of transverse waves, and his faulty beliefs about the *constitution* of those waves are irrelevant to his analysis. This, of course, is why his work has been so successfully encapsulated in contemporary physics.[30]

28. Even this claim might be challenged. One can argue, with some plausibility, that, on at least some occasions, Fresnel thought of the ether as "whatever it is in which light waves propagate," and such tokens could be construed as referring to empty space. I shall not pursue this line of response.

29. Thinking of light as a transverse wave, he concludes that the medium must be an elastic solid, a conclusion in which he is followed by subsequent nineteenth century ether theorists (see Schaffner 1972). Interestingly, the one place in which Fresnel genuinely puts the ether to work is in discussing problems of aberration, brought to the attention of optical theorists by the eighteenth century astronomer James Bradley. Fresnel regards the problem of aberration as one of understanding the interaction between light waves in the ether and moving ponderable matter, and, by making assumptions about ether-matter interactions (specifically by assuming a partial ether drag), he reaches results that agreed with the observations of the time. (See Schaffner 1972 20–29 for lucid discussion.) This is one instance in which Laudan's thesis is sustained: the problem-solution is successful and yet, by our lights, some of the vocabulary employed in it is nonreferential. Interestingly, he does not cite this aspect of the ether's apparent successes, perhaps because this particular achievement is so minor in comparison with the major accomplishments in treating interference, diffraction, and polarization.

30. Notice that, as in Chapter 4, I am committing "whig history" in one sense: using contemporary understanding of nature to identify how Fresnel's language is connected to the world and to see him as referring successfully. (Paul Churchland has suggested to me that "whig" here is an acronym for wildly hopeful insidious gerrymandering. On a correct understanding of reference, I deny that there is any need for trimming or forcing.) I am not engaged in the whig enterprise of forgetting or even downplaying Fresnel's beliefs about

The centerpiece of the prizewinning memoir is the juxtaposition of Fresnel's mathematical treatment of the propagation of transverse waves with the results of his experiments on diffraction. At this stage of the writing, as I noted earlier, explicit references to the ether are very much in the background. Fresnel talks about the intensity of the light at various positions on his screens—presenting numbers that record the intensities he has observed—and it is surely implausible to contend that his belief about the ether is so pervasive that these references are fixed through descriptions concerning the oscillations of ether molecules! By simple extension of this point, I suggest that his tokens of 'light waves' in these passages have their reference fixed to the waves whose combined propagation produces the intensities specified in his mathematical calculations, whatever the medium (if any) in which they are propagated. In other words, I propose that we treat Fresnel's language just as we treated Priestley's discourse in the last chapter (see Section 4), accepting *HR*.

If *HR* is correct then Laudan is wrong about this example. Let me now offer a more general strategy for coping with the broader class of cases to which the Fresnel case belongs. Recall that I distinguished earlier between those parts of the practices of nineteenth century physicists in which successful explanations and predictions were provided by schemata that were taken to presuppose the existence of ethers (of which Fresnel's work is an outstanding example) and the detailed attempts to formulate the principles that govern the supposed ethers. There is no doubt that nonreferential terms are centrally involved in many of the latter discussions. However, it is also clear that decades of work by extraordinarily talented scientists produced no set of principles adequate to all the functions that the ether was supposed to fulfil. An important by-product of these investigations was the provision of mathematical descriptions of elastic solids. (In this work, of course, we have explicitly idealized stories that are grounded in the properties of actual elastic solids.) Otherwise, the nineteenth century ether theories—the serious attempts to study the ether and its properties—were frustratingly unsuccessful and controversial.[31]

While Laudan's invocation of ether theories casts no shadow on the ref-

---

the ether. Those should be given their due but should not be seen as some sort of creeping rot that invades everywhere. Laudan's penchant for tarring past theories as a whole with local misconceptions seems to me to be as unbalanced as neglecting every aspect of the past that does not fit within contemporary views of nature. Moreover, it signally fails to explain how the textbook presentations of historical figures—the whig history of the science text—is ever possible in the first place.

31. Kenneth Schaffner adeptly chronicles the struggles in his (1972). There were occasional successes—witness Fresnel's "ether drag" hypothesis and the solution of the problem of aberration (see footnote 28)—but the overall picture is one of highly sophisticated theorists proving that yesterday's account of the ether won't do (because it is dynamically unstable, unable to yield some one of the properties required of the ether, and so forth). Interestingly, there were always doubters, such as George Airy, impressed with Fresnel's mathematical account of wave propagation but agnostic about the medium in which the waves were propagated.

erential status of those terms that are frequently used in successful schemata, we do learn something important from the example. Distinguish two kinds of posits introduced within scientific practice, *working posits* (the putative referents of terms that occur in problem-solving schemata) and *presuppositional posits* (those entities that apparently have to exist if the instances of the schemata are to be true). The ether is a prime example of a presuppositional posit, rarely employed in explanation or prediction, never subjected to empirical measurement (until, late in the century A. A. Michelson devised his famous experiment to measure the velocity of the earth relative to the ether), yet seemingly required to exist if the claims about electromagnetic and light waves were to be true. The moral of Laudan's story is not that theoretical positing in general is untrustworthy, but that presuppositional posits are suspect.

Laudan recalls, with considerable rhetorical effect, James Clerk Maxwell's remark that "the aether was better confirmed than any other theoretical entity in natural philosophy" (Laudan 1984 114; the formulation is Laudan's not Maxwell's). Although we can understand his claim, based as it was on the multiplicity of phenomena to which schemata appealing to wave propagation had been successfully applied, Maxwell was wrong. The entire confirmation of the existence of the ether rested on a series of paths, each sharing a common link. The success of the optical and electromagnetic schemata, employing the mathematical account of wave propagation begun by Fresnel and extended by his successors (including, of course, Maxwell), gave scientists good reason for believing that electromagnetic waves were propagated according to Maxwell's equations. From that conclusion they could derive the existence of the ether—but only by supposing in every case that wave propagation requires a medium. Thus the confirmation of the existence of the ether was no better than the evidence for that supposition.

Contrast the ether with the working posits of theoretical science that are referred to and characterized directly in successful schemata: atoms, molecules, genes, electromagnetic fields, and so forth. In some cases, we even devise numerous techniques for measuring the magnitudes of quantities that these entities possess or for representing the entities in various ways. If Laudan's story has an antirealist moral, it is that the presuppositional posits of contemporary science may not exist.

Laudan's use of the pessimistic induction from the history of science to discredit the claims to reference and truth of current science and to undermine a realist account of progress is the most sophisticated that I know. I have been arguing that it is not sophisticated enough. It leaves untouched the central realist claim. A finer-grained look at the history of science shows that where we are successful our references and our claims tend to survive even extensive changes in practice and to be built upon by later scientists, giving us grounds for optimism that our successful schemata employ terms that genuinely refer, claims that are (at least approximately) true, and offer views about dependencies in nature that are correct.

5. Settling for Less

Recent philosophy of science contains a number of positions that oppose the claims of my account of progress, by offering a different conception of the aims of science. This is most evident in the work of Bas van Fraassen, whose ideas and arguments I now want to consider.

Van Fraassen takes realism to be a thesis about the aims of science not about the achievements of science. Realists, he suggests, think that the aim of science is to provide us with a literally true story of the world (1980 8). The account of progress that I have outlined is committed to views about both aims and achievements.[32] I have not simply contended that the sciences aim to arrive at concepts that single out natural kinds, schemata that correctly capture objective dependencies, significant statements that are true, but that various parts of the sciences achieve these aims.[33] Avowals of aim without achievement ring somewhat hollow, and, if it were conceded—on the basis of the pessimistic induction, for example—that realist aims of science were never attained and are even unattainable, that concession could be used to persuade us to settle for less (see Laudan 1984, and the discussion in Section 6).

Van Fraassen's characterization of scientific realism thus strikes me as too weak because realists ought to advance theses about what the sciences succeed in doing. It is also too strong. The *aim* of arriving at a literally true story of the world might not only be something that is unattainable, but something we can recognize ourselves as being unable to attain. Van Fraassen's realist seems to have a mad passion for details—all the details. But, since human beings do not have cognitive abilities like those of a Laplacean demon, who would, we presume, be able to identify the structure in a complete description of nature, the realist goal that I have set for science, that of fathoming the structure of nature, requires something different. Our primary tasks are to achieve a language that recognizes natural divisions and a set of explanatory schemata that pick out dependencies. To the extent that we can then advance true answers to significant questions, we aim to do so. But we know, on the basis of understanding our own cognitive limitations, that we will sometimes only be able to determine the approximate values of magnitudes, and that in dealing with complex systems we shall have to provide an idealized treatment if we are not to lose sight of the structure of the situation. Sober realists who

32. In this, my approach is akin to the positions of Hilary Putnam and Richard Boyd, whom van Fraassen cites, who take realism to involve claims about actual attainments. See, for example, (Putnam 1975 69–75).

33. Van Fraassen thinks that the "naive statement" of realism—which supposes that "today's theories are correct"—lumbers the realist with something too strong (1981 7–8). But the trouble here isn't with claims about correctness. Rather it lies with the attribution of truth to *whole theories*. Where van Fraassen qualifies the naive statement by retreating from commitments to achievements, I make it more cautious by specifying the attainments more narrowly. See the discussion in Section 6 of the last chapter.

understand human cognitive limitations will see that the *whole* truth is not something that we want and that, in some cases, the attempt to achieve *truth* (rather than approximation or idealization) would interfere with our primary aims.

The debate between van Fraassen and his realist opponents is an internal dispute within the broader controversy about realism discussed in Section 2. Van Fraassen is willing to allow the possibility of arriving at some kinds of true beliefs. His opposition to realism is based on maintaining that the domain over which we can reliably generate true beliefs is limited to observables: van Fraassen is a local antirealist for whom the feature U that marks the limit of our defensible claims is the property of referring to unobservable entities. He suggests that the aim of science should be to achieve theories that are *empirically adequate,* that deliver the truth about observables.

Once again, I think that van Fraassen's formulation skews the issue in an important way. Let us ask *which* truths about observables we would want an ideal science to deliver. All of them? Surely not—for that would bury what we really want to know in a clutter from which our limited cognitive systems could not retrieve the desired information. The significant ones? But, in this case, what does significance amount to? It is far from clear that van Fraassen's framework allows him to offer the type of account of cognitive significance that I sketched in the previous chapter.

I shall not pursue this line of inquiry, because I think that there is a fundamental difficulty with van Fraassen's project of limiting science to the observable. Why should we settle for less than the delineation of natural kinds, the provision of correct schemata, the answering of significant questions? Van Fraassen's objections to these apparently desirable but ambitious goals would seem to be that they are *too* ambitious. Nothing in his writings suggests that it would not be a good thing to know about the behavior of unobservables. The line of argument appears to be that these goals are beyond our epistemic reach.

Given the discussion of the pessimistic induction in the previous section, this is a species of argument we should recognize as exactly to the point. My general account of scientific progress takes the following form: we would like our components of practice to achieve the qualities X, Y, Z, and so forth; if our account of ourselves shows that we are limited in attaining one of these qualities in certain cases, then we should seek approximations or idealizations that advance the overall project of fathoming the structure of nature. If van Fraassen were able to show that our cognitive limitations prevent us from any knowledge of the unobservable, then the idea of scientific progress that I have delineated would be unfeasible and some modification of it in the direction of the antirealist aim he advances would be motivated. But if there is no reason to believe that our cognitive limitations are so great that they bar knowledge of the unobservable, then there is no reason to abandon the account of progress I have outlined. Van Fraassen's argument is of the right sort because considerations of human cognitive limitations should be used in formulating goals that are attainable for us. Whether it is a cogent critique of the account of progress I have proposed depends on whether our cognitive limitations are as van Fraassen maintains.

Several of van Fraassen's critics have noted that, whereas he allows our ability to arrive at reliable generalizations he frowns on any attempt to offer inferences about the unobservable.[34] They have asked, quite reasonably, how this difference in treatment is to be motivated. Van Fraassen's response to such criticisms is to admit that the belief that theories are empirically adequate invariably involves risks. He denies that his willingness to take the risks commits him to belief in the truth of the theories, "since it is not an epistemological principle that one might as well hang for a sheep as a lamb" (1980 72). But now we should surely ask whether there is a difference between the two kinds of epistemic risks sufficient to warrant the conclusion that our knowledge is restricted to the observable. Or, to pursue van Fraassen's comparison, we need to see why we can make off with the lamb but are likely to be caught if we try for the sheep.

There is a celebrated argument, due to Grover Maxwell (1963), that there is no principled way to distinguish observables from unobservables. Van Fraassen is concerned to rebut this line of reasoning. He concedes that the distinction between observables and nonobservables is vague—as Maxwell rightly pointed out, there is a continuum of cases in which we achieve visual representations of ever smaller objects by interposing pieces of glass (in ever more sophisticated combinations) between the objects and us. However, van Fraassen contends that there are clear examples of unobservables and clear examples of observables. He tests for observability by relativizing to our actual cognitive equipment and to the activities in which we can engage. The moons of Jupiter count as observable because astronauts could go there and look. Electrons do not count as observable because, do what we may, there is no way for us to achieve an unaided visual representation of an electron (1980 16–17).

Now there are many kinds of things that human beings have never actually seen, and which we believe to exist or to have existed. Nobody has ever had unaided observation of a dinosaur or of the members of countless other species which inhabited our planet before we evolved. Are these organisms unobservable or simply unobserved? Perhaps we can think of ourselves, in the style of van Fraassen's thought experiment, as transported back in time (as we might be transported in space to Jupiter for the close-up view of the moons) and provided with an opportunity to confront a Triceratops. But is time travel *physically* possible? If it is not, then it appears that van Fraassen's test will not allow this type of explanation of the observability of dinosaurs. For, if we were to allow that X's are observable when human beings can obtain unaided observations of X's by processes that involve suspension of physical laws, then observability would be easy. Electrons would count as

---

34. See many of the essays collected in (Churchland and Hooker 1984), particularly those by Paul Churchland, Ronald Giere, Gary Gutting, and Alan Musgrave, and also (Railton 1989).

observable if we permitted the Lockean fantasy that "microscopical" versions of ourselves could wander among them—but, of course, that fantasy suspends most of what we think we know about the microstructure of the world.[35]

Van Fraassen's test for observability must demand that there be physically possible processes that could lead to unaided observation. But now the organisms of the past are in serious trouble. Given the physical impossibility of time travel, the only way to conceive of the possibility of unaided observation of them is to suppose that the course of evolutionary history had been different, that human observers had evolved in time to observe each departed species. There are two kinds of problems with that scenario. First, on any conception of species that takes particular species taxa to be constituted by position in the phylogenetic tree—that is, that requires that humans have the evolutionary history we actually have—it will be impossible for us to observe our ancestors.[36] Second, and more prosaically, the environmental conditions required for some extinct species to flourish are incompatible with our survival, even for short periods. I conclude that many extinct organisms, like electrons, genes, the internal structures of stars, and numerous other entities of science, will count as unobservable.

Discussing the merits of far-fetched scenarios for bringing us into contact with extinct organisms strikes me as faintly ludicrous, a venture born of a misguided test for judging our epistemic access to various aspects of nature. Paleontologists think they know a great deal about the past history of life. Why should the decision about whether their claims are true or merely empirically adequate turn on recondite scenarios which, even if possible, are never likely to be actualized? Philosophy-of-science-fiction can be entertaining, but thought is better directed to considering the grounds for believing that there were past organisms like those that figure in paleontological reconstructions.

From Cuvier on, paleontologists have assembled numerous claims about the extinct species that once inhabited our planet. Deleting that entire body of claims from our science would improve our epistemic position in one obvious respect. If P is the conjunction of the statements about extinct organisms belonging to the consensus practice of paleontologists and O is the conjunction of all the accepted statements about observables that occur in the consensus practice of any science, then, as a simple matter of probability theory, $P + O$ is more risky than O. (The probability of $P + O$, conditional on any evidence that does not entail P, is less than the probability of O, conditional on that evidence.) We could lower our risks further by deleting conjuncts from O itself. The sheep that van Fraassen worries about pursuing is $P + O$; the allegedly attainable lamb is O.

Claims about the existence of extinct organisms are sometimes founded

35. Grover Maxwell's defense of the observability of electrons is in the Lockean tradition of fantasizing about "microscopical eyes." Van Fraassen sensibly regards this as a trick (1980 17).

36. The view that species are individuated by their positions in the phylogenetic tree is defended by Michael Ghiselin (1974) and David Hull (1978).

on the collection of fragments, bits of tooth, bone, or shell. Occasionally, paleontologists are lucky and find whole specimens, mammoths buried in the frozen tundra or ants encased in amber. But it is important to see that the hypothesis that the specimens we now find, whether fragmentary or complete, are remnants of previously living organisms *might* be wrong. Although we view the teeth and shells as traces of past life, there could be some alternative explanation of their presence. Should we then be cautious and view ourselves as having no knowledge of any unobserved organisms who inhabited the earth before humans were here?

The natural response: give us a rival explanation, and we'll consider whether it is sufficiently serious to threaten our confidence in P. By accepting P we are able to make some predictions about the kinds of fossils we would expect to find and to give many explanations of the presence of the fossils we do find. (A beautiful example of the former sort is the prediction made by Wilson, Brown, and Carpenter (1967) of the characteristics of the ancestral form of living ants—subsequently, and quite unexpectedly, largely confirmed through discovery of a rare specimen encased in amber.) It is possible that all these successes are *merely* apparent. But to take that possibility seriously we need some alternative explanation of our apparent success.

The last-ditch defense of those who hoped to resist the nineteenth century paleontological account of the history of life was to declare that the apparent fossils were a test from God (or a temptation from the devil). This provides a rival explanation for the presence of what we took to be traces of past life. Outside philosophy classes it typically provokes incredulity—incredulity based on the judgment that our world contains animals that die and leave teeth, bones, and shells behind, but that it does not contain gods and demons who behave in the envisaged ways. We use our general scientific picture of the world and our relation to it to dismiss any rival explanations of fossil findings as implausible.

If that strategy seems suspect, then it is worth considering our beliefs about observables. Science provides us with a picture of how we relate to the bits of tooth, bone and shell. That picture proposes that our beliefs are generated and supported through complex causal interactions with teeth, bones, and shells. Imagine that a philosopher now becomes selectively worried, adopting the *outline* of the story and accepting the idea that beliefs are psychological states with complex causal antecedents but denying that our standard picture of causation is correct. Perhaps we are deceived, he suggests; there are no bones, shells, and teeth, only a malevolent demon who works upon us. The style of argument is van Fraassen's, for, in both cases, the suggestion is that we abandon various risky beliefs since there are possible rival explanations for what we take to be occurring.[37] Both arguments should

---

37. Van Fraassen is aware of this line of objection and concerned to answer it (1980 72). He formulates it as the thesis that statements about sense data would be epistemically more secure than claims about external objects and then responds by contending (correctly) that sense data are "the theoretical entities of an armchair psychology that cannot even rightfully claim to be scientific" (1980 72)—and, moreover, discredited by major programs

be resisted. However, it is impossible to rebut them without taking for granted our normal standards for assessing the possibility of rival explanations, standards that are grounded in our picture of the world and our relation to it.

With respect to extinct organisms, genes, and subatomic particles, confidence is based on the idea that there are no plausible rival explanations of the phenomena.[38] If van Fraassen's maxim is "Be agnostic whenever you might go wrong," then the agnosticism he recommends is far more sweeping than he suggests. (In light of the discussions of earlier sections, I claim that we have no motivation for adopting that maxim.) If it is "Be agnostic whenever there is a plausible rival account of nature and our relation to it," then the maxim is thoroughly sensible. *But the standards of plausible explanation have to be set by our scientific picture.* Once these standards are invoked, it appears that there is no basis for distinguishing our commitments to observables from those to extinct organisms, genes, or subatomic particles.

I shall conclude this discussion of van Fraassen's provocative views by considering briefly a counter that he gives to an argument having some affinities with mine. Hilary Putnam (1975) argued that claims about unobservables are needed if we are to explain the success of science. After provisionally accepting the demand for a scientific explanation of the success of science, van Fraassen continues:

> Science is a biological phenomenon, an activity by one kind of organism which facilitates its interaction with the environment. And this makes me think that a very different kind of scientific explanation is required.
>
> I can best make the point by contrasting two accounts of the mouse who runs from its enemy, the cat. St. Augustine already remarked on this phenomenon, and provided an intensional explanation: the mouse *perceives that* the cat is its enemy, hence the mouse runs. What is postulated here is the 'adequacy' of the mouse's thought to the order of nature: the relation of enmity is correctly reflected in his mind. But the Darwinist says: Do not ask why the *mouse* runs from its enemy. Species which do not cope with their natural enemies no longer exist. That is why there are only ones who do. (1980 39)

By the same token, van Fraassen suggests, the explanation of why the theories of current science are successful lies in the fact that only the successful ones survive the winnowing of scientific ideas.

This passage raises slippery issues. Darwinians want not only to claim that

---

in twentieth century philosophy. But the parallel to van Fraassen's argument does not assume that there are sense data in anything like the traditional sense. It appeals to a small part of our ordinary scientific lore, the claim that we are subjects who can be in psychological states as the result of interactions with something else, and asks us to remain agnostic about the causes of those states. There are numerous different ways to formulate this agnostic position, some of which commit themselves to very little indeed. However, the important common theme is that the conclusion of the parallel to van Fraassen's argument ought to lead us to say less about the world (talking, perhaps, in terms of the appearances of things) rather than to introduce special posits (sense data).

38. As Nancy Cartwright trenchantly puts it, "If there are no electrons in the cloud chamber, I do not know why the tracks are there" (1983 99).

successful organisms are those that leave descendants, but also to investigate those characteristics that promote reproductive success, carrying out the inquiry at whatever level of generality is possible. Thus we can distinguish (i) current species and those that have endured for significant periods in the past; (ii) the possession of high Darwinian fitness, borne by those organisms that have characteristics that dispose them to survive and to reproduce most successfully; (iii) the generic characteristics that endow organisms with high Darwinian fitness (e.g., ability to avoid predators); and, finally, (iv) the particular ways in which organisms belonging to particular species achieve these generic characteristics. Similarly, in the case of scientific theories, we should separate (i) current theories and those that have endured for significant periods in the past; (ii) the possession of high propensities for espousal by scientists, borne by those theories that have great predictive and explanatory power; (iii) the generic characteristics that endow theories with great predictive and explanatory power; and, finally, (iv) the particular ways in which particular theories achieve these generic characteristics. Van Fraassen suggests that the presence of certain species is to be explained by noting that the organisms belonging to them have high fitness in the relevant environments—and this is indeed a relatively shallow level of Darwinian explanation, corresponding to the understanding of (i) in terms of (ii).[39] By analogy, we could explain the endurance and current presence of theories which appear successful by remarking that those which did not have high explanatory and predictive power tended to be discarded—a relatively shallow level of explanation that again explains (i) in terms of (ii).

This should not be the end of the story. Darwinians want to know, at whatever level of generality is possible, what kinds of organism-environment relationships confer reproductive success—they want to explain (i) and (ii) by moving on to (iii). Similarly, Putnam's demand is to ask what it is about those theories that are explanatorily and predictively successful and that posit unobservables that makes for their explanatory and predictive success. The answer given by the realist is that such theories succeed in this way because they fasten on aspects of reality. If they did not, it would be "a miracle" that they were so successful. Now that answer can be challenged—as it is challenged by the pessimistic inductivists—by claiming that highly successful theories frequently introduced posits that we no longer recognize. But, as we saw in the previous section, that challenge can be turned back. I thus suggest that a parallel to the thoroughly Darwinian strategy of trying to elicit the general characteristics of the organism-environment relation that confers

39. To characterize this explanation as shallow is not simply an aesthetic judgment. The simple practice of noting that members of enduring species had high fitness (healthy propensities to survive and reproduce) is close to the caricature view of Darwinism as "tautologous," explaining survival in terms of itself. The explanatory power of this practice derives from its comparison of actual outcomes with probabilistic expectations (see Brandon 1980) and its denial that other factors interfered to separate the actual from the expected. Real evolutionary biologists—unlike van Fraassen's "Darwinist"—typically go further, asking for the sources of the high fitness.

reproductive success is that of trying to elicit the theory-nature relation that confers explanatory and predictive success: (iii) is used to account for (i) and (ii).

Finally, we may probe in more detail, looking at the precise ways in which members of a particular species survive and reproduce so well, noting, for example, the traits of mice that enable them to detect the presence of cats. Analogously, our inquiries into the successes of particular sciences may proceed as far as (iv), explaining the details of the explanatory and predictive successes of paleontology and genetics by showing how references to extinct organisms and to genes play a crucial role in the schemata of these practices and how no explanation of the resultant success rivals that of the simple suggestion that there were once organisms that are now extinct and that there are genes.

6. Goals in Flux?

Let me now take up a different type of concern about my account of progress. On my conception, the cognitive goal of science is to attain significant truth—where significance is understood in terms of charting divisions and recognizing explanatory dependencies in nature—insofar as it is possible for beings with our limitations to do so. This goal is independent of field and time, independent of how we think it might be achieved. By contrast, Laudan has argued that there is no single set of aims that holds for all sciences and for all times. Instead, he proposes a *reticulational* model of scientific change in which not only theories but also "methodologies" and "axiologies" can be debated and modified.[40]

Laudan and I share the idea that the state of a science at a time is multidimensional, but our views about what I call *practices* are importantly different. Ignoring substantial (but currently irrelevant) points about the relations between the individual and the social, my picture of a consensus practice distinguishes the following: (a) the goals of science and the actual best ways of achieving those goals, (b) the implicit characterization of goals in the specification of significant questions and the formulation of principles deemed to govern appropriate ways of achieving the goals in the methodological statements of the practice. Scientific practices contain (b), but (a) stands outside practice. On my account, the goals of science do not change over time—although scientists may offer different ideas about subgoals in the light of their beliefs about the world. For Laudan, by contrast, the counterpart of a practice contains four components, *theories, ontology, methodology, and axiology* (where the latter consists in the specification of a set of aims). In

40. This position is elaborated in his (1984), and there are some tensions between this later work and the view of progress in terms of puzzle solving advanced in (Laudan 1977). I shall not explore possible ways of resolving the tension, since I am primarily interested in the challenge posed by Laudan's claims about variable aims.

the present discussion it will not be necessary to consider the ontology of a science, so I shall formulate Laudan's position by attributing to him the view that science consists of a sequence of *L practices,* each of which can be represented as a triple $\langle T, M, A \rangle$. Apparently, for Laudan, there are no epistemic aims except those represented within the L practice. Thus he has no counterpart for (a).

Laudan would like to use his reticulational model to understand how the sciences can change *rationally,* and, in particular, how there can be rational changes in all three components of L practice. His insight is to see that the rationality of a transition from $\langle T, M, A \rangle$ to $\langle T', M, A \rangle$ (say) is judged by the available evidence and *whatever remains constant throughout the transition* (1984 chapter 3). He envisages the possibility that a sequence of such rational transitions might culminate in changes in methodologies and axiologies as well as in theories.[41]

But there is an obvious question: what are the principles that govern rational L practice transitions? The fact that methodology and axiology are mentioned *in* the picture might blind us to the point (akin to one made by Lewis Carroll (1896)) that there has to be a methodology *outside* the picture. Laudan plainly wants to divide changes in L practice into two types, those that are rational and those that are not. Hence there must be statements of the forms

$\langle T, M, A \rangle \langle T', M, A \rangle$ just in case $T' \ldots T$ given M, A.

$\langle T, M, A \rangle \langle T, M', A \rangle$ just in case $M' \ldots M$ given T, A.

$\langle T, M, A \rangle \langle T, M, A' \rangle$ just in case $A' \ldots A$ given T, M.

where, in each case, $L\ L'$ means that the change is rational. Laudan does not say a great deal that enables us to fill in the blanks in these statements, but it is clear that, in general, the rationality of changing one L practice to another involves remedying various kinds of inconsistencies among elements. In the special case of changes of aims, corresponding to my third schema, he does propose that we revise aims in the light of theories and methods so as to devise substitutes for goals that come to be regarded as unattainable.[42]

There are several worrying features of this reticulational model: for instance, the apparent difficulty of saying how, in a clash between theory and methodology, we should decide whether to revise theory or to change methodology, and the seeming possibility that the state of L practice at the end

41. Indeed, this is supposed to have occurred at various points in the history of science. See the discussion of George Lesage's defense of the "method of hypothesis" and its impact on Victorian views about the goals of science (1984 59).

42. See (Laudan 1984 51–53). There is a second constraint on the advocacy of a set of goals, which receives more prominence in (Laudan 1989), namely the requirement to "preserve the canon." Here the scientist is supposed to espouse a set of aims A that will fit with the "recognized past achievements of science." It is interesting to ask whether this demand is itself grounded in some enduring, context-independent goal for the enterprise— and, if not, why it should be imposed.

of a sequence of changes might look with disfavor upon some of the transitions that generated that L practice (see Doppelt 1986). But the fundamental concern that I have here is with the status of the principles that result when the correct substitutions are made in the preceding schemata. What makes the principles of rational transition among L practices the *right* principles?

Since it is clear that Laudan regards rational transitions among L practices as relieving inconsistencies, and that he does not think that every way of remedying inconsistency is rational, the demand is for understanding why relief from inconsistency should be sought and why it should not be sought in the obvious fashion of hacking away at our L practices until they are so minimal as to avoid the contradictions. The obvious answer is that there are enduring goals that are not represented in Laudan's official picture and that the rules of rationality formulate good strategies for attaining these goals (see the discussion of the next chapter). We eliminate inconsistency because we want our claims to be true and we do not do so by means of some scorched earth policy because we want as much significant truth as we can get.[43] But if Laudan were to offer this explanation, his thesis that the goals of science vary across time would be reformulated so that it was no longer at odds with the picture of progress I have offered. All that would vary would be the formulation of *derivative* goals, and that type of variation is captured in my framework by the conception of shifts in the notion of significance. On the other hand, if Laudan does not explain the status of the principles of rational L practice transition by appealing to their promotion of enduring goals, it is unclear that he has any basis for supposing that these principles have any special quality. Anyone is free to characterize a set of transitions with some honorific term, but the significance of doing so needs to be explained.

The argument so far is intended to show that if there is to be *rational* change in scientific axiology there have to be some enduring goals that provide a basis against which judgments of rationality make sense. Once this conclusion is reached, there is a natural explanation of apparent variation in the goals of science. Scientists seek to present true answers to significant questions, insofar as they are able. As their views about objective dependencies

---

43. Laudan has suggested to me that empirical adequacy would do as well as truth in grounding the demand to eliminate inconsistency with minimal mutilation. However, this suggestion is both problematic and beside the point. First, the demand of empirical adequacy will not require us to eliminate *all* inconsistencies. Assuming that there can be rival empirically equivalent theories which make incompatible theoretical claims, the search for empirical adequacy, rather than truth, will not require us to choose between them. Second, even if it were true that empirical adequacy would underwrite the demand, this would not obviate the need for some enduring aims—possibly disjunctive—to stand outside the sequence of L practices and ground the principles that govern rational adjustment. The point here is fundamental. If the aims that are represented within the L practices are indeed revisable, it cannot be in virtue of *them* that the principles of rationality that govern the adjustment of the components of L practices obtain their force. This is fundamental because once the commitment to *some* enduring aims is established, I believe that the account of those aims which I offered in the last chapter is both natural and defensible by the historical record.

change, what they take to be significant will shift. Moreover, in light of the statements they accept, they will view certain kinds of questions as admitting exact answers, others as approachable in terms of idealizations or approximations, and yet others, perhaps, as unanswerable by human beings. Changes in the statements accepted will be reflected in these appraisals, leading to differences in the formulation of derivative aims.

The explanation I have just sketched should be familiar from discussions of ethical relativism.[44] Apparent differences in moral judgments may stem from differences in belief against a background of shared fundamental values. This, I suggest, is exactly what occurs in the history of science. Laudan's reticulational model is driven by the idea that there are genuine modifications of fundamental aims of science from epoch to epoch. On closer view, I claim, changes in formulations of the aims of science can be understood as expressions of the *enduring* goal of discovering as much significant truth as human beings can in the light of *changing* beliefs about what is significant, what nature is like, and what the nature of our relation to nature is.

Laudan's principal example of axiological shifts in the history of science concerns an alleged oscillation between demanding that theories be (at least approximately) true and demanding merely that they "save the phenomena" (1984 44–66, 1988). Despite the fact that he is an adept chronicler of disputes between those who formulate these aims—particularly in the context of eighteenth century physical science—Laudan misses a simple explanation of the evidence he cites. It is hard to believe that *any* of the major protagonists in these debates would have dissented from the goals that were identified in the last chapter. All would have hoped to formulate explanatory schemata that captured the dependencies of phenomena, to pose and to answer questions picked out as significant by such schemata *to the extent that the formulation of the schemata and the answering of significant questions were possible for human beings*. The dispute is not about desirability but about attainability. Those who settle for less, arguing that hypotheses should "save the phenomena," do so not because they think it undesirable to achieve true (or approximately true) hypotheses but because they think it beyond the power of the human scientists to discriminate the truth from the rival possibilities.

7. Social Interference

It should be apparent by now that my account of progress is not committed to the idea that truth is easily won nor to the view that the facts "force themselves upon us" if the conditions are right. Achieving the truth is tricky, but not uniformly so. Moreover, the history of science cautions us against writing off any domain of investigation as inaccessible. Given the finiteness of our collective lifespan, the boundedness of our distribution in space, and the limitations of our cognitive systems, it is highly likely that there are some

---

44. See Brandt (1959), and the brief discussion of the last chapter.

aspects of the universe that we shall never be able to fathom. However, it is not clear that we can come to know which these are without an enormous amount of effort. Casual attempts to write off certain phenomena as beyond the reach of science are typically premature. After all, who would have thought that we would have come to know as much as we do about the composition of matter or the history of life? Recall how many eminent biologists before Darwin despaired of ever understanding the origins of species, the "mystery of mysteries."

Contemporary work in the sociology of science embodies a vigorous critique of the confident attitude that I have adopted. It would be simple to respond to that critique by pointing out that it assumes that views of science which suppose that realist goals are set and attained inevitably embody a naive epistemology on which the human mind is transparent to nature. Barry Barnes presents the approach he hopes to undermine as follows:

> Although it would be generally agreed that biasing factors occur, in different degrees, in most institutional contexts, and hence that most belief systems embody some degree of error and distortion, the natural sciences are generally accepted as true and undistorted bodies of knowledge; their methods as impartial, unbiased models of investigation. Thus, science can function as a model of how we would be able to orient ourselves to the world in the absence of our psychological biases and social prejudices. (1974 5)

Similar suggestions that one can only maintain that science aims at and attains truth by neglecting the complexities of the processes that give rise to belief and adopting a simple-minded epistemology can be found in a justly famous study of daily laboratory practice.

> At the onset of stabilisation [roughly, the point at which a statement becomes accepted within the relevant community: PK], the object was the virtual image of the statement; subsequently, the statement becomes the mirror image of the reality "out there." Thus the justification for the statement TRF is Pyro-Glu-His-Pro-NH$_2$ is simply that "TRF *really is* Pyro-Glu-His-Pro-NH$_2$." At the same time, the past becomes inverted. TRF has been there all along, just waiting to be revealed for all to see. The history of its construction is also transformed from this new vantage point: the process of construction is turned into the pursuit of a single path which led inevitably to the "actual" structure. (Latour and Woolgar 1979/1986 177)

The quick response is simply to slough off the epistemological ideas that these writers attribute to their opposition and then to proceed with one's business.

The fundamental realist thesis is that we arrive at true statements about the world. That thesis does not imply that we have unbiased access to nature, merely that the biases are not so powerful that they prevent us from working our way out of false belief. Nor should realists be seen as embracing a curious kind of positivism: the justification for belief that p is to be p (a schema that some epistemologists once held for foundational beliefs—"I have a sensory experience of red"—but which has been universally abandoned and was never maintained for beliefs about the structure of molecules!).[45] Intelligent realists will claim that the justification for belief about the structure of TRF is a (very complicated) process involving interactions between scientists and nature, and that the belief is justified because processes of this general sort regularly and reliably generate true beliefs.[46]

Showing that the sociological attack on realism works by foisting upon realists naive epistemologies that they do not adopt and should not want is quick but unsatisfying. There are deep and important insights in recent discussions in the sociology of science, and these are missed unless realists meet the fundamental challenge. Is there an account of our relationship to the realists' supposed "independent reality" that will both accord with the detailed accounts we have of the social lives of scientists and support the claim that we gain knowledge of this "independent reality"?[47]

According to the picture developed in Chapter 3, changes in consensus practice result from modifications of individual practice, in accordance with rules for consensus formation. Those changes in individual practice, in turn, result from two types of process: conversations with peers and encounters with nature.[48] The encounters with nature involve the impinging of stimuli on cognitive systems whose prior states play an important role in determining the new state of the system. Thus there is no suggestion that experience is raw or unbiased. Those prior states are partially caused by conversations with peers, earlier encounters with nature, and, ultimately, the training process and even the earlier intellectual ontogeny of the scientists. The deep point of the sociological critique is that the social forces that operate in this modification of practice—the rules for consensus shaping, the conversations with peers, the training process and broader socialization within a larger community—may be sufficiently powerful that the effects of nature are negligible. If correct, this point would undermine my account of progress from within, accepting the idea that we can use our scientific picture of nature and our relation to it to appraise our beliefs about nature, and maintaining that when the appraisal is done it reveals that relationship to be too tenuous to deliver true beliefs.[49] The persistence of our judgments about nature—the persistence

---

45. See (Schlick 1934/1959) and, for a review, (Chisholm 1966). Foundationalist theories of knowledge, which supposed that certain statements are self-warranting, were always controversial, and, more often than not, associated with antirealism (specifically phenomenalism) rather than realism (see Ayer 1952). For sustained critique, see (Sellars 1963 essays 3 and 5).

46. The locus classicus for this approach is (Goldman 1986). The general point about the need for introducing appeals to the psychological causes of belief into accounts of justification is made by Gilbert Harman (1971), Alvin Goldman (1979), Hilary Kornblith (1980), and me (1983a chapter 1). For further discussion, see Chapter 6.

47. This is surely the point at which one should engage the arguments of Barnes and of David Bloor, both of whom hold that there is an independent reality, but that our interactions with it radically underdetermine the judgments we can make about it. See (Barnes 1992) and Bloor (1974/1991 especially 31–40).

48. Here I oversimplify. Recall that this is not an exclusive and exhaustive dichotomy.

49. Chapter 8 will explore this potentially worrying predicament in considerable detail,

I have used in earlier sections to generate optimistic conclusions—would then be seen as a consequence of the inertia of social systems.

The challenge may be presented in either a radical or a conservative way. On the conservative version the practice of a science would be divided into two parts, a "framework" that remains unaffected by any interactions with nature and a "filling" that adjusts to nature. The radical way of elaborating the challenge would be to contend that nature has no input at all into our scientific practices so that all modifications of consensus practice result from social forces. Both variants can be developed to lead to interesting empirical claims about changes in science.

Let us start with the radical challenge, and with an extreme version of it. Social forces might be so strong at the level of individual practice that precisely the same changes in individual practice would occur whatever stimuli impinged upon the individual scientist. This formulation is surely false. Scientists who are apparently subjected to the same social forces disagree because they have engaged in different encounters with nature. We could deliberately contrive for members of the same research team—perhaps the closest approximation to the social equivalent of monozygotic twins we can reasonably expect to get—to receive different results from runs of the same experiment and test the claim that differences in stimuli will cause differences in modifications of practice.

A far more serious suggestion is that even if social forces do not constrain individual practice in the extreme fashion just bruited, they do make it impossible for the different effects of different stimuli to be absorbed into consensus practice. Those who only *hear* about challenging results will dismiss them as poorly done, and, if the challenger insists on them, he will lose whatever authority he has and may even be driven out of the community. It is not difficult to devise general models of scientific communities, in which those who have greatest authority do not engage in experimentation and in which there is a pronounced tendency among all members to defer to authority, and which show how a community's consensus practice could be completely independent of any input from nature. (Such models are explored in Chapter 8.) This version of the sociological challenge represents a clear theoretical possibility. The question is whether that possibility is actualized in science.

The most plausible versions of the thesis that consensus practice is socially determined are those that combine the idea of the last paragraph with a more conservative approach. Divide the consensus practice of a community into two parts: framework and filling. Elements of the framework are impervious to modification through encounters with nature, either because the framework is never modified in *individual* practice or because modifications of the framework in individual practice are unable to spread throughout the community so as to be absorbed into later consensus. Elements of the filling, however,

---

attempting to understand the dynamics of introduction and diffusion of ideas that might challenge parts of consensus practice.

can be modified as a result of causal interactions with nature and the modifications can spread through the community to enter consensus practice.

I hope that it is now apparent that the sociological critique is both intelligible and difficult to assess. The issue that it raises should recall a familiar dispute. Organismal phenotypes are the result of an interaction between the genotype and the organism's environment. Disentangling the relative contributions of the two has proved highly problematic. Especially with respect to those behavioral traits that most fascinate us (tendencies to sex and violence) there has been a pronounced tendency to gravitate to one of two polar positions. Either it is proposed that the genes exert an iron grip (so that no matter how the environment is varied the same phenotype will result) or it is suggested that the genes provide no constraints at all (for any given genotype we can produce the phenotype we want by varying the environment). As soon as the extreme positions are articulated, they are quickly disavowed: everybody agrees that there are two important determinants. But there is still room for dispute about relative importance, about the extent to which variation in phenotype is limited by genotype.[50]

In the present case, too, there are two types of factor which prove hard to disentangle. The practices of individuals respond both to stimuli from external, asocial nature and to the remarks made by teachers, friends, colleagues, and adversaries. The consensus practice responds to changes in individual practices that are mediated by the social structure of the community. The position that the sociologists attack is the extreme suggestion that social determinants make no difference: given the same inputs from asocial nature there will be the same modifications of consensus practice, no matter what the social structure. They counterpose the extreme view that inputs from nature are impotent, or at least impotent with respect to the framework of consensus practice: given the same social structure, there will be the same modifications of consensus practice, no matter what the inputs from asocial nature, and, in particular, the framework of consensus practice will remain unaltered. I claim that there is a vast, unexplored middle ground between these extremes.

What does a realist account like mine need? It is not necessary to assume the thesis of social transparency that traditional philosophy of science presupposes and that sociologists rightly attack.[51] Nor is it enough simply to deny

50. I have discussed these issues at some length in my (1985b). See also (Hanna 1985) and (Sober 1988).

51. The thesis of social transparency is simply the claim that ultimately all the modifications of individual practice (or, in traditional terms, changes in belief) produced by interactions with other members of the community can be seen as directed by the experiences that gave rise to cognitive changes in those who transmitted information (or in people who transmitted information to them or . . .). Thus we can think of a cognitive subject as a kind of epistemic octopus whose tentacles reach out through the lives of others to obtain vicarious experience, and we can simply explore the modifications of cognitive state in light of this extended experience. At the back of this picture seems to lurk the notion of some primal state in which beliefs are simply induced by nature without any social interactions. I think that the sociologists are correct to disavow any such primal state and

the sociologists' polar antithesis. The relevant issue is whether, given the actual social structures present in scientific communities, the input from asocial nature is sufficiently strong to keep consensus practice on track. (Similarly, in discussions of gene-environment interaction, the appropriate questions concern whether, given the genotypes actually present, modifications of environment can be sufficiently strong to produce modifications of phenotype.)

The issue can be framed precisely as follows: Let us suppose that there is a community of scientists with a consensus practice C_1, a distribution of individual practices, and a particular social structure. Let C_2 be a practice such that, given the way the world is, the shift from C_1 to C_2 would be progressive in any of the senses articulated in the last chapter. The *facilitating encounters* for the shift from C_1 to C_2 are those sequences of stimuli (if there are any) that would lead the community from its initial state to a subsequent state in which the consensus practice is C_2. My account of progress does depend on the idea that, with respect to the kinds of consensus practices, distributions of individual practices, and social structures actually found in science, there are facilitating encounters actually available for scientists to make progressive shifts. *Ideally,* I would like to contend that this obtains for every facet of scientific practice, so that there is no inviolable framework that input from nature cannot displace. However, a conservative version of the sociological critique would be compatible with the partial applicability of my conception of cognitive progress.

So far I have merely tried to arrive at a clear view of what is at stake. Let us now return to the claims and arguments of skeptical sociologists. Latour and Woolgar appear to commit themselves to the extreme position that I have delineated:

> We do not use the notion of reality to account for the stabilisation of a statement . . . because this reality is formed as a consequence of this stabilisation.
>
> We do not wish to say that facts do not exist nor that there is no such thing as reality. In this simple sense our position is not relativist. Our point is that "out-there-ness" is the *consequence* of scientific work rather than its *cause.* (Latour and Woolgar 1979/1986 180–182)

I interpret this passage as recognizing that people (scientists included) engage in causal interactions with something beyond themselves and the social sphere. Their beliefs, or more generally consensus practices, are not affected by these interactions: the fact that TRF has a certain structure does not play a causal role in the genesis of the belief that it has that structure; rather, because we believe, as the result of a complex social process, that TRF has that structure the fact that TRF has that structure is *constructed.* The thesis of the social construction of facts is thus the claim that the acceptance of statements as firm parts of consensus practice is to be explained in a particular

---

to reject any analysis of the grounds of cognition that attempts to reduce the social inputs to complexes of asocial interactions with nature.

fashion, a fashion that makes no reference to the constraining power of stimuli from external, asocial nature.[52]

Why should anyone believe this? Latour and Woolgar offer a detailed account of the social exchanges that preceded the endorsement of the claim that TRF is Pyro-Glu-His-Pro-NH$_2$. That account can obviously raise serious questions about the wisdom of ignoring social processes and social structures in an investigation of scientific change. It can challenge the confidence of those who think that the modification of individual practice and of consensus practice will be affected only by the stimuli received from external, asocial nature. Latour and Woolgar thus make a plausible case for supposing that the extreme opposite view about the modification of consensus is mistaken. But this is a long way short of their conclusion.

To defend the thesis of the construction of facts, as I have interpreted it, one would have to show that the encounters with nature that occurred during the genesis of belief about TRF played no role. *However those encounters had turned out the end result would have been the same.* Since Latour and Woolgar do not describe the actual experiments in any detail, they provide no reason for supposing that that is so. Once again the comparison with nature-nurture debates proves valuable. Those who believe that genotypes exert powerful constraints on phenotypes buttress their case by showing how similar phenotypes result under wide variety in the environment. To defend Latour and Woolgar's thesis, it is not enough to revel in the social complexities of the processes that stand behind changes in consensus practice—although that is a welcome corrective to the simplifications of some traditional thinking about scientific change. It must be shown how the outcomes would remain invariant (relatively invariant?) under modifications of the input from nature.

There is an ideal experiment. Take a large number of populations of scientists, separate them so that no interpopulation communication is possible, start each in the state of the community of molecular endocrinologists at the beginning of Latour and Woolgar's story—same consensus practice, same distribution of individual practices, same social structure. Expose each population to different inputs from nature, perhaps by simply having them run their assays on different substances (all but one community analyzes pseudo-TRF). How much variation in consensus practice do you get (say at the time that the community with actual TRF reaches consensus)? If the consensus

---

52. In Latour's current work (1989a, 1992), he proposes a more thoroughly symmetrical treatment. We are not only to eschew appeals to any fixed asocial nature in explaining the presence of scientific belief but to disavow explanations in terms of a fixed social setting. Reality and society are both equally matters for "negotiation." This position has an admirable formal symmetry, but I find myself quite at a loss in understanding what resources are left for understanding the genesis and modification of scientific cognitive states. My own approach seems to be the mirror image of Latour's: both the social context and nature exist independently of the subject and the subject's state can change in response to either or both; moreover, social relations and nature itself can alter in response to the subject's activities. Traditional philosophy of science gives priority to the impact of nature; interest-oriented sociology of science emphasizes the prior social context. I propose to explore the relative contributions of two forces. (See Chapter 8.)

practices are all the same, then there would be powerful support for Latour and Woolgar's thesis of social determination and the construction of facts. If they vary, with the variation perfectly correlated with the substances presented, then there would be powerful support for the old-fashioned idea that, at least for this kind of social arrangement and this kind of statement, input from reality is strong enough to keep science on track. Other results would testify to an interaction between the social factors and inputs from asocial nature.

We cannot perform the experiment. But there is a common feeling that we know in advance how it would turn out. Surely, we think, there would be *some* variation: *some* of the groups would adopt a *different* belief about the structure of TRF or would at least fail to adopt the belief that TRF is Pyro-Glu-His-Pro-NH_2. Moreover, we expect to find some kind of correlation between the substances presented and the beliefs about structure that result, so that the realist idea that beliefs covary to some extent with the state of nature would be sustained. I find it implausible to think that, with respect to this example, our judgment is entirely self-deception. The experiment we envisage is too close in kind to naturally occurring situations in which people who encounter different parts of nature form different beliefs.[53]

Let us consider some obvious objections. First, it may be suggested that my formulation of the issue, with its pronouncements about what would have happened had the world been different, is suspect. I am sympathetic to the complaint that consideration of "ideal experiments" is somewhat artificial, but it is important to see that such consideration crystallizes what is at stake in the sociological critique. It is not enough simply to observe that there are social processes that occur when consensus practice is shifted: the thesis is that the character of the shifts is determined by these processes and not by inputs from nature. That thesis immediately makes counterfactual claims. It implies that differences in interactions with nature would not systematically be reflected in the modifications of consensus practice. Talk of "ideal experiments" and their consequences presents the counterfactual claims precisely.

Of course, we predict the outcomes of these "ideal experiments" by relying on our judgments about what actual figures do in a variety of actual situations. My formulations should appear less contrived if we recognize their kinship with everyday knowledge about how people in different social contexts and in different relations to nature behave, and how this knowledge serves us as the basis for making counterfactual predictions: judgments of the forms "If

53. It seems to me that Latour and Woolgar, like many contemporary sociologists of science, overestimate the power of the point that observation is theory-laden. Building on the work of Hanson, Kuhn, Feyerabend, and Hesse, they advance what Michael Bishop aptly calls "the expectation argument" —because perception is theory-laden we see just what our theoretical commitments would lead us to expect (Bishop forthcoming). But the expectation argument is a gross hyperextension of what philosophers and psychologists are able to show. Recall Kuhn's eminently sensible conclusion that anomalies emerge in the course of normal science: Roentgen did not expect to discover X-rays (Kuhn 1962/70 57–59).

S had not seen *X* then *S* would not have believed that *p*" and "If *S* had not been in a society of type *Y* then *S* would not have believed that *q*" are hardly alien. The suggestions about the outcomes of "ideal experiments" that I have offered simply introduce such forms of judgment into reflection about episodes of scientific change.

In some instances there are theoretical reasons for worrying that encounters with nature have no power to constrain consensus practice. Sociologists of science have been much influenced by the thesis of the theory-ladenness of observation and by the thesis of the underdetermination of theory by evidence.[54] They have also been much concerned with unsettled controversies, in which, because we do not know how to describe the relations between the participants and external nature, we are not able to use claims about the differences in inputs from reality to explain the variation in belief. But to recognize the methodological usefulness of ignoring scientists' encounters with nature in certain types of analysis, should not lead us to the far more ambitious claim that the world never has an impact on our beliefs. The inferential gap is bridged by taking it for granted that there are alternative ways of responding to any possible sequence of encounters with nature, so that the choice between rival modifications of consensus practice must be determined by social forces.

A sociologist of science might contend that the (functional) cognitive propensities present in mid–nineteenth century naturalists yield no resolution of the question whether organisms are related by descent with modification, given any possible sequence of encounters with nature that those naturalists might have had, so that the actual resolution of that issue had to be effected by social causes. That contention cannot be supported simply by pointing out that there is an interesting network of social relations among the early defenders of minimal Darwinism. The issue revolves around the possibility of identifying a reliable process of reasoning that could generate the switch to minimal Darwinism, given the stimuli that the naturalists actually received. In Chapter 2 I tried to show that this is indeed possible, and that the process is encapsulated in the long argument of the *Origin* (improved through the social interactions of those who debated the *Origin*). I now suggest that this serves as a paradigm for responding to sociological critiques. The first task must be to formulate those critiques as I have done, in terms of "ideal experiments" and the outcomes that both realists and their critics expect. Next one can try to bring to bear empirical information about people's actual behavior in different contexts on study of the conditions of decision. If it is suggested that general considerations—for example that of underdetermination of theory by evidence—reveal that the scientific decision under scrutiny could not have been determined through response to nature, then one must consider whether or not there are available, reliable processes that would

54. Both the theory-ladenness thesis and the underdetermination thesis will be discussed in Chapter 7. There is an obvious sense in which my reply to the arguments of Latour, Woolgar, and especially Barnes and Bloor, is incomplete until something has been done to defuse their claims about the radical underdetermination of our beliefs by all our encounters with nature.

deliver a verdict. Here there can be no substitute for looking at the historical details of the circumstances under which decisions were made.[55]

8. Metaphysical Options

Realism is sometimes a modest doctrine. Much of the debate about scientific realism in recent years has concerned a minimal version, the thesis that the terms of successful scientific theories refer to entities that exist independently of us. This version, *entity realism,* hardly seems worth the fuss. Unless we are quite wrong about the simple (Lockean) idea that we interact with something independent of ourselves, then to the extent that references of our terms are fixed through causal interactions, it is hard to see how we could *not* be referentially successful. Of course, this is quite compatible with the thesis that we know nothing at all about the entities to which we refer and, therefore, seems of little comfort to a realist.[56] Entity realism deserves Michael Devitt's gibe that it is antirealism with a fig leaf (Devitt 1984 15).

I have spent most of this chapter replying to objections to a much stronger version of realism, one that insists that many of the claims we make about nature are true. Traditional critiques of realism are typically directed at this stronger version (or particular articulations of it). However, the point of the enterprise is to guard against challenges to the view of progress that I advanced in the previous chapter. That view appears to be committed to an even bolder version of realism, so that even if I have been successful in turning back the complaints that have figured in recent discussions, the immodesty of my position requires additional defense.

On my account the primary modes of scientific progress are conceptual and explanatory progress. Both involve notions that may make modest realists queasy. Is it reasonable to believe that there are natural kinds that exist independently of human cognition? Or that there are objective dependencies among the phenomena? Affirmative answers to these questions generate

---

55. The most challenging version of a sociological critique of the type of realism that I favor is, in fact, extraordinarily rich in historical detail. In their (1985) Steven Shapin and Simon Schaffer argue that the debate between Boyle and Hobbes over the "experimental form of life" was only settled through "the acceptance of certain social and discursive conventions, and that it depended upon the production and protection of a special form of social organization" (1985 22). Because they are so convinced of the power of underdetermination arguments, Shapin and Schaffer develop their detailed history in a very particular way, not focusing on the gritty details of the encounters with nature or the complexities of the reasoning about a large mass of observations and experiments, but making very clear the filiations to the social issues that concerned the protagonists. As I shall suggest in Chapter 7, cases of putative underdetermination need to be handled extremely carefully, and, when that is done, Shapin and Schaffer's thesis becomes implausible. But, precisely because of the richness of their study—and because of its many insights—a complete reply would require a much longer discussion than this book can contain.

56. As noted in the last section, minimal realism is favored by several sociologists of science. See, for example, (Barnes 1992).

strong realism, a position that conceives of nature as having determinate "joints" and mind-independent causal structure.

Many realists would prefer not to go so far. Ever since Hume, philosophers have faced the challenge of explaining how we are in a position to gain evidence for statements involving a family of notions—statements that identify causal relationships, statements that talk of objective explanatory dependence, statements that assert that a particular set of objects is a natural kind, statements that talk of natural necessities. The root problem seems to be that we have no semantical account of such statements that will fit into an epistemological account.[57] Those theories that offer reductionist analyses in terms that fit neatly into empiricist epistemologies—for example, the efforts of philosophers from Mill to Hempel and John Mackie to make causal claims epistemologically tractable—face apparently insuperable difficulties in doing justice to the concepts they are trying to analyze.[58] Analyses that seem to do better—for example, approaches in terms of possible worlds (Lewis 1973a, 1973b)—raise obvious epistemological concerns. How do our encounters with actual nature provide information about the nonactual?[59]

Strong realism accepts the metaphysical commitments and tries to develop a response to the epistemological worries. Concerns about the possibility of knowing about kinds and causes derive, it is suggested from an oversimplified empiricist epistemology.[60] Empiricists worry that we can never know about causes because our basic perceptual knowledge is limited to identifying certain kinds of attributes. We can see that the first billiard ball moved, that it hit the second, and that the second moved after the impact—but (so the story goes) we cannot see that the impact caused the motion of the second ball. Strong realists can maintain, with some plausibility, that the epistemological presuppositions of such worries about causation are unfounded. Perception

57. The general difficulty is exemplified in the dilemma posed by Paul Benacerraf (1973) for theories of mathematical truth and mathematical knowledge. Benacerraf argued cogently that those accounts of mathematics that most obviously succeed in handling the semantics of mathematical statements, providing an unproblematic account of mathematical truth, strongly realist (Platonist) proposals that take mathematics to describe a mind-independent realm of abstract objects, bring epistemological problems in their train. How do we come into epistemic contact with the alleged objects, since causal contact is apparently denied us? By contrast, those approaches that seem to dissolve the epistemological mysteries (for example proposals that mathematics consists in recognizing the consequences of stipulations that fix the meanings of mathematical terms) have trouble in developing an account of mathematical truth. How does stipulation yield any type of truth? In general, realists find it easier to define and explain semantical and metaphysical notions but then face problems in understanding how we know what we seem to know.

58. See (Hempel 1965, Mackie 1973) for expositions of empiricist programs. Incisive critiques can be found in (Earman 1986 chapter V, Lewis 1973a, and Salmon 1984, 1989).

59. The problem was explicitly posed, and linked to Benacerraf's dilemma, in (Mondadori and Morton 1976).

60. David Armstrong responded to criticisms of Wesley Salmon's attempt to understand our knowledge of causal claims (at a symposium on (Salmon 1984) held in 1985) by declaring forthrightly that the epistemological problems were unworrying because "you can just see causation."

is not simply the registering of information in a hitherto blank mind. Rather, as I suggested in Chapter 3, PI beliefs come about through the activation of underlying propensities, and it might be possible to contend that we have cognitive propensities that regularly lead us to form the PI belief that A caused B against the background of a large class of cognitive states when we confront a situation in which A caused B. This is obviously vague, but strong realists can aim to remedy the vagueness, work out the details, and conclude that there is nothing epistemologically problematic about attributions of causation, natural kinds, or objective dependencies.[61]

Strong realism is not the only way of buttressing my talk of adequate concepts and objective dependencies. An alternative (Kantian in spirit; see Kitcher 1986) is to link the notions of natural kind and objective dependency to *our organization* of nature. Consider the cognitive desiderata of describing the world so that we can formulate generalizations, and, more generally, explain phenomena that are classed together in similar ways. I have suggested elsewhere (Kitcher 1981, 1989; see also Friedman 1974) that scientific explanation consists in achieving a unified vision of the phenomena. We can conceive of the sequence of practices in a field as attempting to modify the language and the set of explanatory schemata so as to achieve the greatest unity among the set of accepted statements. Unity, in this case, is, to a first approximation, understood in terms of generating the largest set of consequences through the use of the smallest number of patterns (schemata). More precisely, let a systematization of a set of statements be a collection of derivations, all of whose constituent statements (premises, conclusions, intermediate steps) belong to the set. Each systematization can be seen as instantiating a set of schemata, the basis of the systematization. The greatest unification of our system of beliefs is obtained when we use a systematization which generates as many conclusions as possible and whose basis contains the smallest number of most stringent schemata.[62]

As I shall suggest in Chapter 7, the growth of scientific knowledge is

61. The most sophisticated attempt to work out an epistemology that will be consonant with strong realism is that of Wesley Salmon (1984). Salmon proposes that a major aim of science is to deliver explanations and offers what he calls "the ontic conception of explanation." According to this conception, explanation consists in revealing underlying causal structure, displaying the mechanisms at work in the world. With admirable intellectual integrity, Salmon faces up to the Humean problem of understanding how we can know that causal relations obtain, advancing a detailed theory of causal knowledge in which the fundamental concepts are those of causal process and causal interaction. Building on ideas of Reichenbach, he suggests that we recognize causal processes through their ability to be marked and identify causal interactions through discerning changes in the causal processes involved. Despite the thoroughness of Salmon's investigations of the issues involved here, I believe that his account of causation is subject to some important difficulties. See (Kitcher 1989, especially section 6, and also van Fraassen 1985, Sober 1987, Woodward 1989).

62. There are many complications that I gloss over here. For more detail, see my (1989 especially sections 4.3 and 7). One point that should perhaps be explicitly mentioned is the possibility that there may be a need to trade off some desiderata against others: we may have to sacrifice stringency or paucity of patterns to increase the size of the set of conclusions delivered by the systematization.

governed by a principle of unification. Modifications of consensus practice are correctly supported by pointing out that they would lead to a system of belief that is more unified.[63] The use of that principle is apparent in Darwin's defense of his minimal schemata (HOMOLOGY and the others that only involve commitment to the notion of descent with modification) and is striking in his correspondence.[64] But even if one accepts the idea that changes in scientific practice can be justified by appeals to the unification they promote, the real task for any alternative to strong realism is to say what explanatory *correctness* amounts to, to explain what is meant by saying that a schema records the objective dependencies in nature. How is that to be done?

There is a line of thought that runs from Kant through Peirce to recent writers such as Sellars and Putnam that can usefully be adapted here.[65] Consider science as a sequence of practices that attempt to incorporate true statements (insofar as is possible) and to articulate the best unification of them (insofar as is possible). As this sequence proceeds, certain features of the organization of beliefs may stabilize: predicates of particular types may be used in explanatory schemata and employed in inductive generalization; particular schemata may endure (possibly embedded in more powerful schemata). The "joints of nature" and the "objective dependencies" are the reflections of these stable elements. The natural kinds would be the extensions of the predicates that figured in our explanatory schemata and were counted as projectible in the limit, as our practices developed to embrace more and more phenomena. Objective dependencies would be those recorded in the schemata that emerged in the limit of our practices.

According to this view the unification of our account of the world is a cognitive desideratum for us,[66] a desideratum that we place ahead of finding the literal truth on the many occasions that we idealize the phenomena. The causal structure of the world, the divisions of things into kinds, the objective dependencies among phenomena are all generated from our efforts at organization. To say that a particular predicate picks out a natural kind is thus to claim that marking out the extension of that predicate would figure in the

---

63. For some of the complications involved in stating this principle of justified change, see section 7 of my (1989). At this stage, since I only aim to sketch an alternative to strong realism, an imprecise version of the principle will be enough.

64. See (F. Darwin 1888 II 13, 29, 78–79, 110, 121–122, 210–211, 240, 285, 327, 355, 362; III 25, 74; F. Darwin 1903 I 139–140, 150, 156, 184).

65. See (Kant 1781/1968, Sellars 1967, Putnam 1981).

66. If this is so, then it is easy to understand the role that appeals to potential and actual unification play in scientific change. Darwin's stress on the unity that his modifications of practice would confer is directly linked to the attainment of a fundamental aim. By contrast, proponents of strong realism cannot take unification as constitutive of causal structure, so that for them there are serious epistemological questions about whether the search for unity will be a reliable guide to the causal structure of the world. This, of course, is reminiscent of Benacerraf's dilemma in the case of mathematics. I have argued the point in some detail in connection with Wesley Salmon's "ontic conception" of explanation. See my (1989 section 8, especially 496–497).

ultimate (ideal) practice.[67] Hailing a schema as correct is predicting that that schema will have a part in the ideal unification of the phenomena.

For my present purposes I am not concerned to articulate the details of this (Kantian) rival to strong realism. Indeed, the project of this book can go forward without choosing between the alternatives. The account of progress I have offered depends on the possibility of making sense of the notions of adequate concepts (or natural kinds) and correct explanatory schemata (or objective dependencies). Unless there are compelling reasons for thinking that both the strong realist and the Kantian strategy for explicating these notions are doomed, then skepticism about my commitment to these notions seems unfounded. I do not see that there are such compelling reasons.

9. Possible Losses

I began by considering a general worry about the coherence of the goal of attaining truth, a worry that is prominent in Kuhn. I shall conclude with a second type of Kuhnian concern, one that challenges ideas of accumulation in science. After offering his traditional philosophical question about truth, Kuhn advances his evaluation "as a historian" of the view that science makes progress by achieving truth:

> I do not doubt, for example, that Newton's mechanics improves on Aristotle's and that Einstein's improves on Newton's as instruments for puzzle-solving. But I can see in their succession no coherent direction of ontological development. On the contrary, in some important respects, though by no means in all, Einstein's general theory of relativity is closer to Aristotle's than either of them is to Newton's. (1962/1970 206–207)

In other places, Kuhn is more explicit about the exact nature of the similarities and differences. He claims that there are important affinities between Aristotelian explanations and explanations in nineteenth century physics:

> As [physics] became increasingly mathematical, explanation came increasingly to depend upon the exhibition of suitable forms and the derivation of their consequences. In structure, though not in substance, explanation was again that of Aristotelian physics. Asked to explain a particular natural phenomenon, the physicist would write down an appropriate differential equation and deduce from it, perhaps conjoined with specified boundary conditions, the phenomenon in question. (1977 26)

The contrast is with the causal-mechanical style of explanation characteristic of Cartesian and Newtonian physics, where, it is supposed, there are "active agents" or "isolable causes."

---

67. Or, given my gloss on how the limiting of practices is to work, in all practices beyond a particular point in a properly developed sequence. Here the notion of proper development of a sequence of practices is partially understood in terms of the attempt to conform to the principle of unification.

Furthermore, Kuhn suggests, there are important differences between the explanatory projects of classical physics and those of Einstein's general relativity. In the eighteenth century, Newtonians found it " 'unscientific' to ask for the cause of gravity" (1977 212). However, "[g]eneral relativity does explain gravitational attraction" (1977 212). So, apparently, an explanatory project that was once rejected has been reinstated.

Kuhn's remarks seem directed against the conjunction of two claims: (1) Physics has made progress from Aristotle to Newton to Fourier to Maxwell to Einstein; (2) progress consists in the accumulation of explanatory strategies. I defend a version of (2): explanatory progress consists in the incorporation of correct schemata, the replacement of incorrect schemata, the extension of correct schemata. I also defend a version of (1): physics made explanatory progress from Aristotle to Newton to Fourier to Maxwell to Einstein. So Kuhn's examples will be at odds with my account if one of the transitions involves the rejection of a schema subsequently recognized as correct, or the incorporation of a schema subsequently recognized as incorrect. It appears that Kuhn's references to the revivals of Aristotelian ideas in the nineteenth century and in the work of Einstein are intended to suggest that the former error has occurred. Let us call such instances *explanatory losses.*

Is there an Aristotelian schema that we can recognize as correct and that was eliminated in the transition to Newtonian physics? Consider first Aristotle's explanations of motion, as cited by Kuhn.

> Stones fell to the center of the universe because their nature or form could be entirely realized only in that position; fire rose to the periphery for the same reason; and celestial matter realized its nature by turning regularly and eternally in place. (1977 24)

The explanations are familiar from Aristotle's writings on motion, and it is not hard to uncover an underlying schema.

NATURAL MOTION

Question: Why does body X move naturally to place P?

Answer:

(1) X is composed of the elements earth, air, fire, and water in the proportions $p : q : r : s$.

(2) The natural place of a body composed of the elements in proportions $p : q : r : s$ is $P(p, q, r, s)$.

(3) Bodies realize their natures most fully when they are in their natural places.

(4) Bodies in natural motion strive to realize their natures most fully.

Hence (5) In natural motion, X moves to P.

Perhaps this schema errs in being generous to Aristotle in that it suggests that there are formulable principles of Aristotelian physics that specify the natural places of mixed bodies.[68] But I offer the schema only as an example of a way in which the explanatory structure of Aristotle's account of motion might be exposed. Plausible amendments will not affect the thrust of my argument.

NATURAL MOTION is abandoned in Newtonian mechanics. But is it correct? No. We do not regard it as formulating the correct dependencies because it wrongly suggests that differences in tendencies to rise and fall result from differences in the composition of bodies out of particular substances. Nothing so specific will do justice to Kuhn's idea that Aristotelian explanation formulates an insight abandoned in Newton and rediscovered later.

Bur perhaps Aristotle can be viewed as mistaken about much of the detail and yet, perhaps, be credited with recognizing something that Newton missed, the anisotropy of space and the relevance of differences in spatial location to differences in motion. Consider

MINIMAL ARISTOTELIANISM (I)

Question: Why does X move to place P?

Answer:

(1) X is initially at place P^*.

(2)   If a body is initially at place P^*, then it will move to P.

Hence (3) X moves to P.

MINIMAL ARISTOTELIANISM (I) is so anemic that I shall take it to be correct. But MINIMAL ARISTOTELIANISM (I) was not abandoned in Newton's account of motion. Consider a Newtonian schema.

GRAVITATION

Question: Why is X at $(x_{1t}{}^*, x_{2t}{}^*, x_{3t}{}^*)$ at time t^*?

Answer:

(1) X is initially at (x_{10}, x_{20}, x_{30}). The only other bodies in the universe are $Y_1, \ldots Y_n$ with initial positions $(y_{110}, y_{120}, y_{130}), \ldots$.

(2)   The mass of X is m; the masses of $Y_1, \ldots, Y_n$ are $m_1, \ldots, m_n$.

(3)   The only force acting is that of gravity.

(4)   The force on a body of mass M_1 from a body of mass M_2 due to gravity is attractive, directed along the line between the bodies, and is of magnitude GM_1M_2/r^2 where G is a constant and r is the distance between the bodies.

(5)   The equation of motion for a body of mass M subjected to a net force F_i along the direction of the ith spatial coordinate is $M\, d^2x_i/dt^2 = F_i$.

---

68. There are general suggestions about the natural places of "mixed bodies" in Aristotle's *Physics,* but nothing as specific as the principle I ascribe. That, of course, is hardly surprising given Aristotle's views about the distortions introduced by mathematization.

(6)   For the $n + 1$ body system consisting of $X, Y_1, \ldots, Y_n$ the equations of motion are . . . . [obtained by computing net gravitational forces dependent on the relative positions of the bodies; there will be $3(n + 1)$ simultaneous differential equations in $3(n + 1)$ coordinate variables and time].

(7)   The explicit motion of X is given by the equations $x_i = g_i(t)$. [Obtained by solving the system of equations (6) and using the initial condition (1)].[69]

(8) $g_i(t^*) = x_{it}^*$.

Not only is GRAVITATION not incompatible with MINIMAL ARISTOTELIANISM (I), it is even an extension of it! The first lines of both schemata formulate the dependency on place, and lines (2)–(7) of GRAVITATION yield a conclusion that is simply taken for granted at step (2) of MINIMAL ARISTOTELIANISM (I).

Of course, I have been cheating. Newtonian explanations only make phenomena of motion dependent on spatial location because the *relative* positions of bodies make a difference to the gravitational forces that they experience. The gravitational field itself is supposed to be invariant over space. So if Aristotelian explanations are going to capture something that Newton missed, then they will have to go beyond the bland formulation of MINIMAL ARISTOTELIANISM I (in which phenomena of motion are seen as dependent, in an undifferentiated way, on position in space) to something that captures the idea that the *gravitational field* is dependent on spatial position.

Let us try to formulate the notion that Aristotle's belief in the anisotropy of space captures an ideal of explanatory dependence that Einstein saw and Newton missed.

MINIMAL ARISTOTELIANISM (II)

Question: Why is the gravitational field in region R, G?

Answer:

(1)   The attributes of space in region R are S.

(2)   The relation between the attributes of space in a region and the gravitational field in that region is given by the function g, where $g(S) = G$.

Hence (3) The gravitational field in R is G.

Obviously this schema is very vague. There are correct versions of it, present in Einsteinian physics and absent in Newtonian physics. Are any such versions also present in Aristotle? That seems to me highly implausible. Aristotle would not recognize the attributes of space that current physics takes to be relevant to the gravitational field. Thus the most we can claim is that Aristotle

69. Of course, this is bluster when $n > 1$, since for three or more bodies the equations are not generally analytically solvable. However, that point is irrelevant to the present discussion.

recognized the dependency of the gravitational field on *some* attributes of space. Even if this is conceived as an insight that generates a genuine explanatory loss in Newtonian physics, it presupposes that Aristotle gave explanations of the characteristics of the gravitational field, and this, I suggest, strains credulity.

Although we could extend the discussion of Kuhn's thesis by considering the abandonment of appeals to formal causes in favor of appeals to efficient causes (or by looking at another favorite example, the alleged loss of the ability to explain what metals have in common in the transition from phlogiston chemistry to Lavoisier's new chemistry[70]), the preceding treatment makes clear the strategy for going further. When the explanatory schemata of the various practices are made completely explicit we find no basis for claiming that later consensus practices have abandoned an earlier correct schema. I conclude that the type of cumulativism to which my account of cognitive progress is committed can survive yet another influential challenge.

6

Dissolving Rationality

1. Good Design

Before the early 1960s, almost all philosophers took the rationality of science for granted. So prevalent was the notion that science is the epitome of human rationality that explicit announcement of allegiance to it was largely confined to popular lectures, introductory books, or preambles to more serious epistemological work.[1] To account for this admirable feature of science it was presupposed that scientists tacitly know methodological rules (jointly comprising "scientific method") which are used to appraise newly introduced hypotheses and theories. Identifying these rules was an important—possibly the most important—task for philosophy of science.[2]

In the next three chapters I shall try to see what, if anything, can be salvaged from the thesis of the rationality of science. I shall begin by returning the notion of rationality to its psychological roots. *Science* is not the sort of thing that uses reasons, good or bad. *Scientists* engage in reasoning. The thesis of the rationality of science ought to be interpreted as claiming that scientists, for the most part or, perhaps, at the most crucial times, base their conclusions on good reasoning. Or perhaps it intends to go deeper and trace the performance of scientists to something in the institution itself that encourages good reasoning.

The discussions of the last three chapters provide a way of thinking about science that enables us to develop this approach. The science of a time is

1. See, for example, Reichenbach (1938), Hempel (1966), and Popper (1959). In all these works, the hard problem is to identify the method that is taken to be at the core of scientific rationality.

2. Many of the consequences of the older traditions become visible in the reactions to Kuhn and Feyerabend. See, for example, Lakatos (1969), Scheffler (1967), Laudan (1978), and, for an especially explicit version, Newton-Smith (1981). An alternative to the implicit psychological hypothesis that scientists know and apply certain special rules is Merton's (1942/1973) suggestion that features of the community of scientists account for the choices made among theories, and thus for the rationality and progressiveness of science. This kind of sociological approach was not developed by the leading defenders of rationalism. However, some proponents of evolutionary epistemology can be viewed as defending the rationality of science without assuming the rationality of scientists. (See, for example, Hull 1988.)

constituted by the collection of individuals engaged in doing science, their relations to one another and to the broader community, their cognitive propensities and their individual practices, the consensus practices of the various fields and subfields. According to the last two chapters, this total system is intended to achieve certain goals. Is it well designed for doing so?

We can make this question more precise by specifying the entity (or entities) whose design we are to investigate, the kinds of goals we bring into consideration, and the standard for good design. The very broadest conception concerns science as a whole and its place in our societies and considers our total set of goals (if, indeed, we can sensibly talk of universally shared goals here). Because I have explicitly foresworn analysis of our overall sense of value, my investigations will be more limited. One question, to be investigated in more detail in Chapter 8, concerns the good design of scientific communities for achieving epistemic ends. There I shall compare different types of social organization and different mixtures of cognitive strategies with respect to their expected generation of progressive sequences of consensus practices.

I conceive of rationality as a means-end notion.[3] Concepts of rationality are generated by thinking of entities (people, groups of people, science as a whole, science and its relations to society) as meeting some criterion of good design (maximization of expectation, expectation of positive modification, high expectation with respect to rival entities) relative to a set of goals (epistemic goals, practical goals, both). Thinking of rationality as a means-end notion is hardly new (Newton-Smith 1981 chapter 7, and, for a lucid, recent presentation, Foley 1988a, 1988b 126).[4] Articulating that familiar idea in the

3. This is, of course, quite at odds with many discussions of rationality, both celebratory and critical. Frequently, rationality is taken to be constituted by a set of rules whose status is independent of their tendency to promote any ends. Explanations by appeal to rationality require showing only that a particular set of statements conforms to the rules. As I shall suggest, this flawed apsychologistic conception of epistemology invites the critiques of David Bloor, Barry Barnes, and others. Moreover, the divorcing of rationality from any tendency to promote epistemic ends fosters relativism. So, for example, the suggestion by P. F. Strawson (1952) and Rudolf Carnap (1952) that the Humean problem of induction can be solved/dissolved by taking inductive logic to comprise a system of principles that is constitutive of rationality is rightly criticized by pointing out that there might be rival systems of principles, espoused by different communities, that were regarded by members of those communities as regulating the adjustment of belief. (Interestingly, this point was made powerfully by a number of philosophers who were concerned with the "foundations of induction." See, for example, (Feigl 1950), (Salmon 1968 33ff.) and (Skyrms 1986 Chapter 2).)

4. Ronald Giere (1988) deliberately contrasts the kinds of evaluative projects of which he approves (thinking about the adjustment of means to ends) with what he views as the "traditional" philosophical conception of rationality (belied by the articles by Feigl and Salmon, cited in the previous footnote.) Richard Foley's (1988a) not only explicates the means-end approach but shows how long and rich is its philosophical history. Giere's contrast thus seems to me to illustrate the way in which philosophical discussions can be influenced by recent fashions, fashions that become canonized as "the tradition."

However, as John Worrall forcefully pointed out to me, a conception of rationality that is solely concerned with the adjustment of means to ends and agnostic about the ends would not capture many important themes that rationalistic philosophers of science have

context of an explicit investigation of the subjects of rationality, the goals and the criterion of good design transforms the debate about the rationality of science.

Recent battles about scientific rationality have not been concerned with the total system, with maximization of the expected progressiveness of a sequence of consensus practices, or anything of that sort. The focus has been on individuals. Here too we can ask whether the performances of individuals are well designed for the attainment of their goals. In the context of scientific activity, individuals count as *overall rational* if what they do meets some criterion of good design for attaining their total set of goals. Conceptions of *epistemic rationality* are generated by restricting attention to epistemic goals. Recall from Chapter 3 that these are of two classes: Individuals have goals for themselves and for the communities to which they contribute. Thus we may inquire whether a scientist's actions and decisions are well designed for promoting progressive consensus practices in the community to which the scientist belongs. This is not the same as asking whether those actions and decisions are well designed for promoting modifications of the scientist's individual practice that would be progressive with respect to either the original individual practice or the consensus practice of the community.

In this chapter and the two that follow, I shall be interested in questions about goodness of design for achieving epistemic goals, some of which relate to traditional problems in the philosophy of science, others of which only come into view when the distinctions of Chapter 3 are appreciated. Philosophers and reflective scientists have devoted much attention to the methods that should be followed and the inferences that should be used in extracting information from nature. I shall discuss such issues in Chapter 7, by focusing on encounters with nature. Good design of such strategies is neither necessary nor sufficient for achieving progressive sequences of consensus practices. This point will be demonstrated in Chapter 8.

So much for an overview of the territory. Let us now descend from the heights and remind ourselves of the type of episode that provokes the claim that science is rational.

2. Illustration

At the 1860 meeting of the British Association for the Advancement of Science, Robert Fitzroy, onetime Captain of HMS *Beagle* and former friend and mess-mate of Charles Darwin, stalked the hall, holding a Bible above his head and intoning the words "The Book! The Book!" Meanwhile, up on the platform, the Bishop of Oxford, Samuel ("Soapy Sam") Wilberforce, engaged

---

wanted to emphasize. To generate a notion of epistemic rationality it is necessary that the ends be epistemic. Thus, in Chapters 4 and 5, I have tried to offer an account of the epistemic ends of science that could be put to work in articulating a notion of epistemic rationality in terms of the adjustment of means to *those* ends.

T. H. Huxley in debate about the merits of another book, Darwin's *Origin*. Wilberforce, so the story goes, had been coached by Richard Owen, who was later to launch another undercover attack on Darwin's proposals by writing an anonymous review.

This scene dramatizes traditional thinking about scientific rationality. There is a cast of five characters (two of whom, Owen and Darwin, are off-stage). Two (Darwin and Huxley) decide rationally in favor of common descent. Three (Fitzroy, Owen, Wilberforce) act irrationally. All five share a large number of beliefs about the natural world, accepting the phenomena about the distributions of organisms, comparative morphology, the fossil record, and so forth, that Darwin reported in the *Origin*. If that information were stored in the declarative memory of a cognitive system with an activated goal to adopt a progressive shift in individual practice and if that system had activated cognitive propensities that were well designed for making progressive shifts, then that cognitive system would modify its state to embrace the claims and commitments of minimal Darwinism. Darwin and Huxley did this: they are therefore rational. Fitzroy, Owen, and Wilberforce did not: they are therefore irrational.[5] We explain their irrationality, perhaps, by suggesting that other, nonepistemic, goals become activated and, in consequence, their cognitive activity is diverted.

What we know of the men's behavior enables us to sketch pictures of their psychological lives. Fitzroy's apparently demented display suggests that the goal of defending certain religious doctrines has become so dominant in his thinking that he is unable to respond at all to Darwin's evidence and arguments. Propensities that might have been activated to reach Darwinian conclusions are immediately inactivated once it is recognized where their application leads. The case of Wilberforce is more subtle and more interesting. Dedicated to a long-term goal of protecting the Anglican church, Wilberforce was favorably disposed to hear that there were flaws in Darwin's argument. Deferring to an eminent authority who told him what he wanted to hear, he decided to engage Huxley in the famous debate. His nonepistemic (pastorally professional) goal was activated, and, in consequence, he applied inferential propensities that made him sympathetic to criticisms of Darwin's arguments. A similar story can be told for Owen, the most interesting of the three. Already disinclined to form beliefs about the relationships among organisms, Owen seems to have been motivated by professional jealousy—the goal of retaining his status of top dog in British biology was activated. When his tendency to suppress propensities that might lead to risky conclusions (a part of his cognitive temperament) was reinforced by his desire to find holes in Darwin's arguments, he activated propensities that led him to scrutinize the minutiae of the case. His carping response to the *Origin* was the result.

These are psychological speculations. I do not claim that they are correct. Indeed, historians would surely want to tell far more subtle stories about the

---

5. Here I use the traditional language. As we shall see later in this chapter, the epithet is only appropriate for at most one of the participants (Fitzroy).

complicated characters I have mentioned. My aim is to illustrate, showing how the traditional idea that there was a distinction between rational and irrational responses to the *Origin* can be translated into my framework. Those who believe in that distinction and its application to divide Darwin and Huxley from Fitzroy, Owen, and Wilberforce are, I think, committed to some cluster of psychological hypotheses such as I have offered. What Darwin and Huxley did must be seen as well designed for the making of cognitive progress. The cognitive systems of Fitzroy, Owen, and Wilberforce, by contrast, should be revealed as defective in some way (by no means necessarily the same way in all cases). Perhaps nonepistemic goals became inappropriately activated, perhaps the inferential propensities they applied were unreliable.

Once again, I intend only to illustrate. The point of the traditional distinction between the rational and the irrational is to mark out features of cognitive systems that are well designed for making cognitive progress. Having reframed the issues in this way, the next task is evidently to try to specify the relevant notion (or notions) of good design.

3. Preliminary Clarifications

Because debates about scientific rationality have often taken for granted a very different perspective from the one I adopt, there is real danger that firmly entrenched ideas about rationality will be carried over in thinking about my conception of rationality as good cognitive design. Accordingly there are some misunderstandings about past attempts to discern rationality in science that it is important to correct and some criticisms of any such enterprise that ought to be answered.

Defenders of the rationality of science are sometimes lumbered with the doctrine that only those who hold true beliefs can be rational. Thus Bruno Latour imagines attempting to divide figures from the past and the present into two classes, one labeled with "derogatory adjectives" (of which 'irrational' is a paradigm) and the other with "laudatory ones" ('rational,' for example). The attempt, Latour claims, is doomed. Descartes initially appears to be on the right side—until we recognize his commitment to bizarre vortices. Newton only lasts among the praiseworthy while we keep his alchemical researches under wraps. So it goes. Latour concludes wryly: "The only way to stop adjectives jumping randomly from one side of the divide to the other would be to believe that only *this year's* scientists are right, sceptical, logical, etc." (1987 191).

Well, of course one can make the enterprise of appraisals of rationality look foolish and unworkable by choosing the wrong units of analysis. Whoever thought that a person would be rational or irrational *throughout her entire career*? Indeed, the tradition aimed at specifying the rationality or irrationality of particular beliefs or decisions, held or made by particular people in particular contexts. The appropriate candidates for division are not therefore *people* (Newton, Descartes) but the beliefs they held or the things they did at particular times (Descartes's belief in the law of refraction in 1637, Newton's belief that gravitational attraction changes inversely as the square of the distance separating the bodies, his belief that isolation of the "green lion" would enable him to turn lead into gold). However, interesting problems must still be addressed. Persistent lapses might betoken serious flaws in cognitive design. When performance becomes sufficiently mixed, what we previously viewed as clear successes may come to seem matters of epistemic luck.

There is a more fundamental error in Latour's critique. The distinction between the rational and irrational should not be conflated with that between the true and the false. People sometimes rationally hold false beliefs and sometimes irrationally hold true beliefs. Rationality is supposed to be an attribute of the processes that generated or sustained belief.[6] Once this point has been established, there is no basis for Latour's ironic conclusion that only *this year's* scientists deserve the laudatory title of 'rational.'

However, in making the character of ascriptions of rationality explicit, I have exposed a point that champions of scientific rationality have not always appreciated, and have sometimes even denied. To say that the beliefs or decisions of past scientists are rational is to advance *psychological* claims. Galileo did various things: he turned his telescope on the night sky, he came to believe that Jupiter has satellites, and he maintained that the Aristotelian account of the heavens is incorrect. We suppose that these actions, decisions, and beliefs are rational only because we think that there is a particular kind of causal explanation for them. We would receive a rude shock if we learned that Galileo was a huckster; that he had no faith in the power of the telescope to reveal facts about celestial bodies or that his confidence was the result of wishful thinking, that he only hastily glanced toward Jupiter and, seeing some spots of light, decided to ingratiate himself with his Medicean patrons; that his attack on Aristotelianism was based on efforts to make a reputation and that his arguments moved him because of the euphony of his Italian formulations.[7]

The point is akin to one that is now familiar in general epistemology. Despite what the logical positivists taught and what generations of their successors have dutifully learned, rationality and justification are not simple matters of logical connection among beliefs.[8] When psychology is left out of

6. This point is often muffled in apsychologistic approaches to epistemology, especially those that insist on the self-authenticating character of some statements (descriptions of sense experience, principles of logic and methodology). According to such approaches, accepting the statements in question (perhaps in the right context, perhaps—with logic— in any context) becomes constitutive of rationality. Nonetheless, *all* twentieth century epistemologists of whom I am aware would reject Latour's blanket identification of truth and rationality.

7. Despite Galileo's interest in and talent for securing patronage (see Westfall (1985) and Biagoli (1990)), the overwhelming evidence is that his actions and decisions stem from cognitive strategies designed to promote cognitive goals.

8. Before the end of the nineteenth century, accounts of human knowledge were thoroughly grounded in explorations of human psychology. The links between epistemology and psychology are very clear in Locke, Hume, and Mill. Even Kant—long assimilated to

epistemology (and it is claimed that epistemology is a development of logic) we are offered the following picture: there are epistemic rules declaring that a subject who has received certain stimuli and who believes that p is automatically justified in that belief (the belief is rational); other rules declaring that if the subject rationally believes that p and if q bears certain logical relations to p (in the simplest case, if q is a logical consequence of p), then, if the subject believes that q, the belief that q is automatically rational. As the last type of example makes especially obvious, these rules may fail precisely because the justificatory connections are never made in the life of the subject. I may rationally believe that p, I may believe that q, and there may be a deductively valid argument from p to q, but *I may never appreciate this argument and base my belief in q on something quite different and irrelevant.* The general moral is that epistemology should be *psychologistic.* Whether or not people are rational in their beliefs depends not simply on what beliefs they hold or how the propositions they believe are logically connected, but also on how their beliefs are *psychologically* connected.[9]

As David Bloor has recognized, there is an important symmetry: rational beliefs and actions, as well as irrational beliefs and actions, require causal explanation.[10] For any state of belief, decision, or action, we can ask for a causal explanation, focusing on proximal psychological causes or on the impact of the social and asocial environments.[11] However, Bloor's thesis is easily

---

twentieth century preferences for banning psychology (see Bennett 1966, Strawson 1966)—used highly abstract description of psychological processes to undergird claims about knowledge (Patricia Kitcher 1990). The role of psychology in philosophy comes under attack in Frege's writings (witness the introduction to the *Grundlagen*), but, as I have argued (Kitcher 1979), the significance of Frege's project can only be understood against the background of his Kantian (psychologistic) epistemology. But that epistemology remained tacit in his writings, and the success of the logical techniques that he introduced inspired his successors to make a thoroughgoing split between epistemological and psychological questions. The divorce is announced in Wittgenstein's *Tractatus* (highly influential on the Vienna Circle, and, through them, on twentieth century English language epistemology): "Psychology is no more closely related to philosophy than any other natural science" (4.1121). (Frege's critique of Husserl's views on arithmetic was also influential in prompting European philosophers to avoid psychology.) Even when epistemology was no longer seen as an extension of logic, this separation lived on until the 1970s (see Kitcher 1992).

9. This seems to me to be the most important consequence of the responses to Gettier's problem about the analysis of knowledge. For elaborations of the argument of this paragraph, see (Harman 1971 chapter 2), (Goldman 1979), (Kornblith 1980), (Kitcher 1983a chapter 1), and, for what is now the classic source, (Goldman 1986).

10. Bloor's insight was originally formulated in his (1974). As will become clear in what follows, while I take him to have offered an important corrective to epistemological positions then current, I believe it to be important to separate versions of the symmetry thesis that are wrongly conflated.

11. It is worth remarking explicitly that the placing of human cognitive states—such as beliefs—firmly in the natural order does not dictate a particular species of causal explanation. Sometimes it appears that the salutary recognition that the presence of a belief in a subject cannot simply be explained by pointing to logical relations among statements is understood as entailing that we need a *social* explanation of the presence of the belief. But there are all kinds of causal processes that initiate or maintain belief, among them some that involve the reflection of logical relations in psychological connections.

overinterpreted. It does not follow that there are no important distinctions among the types of processes that generate and sustain beliefs, decisions, and actions. Only when rationality is construed in the faulty apsychologistic fashion does Bloor's insight pose a threat.

Many sociologists of science view the symmetry thesis as checkmate for the champions of rationality. Once it has been conceded that beliefs, actions, and decisions have causal explanations, then they see no point in trying to distinguish among the forms that these explanations may take. Resurrecting rationality, they suggest, is idle moralizing. Latour formulates the idea with characteristic panache. After contending that appraisals of rationality can typically be reversed, he concludes:

> There are only two ways to get out of this situation. One is to use derogatory and laudatory adjectives and their accompanying adverbs whenever it suits you. 'Strictly logical', 'totally absurd', 'purely rational', 'completely inefficient', thus become *compliments or curses*. They do not say anything more on the nature of the claims being so cursed or complimented. They simply help people to further their arguments as swear words help workmen to push a heavy load, or as war cries help karate fighters intimidate their opponents. This is the way in which most people employ these notions. The second way is to recognize that these adjectives are so unreliable that they make *no difference* to the nature of the claim, each side of the divide being as rational and as irrational as the other. (1987 192)

I shall argue that Latour's suggestions disguise both the serious purpose and the genuine difficulties involved in appraisals of rationality.

Focus for the moment on the simplest kind of cognitive progress, that of finding a true answer to a significant question. Imagine that various subjects have all the information needed to generate belief in the correct answer: there is an inferential process that could lead any of them from items in declarative memory to a state of belief in the correct answer and all of them have the propensities required to undergo this process. Some of them activate the right propensities and achieve the true answer. Others activate propensities that are very unlikely to generate true answers (for example, suppose that they lexicographically order the alternatives and choose the eleventh), and they come to believe incorrect answers. There is a distinction to be drawn here. Some undergo processes that reliably generate true beliefs, while others undergo processes that have a very small chance of yielding true beliefs.[12]

The point here is very simple. People can make cognitive mistakes, per-

---

12. Simple examples are readily generated by thinking about situations in which students (from very young children to apprentice scientists) are tested. Of course there are many instances in which students arrive at the right answer but are properly corrected for having engaged in reasoning that is unreliable (likely to generate false answers, and only successful on the present occasion as a matter of luck). While these examples are distant from the more complex situations of scientific debate, so that there is a genuine issue about whether or not the distinctions of good from bad cognitive design carry over to those contexts, they do demonstrate the possibility of honoring the idea that beliefs are part of the natural order and simultaneously pursuing normative questions.

ceiving badly, inferring hastily, failing to act to obtain inputs from nature that would guide them to improved cognitive states. Nobody who accepts the symmetry thesis, who believes (rightly) that changes in cognitive state have causal explanations, should go on to infer that there is no basis for distinguishing the types of processes that generate and sustain beliefs, actions, and decisions. Some types of processes are conducive to cognitive progress; others are not.

From this perspective Latour's dismissal of talk about rationality disguises the hard problems by suggesting that we can and do make judgments about rationality in any way we like. He thinks that we can vary our perspective on Galileo, seeing him alternatively as "courageously rejecting the shackles of authority, arriv[ing] at his mathematical law of falling bodies on purely scientific grounds," or as "a fanatic fellow traveller of protestants [who] deduces from abstract mathematics an utterly unscientific law of falling bodies" (1987 191, 192). However, if our aim is to do more than make up stories about Galileo and we proceed with the approach to rationality I have been recommending, then we face two difficult projects. First we must identify the processes that underlie Galileo's beliefs and decisions. Second we have to assess the ability of such processes to generate and sustain cognitively valuable states. Both endeavors are error-prone. But there is a fact of the matter: either Galileo's reasoning exemplified a strategy likely to promote cognitive goals or it did not.

Why should we engage in such difficult endeavors? Since Bacon and Descartes, explicit methodological reflection has been designed to improve human cognitive performance, and we look to the apparent successes of the past for guidance.[13] So the (fallible) enterprise of evaluating the rationality of our predecessors does not simply express our penchant for grading people, awarding gold stars here and black marks there. Methodologists hope to analyze successful scientific reasoning in order to integrate those cognitive strategies that promote progress within the science of the present. Psychological research not only demonstrates that subjects have difficulty assimilating certain types of cognitive strategies (appealing to base rates, for example) but also shows that successful strategies can sometimes be absorbed into their cognitive lives (Holland et al. 1986 260–271). The delineation of formal rules, principles, and even informal canons of reasoning, when supplemented by an appropriate educational regime, can thus make people more likely to activate propensities and undergo processes that promote cognitive progress.[14]

---

13. Here, and elsewhere in Chapters 6, 7, and 8, I emphasize the *meliorative* project of epistemology. However, as both Peter Godfrey-Smith and Kim Sterelny have pointed out to me, epistemologists are rightly also interested in *understanding* the extent to which human cognitive performance is successful. Given my account of cognitive goals in Chapter 4, it would be ironic if I were to slight the significance of understanding, whether or not it leads to improved performance. Thus my claims about the meliorative project should be seen against the background of a broader epistemic enterprise.

14. Consider one of the canonical problems for which Bayesian analysis is useful, the computation of risks in cases where a medical diagnostic test has a known rate of false

Methodological reflection on the history of science is important to the practice of science, and it is hardly surprising that practicing scientists often scour the recent past for clues about strategies of reasoning, acting, and decision making that might be employed to promote their goals. In some sciences, courses in methodology form an important part of training—sometimes explicitly endeavoring to correct the errors that earlier workers in the field were liable to make (see, for example, Sokal and Rohlf 1981).[15] Dividing figures of the past is therefore not a matter of complimenting and cursing, but a preliminary to difficult empirical study of how to improve human cognitive performance.

Let me take stock of likely objections. First, I have taken for granted the idea that there are *some* distinctions among people's beliefs. We cannot sensibly divide processes and propensities into those that are likely to promote cognitive progress and those that are not, without prior division of those beliefs that are true and significant from those that are either false or insignificant. Objections to the coherence of this division or to our ability to draw it will therefore affect my treatment of rationality. However, I have attempted to forestall objections of this kind by offering, in the previous chapters, a relatively detailed account of scientific progress and a response to broad criticisms, worries about truth, or about our ability to identify truth. My replies are intended to defend presuppositions of the approach to scientific rationality undertaken here.

Second, historians of science are likely to wonder whether appraisals in terms of good design for cognitive progress inevitably involve a "chauvinism of the present" that rides roughshod over the thoughts and inferences of our predecessors. I hope that my discussion of historical examples later in this

---

positive results. If you are given a test for a rare disease that gives a small rate of false positives, and if you test positive then your risk is correctly assessed by employing information about the base rate and using Bayes's theorem. If you ignore the base rate and follow some non-Bayesian process of reasoning you will almost certainly arrive at erroneous conclusions about your risk. Granting that people do not *naturally* perform the Bayesian computation, the significant points are that that computation improves their decision making and that they can be taught to do it.

15. It is important to distinguish here between the "nuts-and-bolts" methodological advice about experimental design and statistical analyses of data that is genuinely useful for scientists, and the more grandiose methodological pronouncements ("Always try to falsify your pet hypothesis!," "Prefer simpler theories!") that are sometimes adapted from philosophy to provide exhortations to the scientific apprentice. These slogans, often at odds with scientific practice, should no more be taken as philosophy's main contribution than should the historical anecdotes that leaven scientific texts be seen as history's gift to science. In both cases, simple stories and simple slogans may provide rough and ready pictures, valuable to the beginner but unneeded (and discarded) by the veteran. However, even the veteran may sometimes be well served by historians and philosophers who can provide nuanced accounts that are relevant to current debates. Recent discussions in evolutionary theory have drawn, sometimes substantially, on historians' reconstructions (see, for example, Gould 1977, Eldredge 1985) and philosophers' attempts to articulate methodological ideas that are directly relevant to current debates (Sober 1984, 1988, Brandon and Burian 1984, Kitcher 1985b, Dupre 1988).

chapter will lay any such concerns to rest. In the meantime, it is possible to explain in general how the thinking and decision making of the past can be evaluated at least partly on their own terms.

To evaluate a scientist's efforts to achieve cognitive goals, we need a view of the scientist's situation, a view of the scientist's capacities, and a view of what would have promoted attainment of the goal. These views should draw on the best available information. Acquiescing in current science, we use contemporary knowledge, including historical knowledge, to make our assessment. Does this mean that we necessarily criticize the scientist for failing to see what we see or for failing to undergo the processes that we believe to promote cognitive progress? Of course not. We use our knowledge of human cognitive systems and of the historical situation to identify the kinds of propensities that would have been present in our subject and thus to delimit the kinds of psychological processes that might have been available to modify the cognitive state. We employ our current understanding of the world to consider the relative abilities of such processes to promote cognitive progress. Cognitively well designed scientists use processes that are relatively good at promoting cognitive progress, within the comparison class of those that are available to them. They do not have to believe what we believe or even to reason as we would reason. They are seen as optimizing (or, at least, doing relatively well) subject to constraints (often constraints from which we are liberated). They do the best they can.

This general explanation shows that the more blatant forms of historical chauvinism can be avoided. We do not have to assume that only those who share our beliefs or emulate our strategies of good reasoning count as rational. But it is perfectly reasonable to wonder whether subtler versions have been circumvented. As we shall see in the next two sections, the notion of good cognitive design needs to be carefully disambiguated before such issues can be resolved.

4. The External Standard

Rationality is only a problem for limited beings. In the context of goal-attaining behavior, the actions of an unlimited being would not suffer a contrast between rational strategies and goal-attaining strategies. In the particular enterprise of aiming at a progressive sequence of cognitive practices, unlimited beings would pursue the trivial strategies: adopt adequate concepts and correct schemata, pose significant questions, accept true answers. The suggestion that we should follow these strategies is unhelpful. Ideals of rationality arise from considering strategies that would best enable subjects with our limitations to attain their goals.

My use of the plural is deliberate. As we think about design for cognitive progress it quickly becomes apparent that there are a number of possible ways to construct standards. There is a set of strategies A available to human inquirers for use in modification of their practice. The aim is to pick out those members of A that do well—optimally, relatively well—in making progressive shifts across a class of epistemic contexts. Epistemic contexts are individuated by the initial practices with which the investigators begin and the stimuli they receive. To do the accounting, we require both a way of measuring cognitive improvement and a criterion for deciding how much expected cognitive improvement is enough. So there are *four* loci at which specification is needed: (i) the set A of available processes, (ii) the set C of epistemic contexts, (iii) the measurement of cognitive improvement, and (iv) the criterion of adequacy.

To make this more concrete, I shall generate one standard of rationality. Let A include all the processes of cognitive modification that have been, are, and will be used by human beings.[16] Let C include all ordered pairs of possible practices and sequences of stimuli that the world will afford human subjects. I shall assume that C is large, but finite. Let the *improvement set* of P be the set of epistemic contexts in which P would yield a progressive shift in practice. The *success ratio* of P is the ratio of the cardinality of P's improvement set to the cardinality of C. The criterion of adequacy demands that the success ratio of a process be the maximal success ratio for members of A. Putting all this together:

> (ES) The shift from one individual practice to another was rational if and only if the process through which the shift was made has a success ratio at least as high as that of any other process used by human beings (past, present, and future) across the set of epistemic contexts that includes all possible combinations of possible initial practices (for human beings) and possible stimuli (given the world as it is and the characteristics of the human recipient).

(ES) is extremely demanding. It makes no concessions to the fact that processes available to subjects at some stages in the history of science may not have been available at others, and it requires that only optimal processes count as rational.

These requirements could easily be relaxed. So, for example, the criterion of adequacy could be adjusted to demand that the success ratio of a process only be in the top k percent of success ratios of processes in A. Alternatively, we might want to partition the set of epistemic contexts —focusing on those that call for a particular species of cognitive strategy—and compare strategies that belong to the same species. (Intuitively, the idea would be to demand

---

16. Ideally, we would like to delimit the strategies that are psychologically possible for members of our species. Those that are actually used comprise a proper subset of these. Moreover, it may well be that there are some strategies that would be excellent and would be employable by us, but which are never actually going to be used. For present purposes my concern is to provide a relatively sharp standard of rationality, so I shall resist the temptation to take the notion of "psychological possibility" or "availability" as well understood, even though there is a danger that the class of strategies on which I focus may seriously underrepresent the target.

that a strategy be optimal for doing a particular kind of job, rather than optimal *simpliciter*.) The set *A* could be indexed to the cognitive resources available at the time of the subject whose rationality we are considering, so that we exclude processes that would have been simply inaccessible to people at that time—processes that only become available through the explicit articulation and inculcation of methodological principles, for example. We might also decide that success ratios over the entire set *C* are not of interest to us, limiting our attention to the kinds of practices that are (in some sense) available to the contemporaries of our subject and to the kinds of stimuli that they might have received. Unlike the earlier proposals for modifying (ES), this one does not necessarily weaken the standard—the set of shifts it would admit as rational does not necessarily include those admitted by (ES). Similarly, if we amend the measurement criterion by taking into account the magnitude as well as the direction of the change in practice, we are likely to obtain a different standard but one that may not be any weaker.[17]

Judgments about the history of the sciences often resolve the vagueness in context by implicitly comparing two rival processes. Simple claims about rationality can avoid fine-grained resolutions by focusing on the differences between the processes undergone by those whom we retrospectively praise and the processes of their opponents. Galileo's modification of his practice counts as rational because *looking at the heavens* rather than *not looking* is, it seems, likely to have a high success ratio in modifying astronomical practice. The details of which processes are available, the epistemic contexts we consider, the ways cognitive improvement is measured seem insignificant beside the gross differences we expect to find. However we do the accounting we think that we shall get the same result. Of course, as the history becomes more refined and it is understood why the telescope might have seemed an unreliable instrument, confidence that we can ignore the vagueness begins to waver.

Which standard is the right one? There is no answer. A methodological ideal would be to describe optimal improving processes for as broad a range of contexts and as large as class of alternatives as possible. (ES) provides a valuable target for the methodologist: pointing to a goal—optimal cognitive design—at which we aim. Where explicit methodology is done, one can thus expect that there will be attempts to meet the exacting demands of (ES).[18]

17. However, there is a reasonable worry that (ES) as formulated is too demanding because of the presence of relatively trivial processes of modification with success ratios close to 1. Consider the possibility of always effecting a small improvement in practice by carrying out some deduction from "safe" premises. This would, presumably, yield minuscule improvements in practice across almost all contexts (the exceptions being those in which rigorous house cleaning was in order). Risky strategies for making major improvements in practice would thus be debarred because they would have lower success ratios. This problem can obviously be mitigated by taking into account the magnitude as well as the direction of the shift, so that we allow for processes that occasionally yield big dividends.

18. Or of some kindred principle of maximization that takes into account not only the fact that processes generate positive modifications but also the extent to which they promote cognitive progress.

For study of the historical development of science, however, (ES) may prove far too demanding. Recall that the problem of rationality arises for limited beings. Rational inquirers are those who do the best they can to promote cognitive progress (more exactly, for present purposes, their own cognitive progress) subject to their limitations. (ES) sees them as having available to them all possible human processes of cognitive modification and as failing unless they modify their cognitive states through undergoing those processes that, as the world actually is, are most likely to generate cognitive progress. That concedes very little to our limitations. In effect, we are recognizing that people are unlikely to attain ideal practices but we ask that their methods of practice modification be ideal. Allowing that we are not substantively omniscient, we demand *methodological* omniscience (or, at least, procedural perfection). Just as the goal of attaining significant truth sets a worthy ideal for human cognition, so too (ES) is a valuable methodological ideal. Just as our beliefs need not be true to be cognitively good, so too subjects may be commended for their modifications of cognitive state, even though they do not measure up to (ES).

(ES) demands too much of historical subjects. It may also require too little. Imagine a scientist in some epistemic context who modifies her cognitive state (and her practice) through a process that fortuitously satisfies (ES). The activation of her inferential propensities comes about because of her prior history of enthusiasm for numerological mysticism. Despite her odd penchant for mumbo-jumbo she was, for once, lucky. (For examples to similar effect, see Goldman 1986 52–53; for alternative efforts at resolution, see Papineau 1988 and Peacocke 1987.)

The obvious way to circumvent this type of difficulty is to apply a version of the external standard at a higher level, requiring not only that the actual process involved in cognitive modification be well designed for promoting cognitive progress but that the higher-order propensities responsible for the production of that process themselves should be likely to generate progress-promoting processes. Have we embarked on a regress? Perhaps. Someone with sufficient ingenuity might be able to construct a (tolerably convincing) example in which the presence of a higher-order propensity meeting the appropriate standard was itself due to some bizarre feature of the subject's causal history which would not have been expected to contribute to cognitive performance but somehow does so in a fortuitous way. I shall not be interested in exploring such recondite possibilities. More significant seems to me the possibility that a scientist might activate a higher-order propensity that does relatively well, when compared with rivals, at generating progress-promoting processes, even though the process actually produced does rather poorly at modifying practice in a progressive way.

Consider two examples in which scientists fail to meet the rigorous requirements of the external standard (not only (ES) but also versions generated by more liberal decisions about the available processes and the criterion of adequacy). One tries to follow a process that we actually endorse. Faced with a problem that allows for Bayesian reasoning, he patiently assembles statistical information, attempts to formulate the right versions of Bayes's theorem and other probabilistic principles that are needed, and computes an answer. Unfortunately, something goes wrong. Perhaps there is a flaw in the collection of statistics (faulty categorization, failure to assess for some type of relevant variety, etc.), or a misapplication of the probabilistic theorems, or even a simple arithmetical error. As a result, the process actually followed has a very high chance of generating false conclusions, so that it fails to meet any of the external standards. The other scientist, faced with a different problem of evaluating a new hypothesis, proceeds by appealing to the methodological traditions of her community. These traditions suggest that new hypotheses are to be assessed by considering a number of qualities (compatibility with evidence, simplicity construed in a particular fashion, ability to yield new predictions, or whatever) and offers a proposal for weighing the relative significance of these qualities. Our scientist follows the traditional procedures impeccably. She arrives at a correct determination of the extent to which the novel hypothesis satisfies the accepted desiderata and makes her evaluation in accordance with the method of weighing different virtues. Unfortunately, however, undergoing that process—following the advice of tradition—is far worse than available courses of reasoning that the scientist considers and explicitly rejects: the dismissed rivals would have been far more likely to issue in cognitive progress.

I claim that there are scientists meeting the descriptions of the last paragraph whose decisions, on the occasions under consideration, are eminently rational. Indeed, I strongly suspect that many of the epistemic performances of scientists, great and humble, are similar to those I have sketched.[19] We are no more immune from methodological mistakes (as judged by (ES) or some other version of the external standard) than we are from substantive error. *Yet our susceptibility is not aptly characterized by declaring us to be irrational—or even not rational.*

Especially in the large changes that have attracted most of the discussion about scientific rationality, scientific decision making is very complicated. There are many considerations that might come into play in assessing the merits of Darwin's proposals—or those of Copernicus, Galileo, Newton, Lavoisier, Dalton, Einstein, or Wegener. Those who attempt to think through the issues may go astray in any number of ways—they may fail to remember or to appreciate certain kinds of relevant information, they may misformulate mathematical problems or miss possible lines of solution, they may fail to be sufficiently imaginative to see how apparent difficulties might be answered, they may make logical or mathematical errors in the middle of complex lines of argument. In consequence, the processes they undergo may not be well

19. Even the most mathematically talented scientists make mistakes in computation or in the formulation of mathematical problems (recall Newton's treatment of the orbit of the moon or consider the history of debates about genetic loads [see Lewontin 1974 for details]). Conversely, some of Darwin's opponents—Owen, Agassiz—adopted an extremely conservative methodological stance which they applied unproblematically to support their opposition.

suited for the generation of cognitive progress. Simply lumping these various kinds of mistakes together with one another and with the extreme lapses of those who serve us as paradigms of irrationality is unhelpful. It is possible to introduce a finer-grained taxonomy of cases by recognizing that propensities for reaching decisions by undergoing certain kinds of processes, some of which would be progress-promoting, can also lead to performances that go astray in various ways. So, for example, my imaginary Bayesian bungler activates a good higher-order propensity, *trying* to resolve a problem by applying Bayesian ideas in a case to which such ideas are appropriate. The goodness of that propensity consists in the fact that the processes it can be expected to produce include some that would promote cognitive progress, even though the actual process generated is not among them. Similarly, in instances of complex scientific evaluation, proponents of rival positions may attempt to assemble evidence, assess arguments, and apply methodological principles in an overall fashion that would promote cognitive progress, despite the fact that their actual performances are flawed. We may applaud the basic propensities while recognizing the foibles of the resultant processes.

Turn now to my second scientist. Evaluating her epistemic performance depends on the details of the story. The appeal to tradition may sometimes be born of inflexible, dogmatic thinking (of this, more later). It may also arise from modesty (or from the fact that not everything can be checked at once). Once we have broken free of the idea that methodological principles can be known a priori—so that one can know, just by thinking hard enough, how to satisfy (ES)—then methodological misinformation, sanctioned by tradition, can be treated in the same fashion as substantive misinformation, encapsulated in hallowed lore. Scientists may tentatively explore a line of reasoning that would, as things stand, promote cognitive progress, drawing back from it because they perceive it as at odds with approved canons. Their rejection may be deliberately based on a sense of the improbability that they will have made a methodological breakthrough that their wise predecessors and contemporaries have missed. Something must be wrong with the apparently appealing argument. So they reason, and, in following traditional ideas, they undergo a process that is far less well designed for promoting cognitive progress.

Let us take stock. I began by approaching scientific rationality from the perspective of the external standard: rational decisions are those that issue from processes that have a high expectation of promoting cognitive progress. But there is no single external standard. Rather, there are many different ways in which we may make the general idea more precise. However, satisfaction of the external standard, even in its most exacting form, does not guarantee that all kinds of cognitive lapses have been avoided, for there may be people who meet that standard fortuitously by following bizarre higher-order strategies. On the other hand, sometimes those who are moved by higher-order propensities that have a high expectation of generating processes that meet the external standard should count as rational, despite the fact that the actual processes that underlie their decisions do not meet the standard.

Doubtless, with more exploration of cases, more varieties of rationality could be introduced, more complications unearthed, yet more intricate criteria proposed.

Enough! The moral of my discussion is not that we should continue to refine our definition of rationality and to seek some complex disjunctive analysis that will cover all the possibilities. Instead I propose to *dissolve* the notion of rationality. There are two contexts in which the notion of rationality functions. The ideal of methodology is to formulate principles of rational inference (or, more generally, rational rules for the modification of cognitive state). Here the external standard plays an important role: just as we would like our practices to attain the ideals delineated in Chapter 4 (to be full of adequate concepts, correct schemata, and significant truth), so too we would like to be able to specify processes satisfying (ES)—and, for good measure, to identify higher-order propensities that would generate such processes. The other context is that of epistemic appraisal. Here there are numerous kinds of cognitive appraisals that we might wish to make, *and no one of them has a unique title to the term 'rational.'* The processes that people undergo vary with respect to their progress-promoting qualities.

The notion of rationality preserves a minimal usefulness in allowing us to discriminate the normally functioning members of our species from those with pronounced cognitive deficiencies. But for the purposes of evaluating the performances of past historical actors, we do better to think in terms of goodness of cognitive design, recognizing both that this varies and that there are many sources of variation. Scientists, past and present, are variously imaginative, thorough, rigorous, bold, clear-headed, open-minded, modest, and so forth. Almost all of them are "rational."

5. The Limits of Tolerance

My approach to rationality accommodates the historians' contentions that many of those who were on the "wrong side" were epistemically virtuous in at least some respects. Is it too ecumenical?

Some of those who play major roles in the scientific debates of the past act in ways that suggest that they are attempting to realize nonepistemic rather than epistemic ends. There is no evidence that the friars who intrigued against Galileo were sensitive to some clever objections to his scientific doctrines.[20] On the contrary, their assessment of his ideas seems to have been thoroughly subordinated to protecting a certain kind of religious life and advancing their own ecclesiastical careers. In similar fashion, Bellarmine's decree of 1616, banning the teaching of Copernicanism, looks like a calculated political move, designed to secure the interests of the church. Part of the traditional idea of

20. For an account of the origins of Galileo's troubles, see (De Santillana 1955); for radically different suggestions, see (Redondi 1987). (Finocchiaro 1989) provides an extremely useful set of primary sources for understanding the Galileo affair.

an opposition between the "rational scientist" and the "prejudiced opponents of science" is best captured by noting that some people do not give the highest priority to impersonal cognitive goals but to certain practical ends (sometimes personal, sometimes impersonal) and that their actions are well designed for achieving these ends.

The label "irrational" is most tempting when behavior that appears to indicate the pursuit of nonepistemic ends is accompanied by professions of devotion to the ideal of cognitive progress. Sometimes people's decision making exhibits deviations from progress-promoting processes that are hard to explain except by supposing that the cognitive goals they explicitly honor are not those that motivate their decisions. The hallmarks of such cases are varieties of inflexibility, blindness or deafness. Thus when scientists continue to defend their assertions by rehearsing the same arguments, even when they have been presented with criticisms and counterarguments that their contemporaries take extremely seriously, when they neither reply to nor even acknowledge such counterarguments, it seems that we must either credit them with insights that are denied to the multitude or else suppose that the conclusions they maintain are too valuable to be risked by engaging in any kind of dialogue.

Consider, in this light, the case of "creation scientists." Ever since this group of critics of Darwinian evolutionary biology achieved prominence, champions of orthodoxy have wanted to label them as "pseudoscientists." The apsychologistic character of twentieth-century philosophy of science influences the formulation of the charge. If creation scientists are pseudo*scientists* that must be because they defend a pseudo*science,* a doctrine that can be distinguished from genuine science by its logical characteristics. Philosophers shift uneasily at this, because one of the great morals of the demise of logical positivism was the difficulty—or, to put it bluntly, apparent impossibility—of articulating a criterion for distinguishing genuine science (Quine 1951, Hempel 1951). Moreover, a sober look at the history of paleontology will reveal that the creationists effectively espouse what was once scientific consensus, not a scientific consensus that was overthrown by Darwin in 1859 but one that began to erode in the early years of the nineteenth century.

The apsychologistic point of view has matters exactly backward. There is nothing intrinsically unscientific about the doctrines—no reason to castigate Thomas Burnet, or others who held them, as pseudoscientists. The primary division is a psychological one between *scientists* and *pseudo-scientists.* The behavior of creation scientists indicates a kind of inflexibility, deafness, or blindness. They make an objection to some facet of evolutionary biology. Darwin's defenders respond by suggesting that the objection is misformulated, that it does not attack what Darwinists claim, that it rests on false assumptions, or that it is logically fallacious. How do creation scientists reply? Typically, *by reiterating the argument.* Anyone who has followed exchanges in this controversy or has read the transcripts of a series of debates sees that there is no adaptation to any of the principal criticisms. One important example among many is the creationist use of the second law of thermodynamics. For nearly twenty years, the major exponents of creation science have been declaring that the second law of thermodynamics is incompatible with the evolution of life. Creationists have been in the presence of people who have given lengthy critiques of their objection and there is substantial evidence that their eyes have wandered over some of the pages on which such critiques have been printed. How has their thinking adapted to these critiques? Apparently not at all, for they make no replies to them and continue to present their ideas in exactly the same ways.

There are limits to proper tolerance. In some cases, epistemic performance is so inflexible that we either view the cognitive systems in question as poorly designed for the promotion of cognitive goals or suppose that the goals that are being activated are not cognitive at all. Where the latter supposition is correct and the subjects in question profess cognitive goals, some form of deception or self-deception is occurring. The category of pseudoscientists is a psychological category. The derivative category of pseudosciences is derivatively psychological, not logical as philosophers have traditionally supposed. Pseudoscientists are those whose psychological lives are configured in a particular way. Pseudoscience is just what these people do.

6. Closing Debate

Dissolving rationality, we have refocused the issues to which that notion was directed in terms of goodness of cognitive design. Let us now turn back to the question of the "rationality of science." In holding science to be the supreme embodiment of human reason, rationalists[21] are advancing claims about the ways in which scientific debates—including those debates that mark times of large transition—have typically, if not invariably, been closed. Their antirationalist opponents offer rival ideas about the closure of those debates. I shall start by showing how the fundamental themes of rationalism and antirationalism can be expressed in my preferred idiom.

The *rationalist model* embodies the following assumptions:

(R1) The community decision is reached when all individuals within the community have independently made the same modification of their practice.

(R2) Each member of the community is moved solely by the epistemic goal of modifying practice as progressively as possible.

(R3) All members of the community are in the same epistemic context: each begins from the same practice and each receives the same stimuli.

21. Here, and in what follows, I use this term to designate those who believe that scientific debates are closed through the formulation and appreciation of decisive reasons.

(R4) While there is debate within the community,[22] those who champion the modification that ultimately triumphs do so by undergoing processes that are well designed for promoting cognitive progress (processes that satisfy (ES) or some such standard); their opponents undergo processes that are less well designed for promoting cognitive progress.

(R5) Debate closes when those who employed the inferior processes rightly modify their cognitive activity so as to undergo the progress-promoting processes; in some instances, a small minority may fail to make this shift and members of such minorities are excluded from the pertinent communities.

Although defenders of the rationality of scientific decision making have not used the idioms that figure in my formulations, I think that Popper, Lakatos, Laudan, Levi, Glymour, Worrall, and various Bayesian philosophers of science all subscribe to the foregoing model. The differences among them center on the accounts they provide of superior and inferior processes of cognitive modification—substantive versions of traditional rationalism take (R1)–(R3) for granted and use (R4) and (R5) as bases for detailed specifications of the kinds of transitions in practice (or, in the traditional language, "theory") that can be expected to promote cognitive progress.

Everybody knows that, strictly speaking, (R1)–(R3) are false. The rationalist model is an idealization, and it neglects complications that are regarded as insignificant. The central thrust of rationalism is that the power of the right kinds of inferences is sufficiently strong to overwhelm effects of interdependence, of nonepistemic goals, or of background variations in practice and stimuli. Hence it is important to show that there is a relation between Copernicus's claims and the (universally) available evidence (as of 1543) that does not hold between Ptolemy's claims and that evidence, a relation that warrants acceptance of the ideas of Copernicus and rejection of those of Ptolemy. (Other major examples involve Einstein and Lorenz in 1905, Lavoisier and Priestley in the late 1770s, Newton and Descartes in 1687, Darwin and his various opponents in 1859.)

The centerpiece of the antirationalist account of major debates in the

22. The extreme version of rationalism supposes that, from the first public introduction of the ultimately successful view, those who adopt it do so because they undergo cognitively superior processes. So, for example, Copernicanism would have been epistemically preferable as of 1543, Darwinism as of 1859, special relativity as of 1905. There are often sustained efforts to show that the enlightened were thinking clearly from the beginning—see, for example, the treatments of Copernicanism by Glymour (1980) and by Lakatos and Zahar (1976). As I shall suggest, this is a rationalist theme that ought to be dropped. Just as rationalists should be happy with the possibility that those who break with tradition to develop completely new views—Copernicus in 1509, Darwin in 1838—take large epistemic risks, so too, they should allow that early adherence by others need not be on the basis of undergoing cognitively superior processes. Once the issue is posed, many erstwhile rationalists may prefer to remain more flexible about the time at which cognitively superior processes become available within the community, maintaining only that this occurs before closure of the debate.

history of science is the denial of (R4).[23] However, antirationalists also offer a different account of individual scientists and the communities to which they belong. So I suggest that an antirationalist model comprises the following claims:

(AR1) The community decision is reached when sufficiently many sufficiently powerful subgroups within the community have arrived at decisions (possibly independent, possibly coordinated) to modify their practices in a particular way.

(AR2) Scientists are typically moved by nonepistemic as well as epistemic goals.

(AR3) There is significant cognitive variation within scientific communities, in terms of individual practices, underlying propensities, and exposure to stimuli.

(AR4) During all phases of scientific debate, the processes undergone by the ultimate victors are no more well designed for promoting cognitive progress than those undergone by the ultimate losers.[24]

(AR5) Scientific debates are closed when one group musters sufficient power to exclude its rival(s) from the community; the subsequent articulation and development of the successful modification of practice absorb all available resources, so that later comparisons can be made between a highly developed tradition and an underdeveloped rival; in this way it is ensured that *later* scientists will be able to defend the victorious transition by undergoing processes that are better-designed for promoting progress than those available for supporting the dismissed rival; if the original decision had gone the other way, the victors (who would have been those who actually lost) could also have mounted just the same later defense.

As in the case of the *rationalist model,* opponents of Rationalist views of scientific change would not employ the idioms of my formulations. Nonetheless, it seems to me that the claims I have presented expose what is crucial in the challenges launched by Feyerabend, Barnes, Bloor, Shapin, Schaffer, Collins, Latour, and all the many historians of science who think that rationality is a philosophical bugaboo.

23. In denying (R4), antirationalists also raise to prominence features of the community of scientists that are explicitly ignored in (R1)–(R3). They should thus be viewed as denying the validity of rationalist idealizations.

24. This thesis encapsulates the popular idea of underdetermination of theory by evidence (see Chapter 7). A more extreme version of (AR4) would claim that the processes undergone by the ultimate victors are *less* well designed for promoting cognitive progress than those employed by the ultimate losers. Thus Feyerabend takes up Galileo's praise of Copernicus for denying reason. However, the ultimate point of Feyerabend's argument seems to be that no coherent general account of progress-promoting reasoning can be given, so I shall suppose that he, like other antirationalists, would be content with (AR4).

Antirationalists typically do not place much emphasis on (AR1)–(AR3). These claims are presupposed in the arguments about main points of controversy: (AR4) and (AR5) versus (R4) and (R5). Now (AR1)–(AR3) plainly provide a more accurate picture of scientific life than do (R1)–(R3), and I absorbed them within my picture of the microstructure of scientific change (see Chapter 3). However adoption of them does not prevent expression of the central point that rationalists ought to want, to wit that, at the time of community decision, the victors base their modification of practice on progress-promoting processes, while their opponents do not, and their relative power is grounded in this difference.

When the two models are juxtaposed, it becomes clear that the struggle is between contrary generalizations. Each side believes that the epistemic situation is constant throughout the period of debate. Rationalists insist that the good guys were epistemically superior from the beginning, and that, after a period of stubbornness, prejudice, and/or irrationality, their opponents saw the light. Antirationalists maintain that none of the protagonists, even those who hold out for the defeated doctrines until the bitter end, is any more stubborn, prejudiced, or irrational than any of the others. There is obvious middle ground. Perhaps the debate *evolves* in important ways.[25] At early phases the processes undergone by protagonists on both sides may be comparable in terms of their progress-promoting qualities. During the debate, through further encounters with nature *and* conversations among peers, the processes underlying the rival claims may be modified, with the consequence that there is a clear difference in ability to meet the standards discussed in previous sections. Antirationalists concede that there is evolution of the epistemic context, for they suppose that the approved modification of practice is *ultimately* defensible by arguments that inevitably appear superior to those that can be assembled for the abandoned rival.[26] However, they maintain that this success is spurious because the presence of such arguments is guaranteed by the initial victory (AR5). There is an important challenge here, one that must be met head on by considering major examples. The crucial thesis of rationalism can be sustained if transitions come about through the accumulation of power on epistemic grounds, through the widespread use of processes that are well designed for promoting progress (in contrast to the processes used by rivals). The crucial thesis of antirationalism is that such processes only become available as artefacts of transitions that have to be made on other grounds.

Plainly, changing the rationalist model to allow for the evolution of debate involves weakening rationalist claims. Does it make dangerous concessions?

25. One important way in which a debate can evolve is if the points in dispute are significantly modified. Later in this chapter I shall discuss an example in which this happens (the course of "the great Devonian controversy"). Here I am concerned with changing argumentation directed at a pair of proposals for modifying practice that remain constant throughout the period of debate.

26. This is because the victors have shaped the culture in which science is done, setting the standards for which kinds of considerations will be taken as decisive.

By supposing that superior progress-promoting processes only become available late in the course of the debate, one raises obvious questions about the early champions of proposals for modifying practice. Their reasoning (more generally, their cognitive activity) cannot be seen as superior to that of their opponents. Yet out of their efforts, indeed perhaps out of interactions with their critics, there emerges a line of argument that enables all members of the community to undergo superior progress-promoting processes.

As I have suggested, and as (AR3) explicitly maintains, there is cognitive variation within scientific communities. Scientists differ with respect to their individual practices, with respect to the information they exchange with other members of the community, in the stimuli they receive from nature, and in their propensities for modifying their cognitive states. I shall suppose that it is possible that, at early stages of scientific debate, the processes that underlie the practices of differently situated members of the community should satisfy some standard of good cognitive design (not necessarily the same in all cases) even though those practices differ with respect to the issues involved in the debate. In virtue of the adoption of those practices and the conversations with peers and the encounters with nature that they generate, the participants receive new stimuli. They modify their practices and the underlying processes in ways that meet some standard of good cognitive design (again, not necessarily the same in all cases). After a number of rounds of debate, matters *crystallize*. Certain kinds of cognitive homogeneity have been achieved, crucial issues and arguments formulated, and the product is a form of reasoning that encapsulates a process that is markedly superior (judged by (ES) and other standards of good cognitive design) both to the processes that sustain the rival proposal(s) for modification of practice and to those that originally underlay the successful change. Internalization of this form of argument by most members of the community builds sufficient power among those who champion its conclusion to dismiss dissenters from further discussion.

Traditional forms of rationalism about scientific change implicitly condemn those who behave in the fashion that I have sketched. Rational scientists with partial evidence are supposed first to assemble all relevant evidence and then to believe exactly what is warranted by the total evidence; if there is a standoff, then the proper scientific stance is agnosticism. Two important questions emerge from the contrast between this picture and mine. First, is good cognitive design always reflected in action to gather more evidence and agnosticism when the evidence is balanced, or are risky cognitive processes sometimes progress-promoting? Second, is it a good thing from the point of view of the community that scientists are sometimes prepared to jump to conclusions, pursuing lines of inquiry that lead to clarification of the issues under debate? I shall examine both questions in the next two chapters and (as may be expected) suggest that the answers to them favor my views rather than those of the tradition.

I have been attempting to motivate a model for the closure of major scientific debates that embodies some of the ideas of rationalism and some of antirationalism. This *compromise model* comprises the following claims:

(C1) The community decision is reached when sufficiently many sufficiently powerful subgroups within the community have arrived at decisions (possibly independent, possibly coordinated) to modify their practices in a particular way.

(C2) Scientists are typically moved by nonepistemic as well as epistemic goals.

(C3) There is significant cognitive variation within scientific communities, in terms of individual practices, underlying propensities, and exposure to stimuli.

(C4) During early phases of scientific debate, the processes undergone by the ultimate victors are (usually) no more well designed for promoting cognitive progress than those undergone by the ultimate losers.

(C5) Scientific debates are closed when, as a result of conversations among peers and encounters with nature that are partially produced by early decisions to modify individual practices, there emerges in the community a widely available argument, encapsulating a process for modifying practice which, when judged by (ES) (and other standards canvassed in earlier sections), is markedly superior in promoting cognitive progress than other processes undergone by protagonists in the debate; power accrues to the victorious group principally in virtue of the integration of this process into the thinking of members of the community and recognition of its virtues.

(C1)–(C3) are, of course, merely (AR1)–(AR3) with new designations. (C4) comes from (AR4) by restricting the antirationalist thesis about the entire debate to the early phases. The decisive break with antirationalism and the revival of a rationalist theme come with (C5), which claims that scientific debates are ultimately closed through the articulation and acceptance of decisive arguments.[27] However, even here there are some concessions to antirationalism. While the process of crystallization, out of which the crucial argument emerges, *may* be governed by transitions in individual practice that promote the attainment of cognitive goals, the model makes no claim that this must be so, allowing for an important role for the nonepistemic goals mentioned in (C2). Moreover, various kinds of social factors may operate in the acquisition of power by the victorious group. The demand is only that the argument emerging when the debate crystallizes should be the *principal* source of power—in other words that in a competition between the social

27. It is quite possible that erstwhile rationalists may regard this as the crucial point and thus view the compromise model as entirely congenial. The model may be viewed as making some rather obvious amendments to classical rationalism *once the apsychologistic approach to rationalism has been abandoned.* So, once convinced of the need to do epistemology in psychologistic terms, a rationalist might regard (R1)–(R5) as obviously flawed and take the compromise model to be the straightforward way of amending them. If so, then, as I have argued in earlier sections, approaching epistemological questions psychologistically is really the crucial step.

factors and arguments leading in a contrary direction, the acquisition of power should be more affected by the arguments. (Social factors may retard a decision but not reverse it.[28])

It may be too optimistic to think that every debate, even every major debate, in the history of science has been resolved in accordance with the compromise model. There may be occasions on which modifications of practice are made prematurely, so that it is only later, after further encounters with nature, that cognitively superior processes are available to support the modified practice.[29] Would-be rationalists can tolerate lapses from the closure of debate by dissemination of superior reasoning, if it can be shown that the lapses do not matter.

Recall the point of insisting on the rationality of science. Those who believe, as I do, that we discover more and more about the world while simultaneously learning how to investigate the world are vulnerable to skeptical challenges to the effect that cardinal tenets of contemporary science are mistaken. Rejecting the possibility of any a priori foundation either for science or for methodology, we can only answer skeptics by pointing out that our current knowledge is the product of a self-correcting process. Our right to this defense is undermined if there are transitions in the history of science which could have been made differently, on the basis of cognitively equivalent processes, and which would have yielded very different contemporary practices. The simplest response to that deep skeptical complaint is to argue that all transitions in the history of science accord with the compromise model. Failing that, we can block the complaint by arguing that the cognitively superior processes would have emerged later, whether or not the premature decision had been made, or by showing that an alternative modification of practice would ultimately have converged on the actual historical sequence of practices. In both instances, there are substantial tasks to discharge. For example, one must defuse the idea, bruited in (AR5), that decisions to modify practice contaminate the reception of further stimuli, so that, once the premature decision has been reached, the scales are automatically tilted so as to ensure its success according to any cognitive criterion, no matter how demanding. Similarly, showing how an alternative decision at some key point would still eventually have generated practices indistinguishable from our own plainly requires ingenuity in spinning historical fiction. Hence it is far simpler to ward off skepticism by contending that transitions in the history of science fit the compromise model. If, that is, the charms of simplicity and historical accuracy can be combined.

In the next two chapters I shall explore individual and community decision making with an eye to seeing how much of this residual rationalism can be

28. Although the *compromise model* has not been explicitly formulated before, I believe that it has affinities with *one* of the positions about scientific change offered in (Kuhn 1962/1970) and developed in chapter 13 of (Kuhn 1977). Certain forms of Bayesianism may also find the compromise model congenial (see, for example, Salmon 1990).

29. Here, as elsewhere, I count an unaltered practice as undergoing a "null" modification.

maintained. Before embarking on that investigation, it will be useful to develop the main ideas of the compromise model in more detail, descending from the abstractions of this section to engage with some illustrative examples.

7. Scientists at War

Scientific debates are rarely resolved in an instant. Most, if not all, of the examples that have figured in rationalist reconstructions of the history of science concern episodes in which over a period of years, even generations, the merits of a particular way of modifying consensus practice are vigorously debated. During these extended debates, the participants undergo innumerable cognitive processes as they struggle to assess the credentials of rival proposals. Neither the proposals nor the arguments offered in their favor are stable. It would thus be highly surprising if there were available to the participants some simple way to reach a rational decision. Those who believe in some clean criterion for distinguishing (say) Copernicanism from Ptolemaic astronomy—the progressiveness of Copernicanism as a research program or the greater bootstrap testability of Copernicanism, for example—can only defend the rationality of the early Copernicans if they can show why it was cognitively preferable for midsixteenth century people to focus the debate on this criterion, ignoring the mass of other factors that occur in discussions of *De Revolutionibus*.

Thinking about scientific controversies is considerably advanced by noting their resemblances to war (or to a complex "war game" like grandmaster chess). In scientific debate, as in warfare, patterns of alliance and opposition may change. Just as in scientific debates the proposals defended and the issues at stake may be altered, so too in warfare the demands and goals of the parties may shift with time. Most importantly, wars are rarely resolved through some single, local incident. There are occasional exceptions—perhaps the Norman conquest of England can be traced to the arrow that killed Harold—but, in almost all cases, individual skirmishes do not decide battles, individual battles do not determine the results of campaigns, and individual campaigns are not always crucial to the outcome of a war.

In major scientific debates, we can also think in terms of skirmishes, battles, even campaigns. Nor is this idiom foreign to the way in which the scientists involved are prepared to talk: Darwin, recall, describes his book on orchids as a "flank movement" against the enemy. Just as the ultimate victors rarely win every battle (and may even, as the Russians against Napoleon, lose all the early battles), so too, the fortunes of scientific debate may be variable. We can expect, then, that scientists who end up on the losing side may still "fight well" (produce good reasoning) and may even win some victories.

I propose to extend the compromise model by drawing on the resemblances between science and warfare, and, in particular, underscoring the idea that scientific controversies are long and complex episodes, in which local points and items of evidence do not determine the final result. The participants in the debate defend proposals for modifying consensus practice by offering a list of characteristics that a practice should have and presenting reasons for thinking that that modification satisfies the desiderata or is likely eventually to satisfy those desiderata. These defenses are attacked by criticizing the desiderata—either by suggesting that important requisites of cognitive progress have been omitted or by contending that the alleged desiderata are inappropriate—or by objecting to the claims about the likelihood of satisfying the desiderata. So there arise *skirmishes* about particular issues. A skirmish is resolved in favor of the attacker if the defender acknowledges both that there is a legitimate desideratum of doing X and that all of the specified avenues for doing X seem to be blocked. A skirmish is resolved in favor of the defender if the attacker concedes either that an alleged desideratum is inappropriate or that the defender has specified a way to meet it. There are a number of intermediate cases, in which the defender proposes one or more possible avenues along which a solution to the problem of meeting the desideratum might be found. These proposals may raise subsidiary questions about the probability with which exploring those avenues will prove successful.

The weapons available for carrying out these skirmishes are an agreed-upon conception of the goals of science, possibly only at the most abstract level (that presented in Chapter 4), possibly involving agreement on derivative goals as well, a consensus on certain elementary forms of argument, and agreement about some statements.

I shall postpone to the next chapter the task of considering how individual skirmishes are decided and how the outcomes of those skirmishes affect the outcome of the debate. Here I shall only sketch some features that appear in many scientific controversies. At early stages of debate matters are typically confused. Each side is able to resolve some skirmishes in its own favor. Confusion dissipates as one party manages to reopen the issues ("terrain") lost in previously unfavorable skirmishes, now engaging in inconclusive skirmishes (by thinking of some hitherto unexplored avenues for meeting the desideratum) or even in skirmishes that are resolved in its own favor.[30] If this successful defense is accompanied by an increase in the number of unanswered attacks, then, just as in war or in chess, there is a clear sense of a position crumbling.

Assuming that the forms of argument have desirable qualities (reliability in generating true conclusions from true premises—see the next chapter) and that the participants accept their common premises on the basis of processes that promote cognitive progress, then the arguments they offer reflect processes of reasoning that cannot be dismissed as *poorly* designed for the promotion of cognitive progress. At early stages of the debate, the processes underlying the claims of each side are risky: each has to admit that certain

---

30. Here, and elsewhere in this section, there is kinship with the ideas of Larry Laudan in his (1977). Skirmishes are fought around what Laudan calls "problems." Both notions deserve further analysis and explanation. My account will be offered in the next chapter.

problems will have to be overcome in ways that they are not yet able to specify. As the debate crystallizes, one side finds avenues which can be explored to find solutions, and there emerges a progress-promoting process which, once made available to all, inclines virtually all members of the community to the pertinent modification of practice. This is analogous to the major battle, or final push, that concludes many wars.

8. Darwin's "Long Argument" Revisited

One obvious illustration of the *compromise model* is the modification of consensus practice in biology from 1859 to 1867. I have already told the story in Chapter 2. Here I shall very briefly indicate how the conditions of the *compromise model* are satisfied.

Consider (C1). Darwin's victory in the 1860s turned on his winning the public support of major members of the biological community: Lyell had finally given the benediction Darwin so eagerly awaited; Owen had been reduced to disgruntled silence. The role of powerful subgroups appears in the introduction of minimal Darwinism within the university curriculum. (C2) is easily supported by recognizing Darwin's desire to receive credit for his ideas, the ambitions of Huxley and Owen, the religious sensibilities of Asa Gray and others. The diversity of practices found among Owen, Sedgwick, Lyell, Huxley, and other participants in the debate, such as Hopkins and Jenkin, underwrites (C3).

To see how (C4) and (C5) are fulfilled in this debate, it is enough to consider the objections of one of Darwin's critics, such as Hopkins, and Darwin's own response to them. As I argued in Chapter 2, Hopkins's methodological objections are by no means silly: the cognitive processes that stand behind Hopkins's querying of Darwinism are not inferior to those that supported the beliefs of many of Darwin's supporters. Furthermore, by forcing Darwin to extend and amplify his reasoning, Hopkins played a valuable role in the generation of a more convincing argument, whose public accessibility in the late 1860s made the case for minimal Darwinism too epistemically powerful to resist.

9. From Copernicus to Galileo

My second example has loomed large in discussions of the rationality of scientific change. The story of the articulation, reception, spread, criticism, and ultimate endorsement of Copernicus's ideas is too long and complicated for me to pursue it in detail here. However, I do want to suggest that major features of the history are handled rather naturally by the *compromise model*.

The principal challenge to rationalism, and consequently a major project for philosophers of science, has been to identify the epistemic advantages of the system presented by Copernicus in *De Revolutionibus* over the late medieval versions of Ptolemaic astronomy, given the evidence available in the early sixteenth century.[31] Apparently, the hope has been to show that Copernican astronomy had such virtues that rational people, fully cognizant of the evidence, would have chosen it rather than Ptolemaic astronomy in 1543. The stage would then be set for condemning those who failed to adopt Copernicanism as ignorant or prejudiced or irrational.

From the account offered in the preface to *De Revolutionibus,* it appears that Copernicus began to develop his heliocentric approach to the planetary motions in the first decade of the sixteenth century. At that stage there was no heliocentric system to enjoy whatever epistemic advantages accrue to the full treatment of *De Revolutionibus.* If we praise Copernicus for his decision to explore an alternative to Ptolemaic astronomy, and to plunge resolutely into the mathematical details, then it must be because the problems of the received system were sufficiently severe to make it worth pursuing a rival. We imagine Copernicus undergoing some process of evaluating the merits of received approaches and the possibilities for improving them, and setting out to try an alternative.[32] It is doubtful that that process was optimally designed for promoting cognitive progress, but it was surely not poorly designed for promoting such progress. I shall take it to have been *adequate.*

In the early 1540s Copernicus acceded to the urgings of his friend Rheticus, allowing the publication of his system, and, on his deathbed in 1543, he received the first copy of *De Revolutionibus.* That book gave the most accurate treatments available of the apparent motions of the planets. It was also at odds with both Aristotelian cosmology and Aristotelian physics. In book I, Copernicus did offer brief defenses of his departures from Aristotle. The technical exposition of the later books was very much in the Ptolemaic tradition. Copernicus employed two of the three main geometrical devices that astronomers had used to represent planetary motions, epicycles and eccentrics. He rejected the technique of equant points. The number of epicycles in his system was comparable to that in contemporary versions of Ptolemaic astronomy (of the order of 50).

Why should anyone have accepted Copernican astronomy in 1543? Perhaps because it was more accurate. But the geometrically sophisticated could see that equal accuracy might be achieved by building a Ptolemaic system that would be geometrically equivalent to Copernicus's. Perhaps because it was simpler. But the relative mathematical simplicity of the heliocentric scheme is not enormously impressive. Copernicus employs fewer epicycles but his system still uses a lot. Moreover, this mathematical simplicity is offset by the abandonment of the "simple" picture of the Aristotelian cosmos. Finally, there are unresolved questions about the possibility of the earth's

31. To speak of "the evidence available" is already to slide over important issues, for different determinations of stellar and planetary positions were made by different observers. Many astronomers of the period inherited a hodge-podge of results from antiquity, to which, of course, they added their own individual findings.

32. For psychological speculations about Copernicus's route to discovery, see (Margolis 1987).

motion, some of which are based on the theoretical claims of Aristotelian physics and cosmology, others that are grounded in the common observations of motion that Aristotelianism attempted to systematize.

In the face of this situation, philosophers who want to defend rationalism are forced to identify subtle epistemic virtues that Copernican astronomy has and Ptolemaic astronomy lacks. Now there are two attractive candidates that Copernicus himself emphasizes: (1) in the Copernican system, there is an immediate explanation for the observed fact that the planets Mercury and Venus always remain within a relatively small angular distance of the sun; (2) the Copernican system is harmonious in the very particular sense that it allows for unambiguous determinations of planetary distances (relative to the earth-sun distance).[33] Rationalist accounts seem to want to calculate the net evidential force of these kinds of considerations and to show that it exceeds that of the contrary suggestions about the impossibility of a moving earth.

I do not trust such accounting. The obvious thing to say about the state of astronomy in 1543 is that both geocentric and heliocentric approaches face problems. Astronomers have to base their choices on assessments of which problems are more likely to be overcome. There are adequate processes of assessment that issue in different conclusions—decisions to try to improve Ptolemaic astronomy (make it as accurate as Copernican astronomy, make it mathematically simpler, find explanations for the behavior of the "inferior" planets, discover ways of unambiguously determining planetary distances), or to improve Copernican astronomy (improve its accuracy, make it mathematically simpler, answer objections to the earth's motion, defend against the criticism that no stellar parallax is observed), or to work out a system with the advantages of the main rivals. One cannot unambiguously declare that one of these processes of assessment is more likely to yield cognitive progress than the others, and that one and only one of the decisions is therefore "rational."

In fact, we can distinguish a number of ways in which midsixteenth century astronomers might frame their enterprise and modify their practice. There are three levels of explanatory commitment. Level 1 (the least ambitious) maintains that the history of ventures in astronomy has shown that the task of arriving at the true motions of the heavenly bodies is (at least at present) too hard, and that astronomy should settle for discovering the best scheme for predicting apparent motions (finding the angular coordinates of heavenly bodies on the stellar sphere at various times). Level 2 aims to advance a scheme that represents the true motions and to use this scheme to predict and explain the apparent motions, but it does not take up the project of explaining why the true motions are as they are. Level 3 (the most ambitious)

---

33. Copernicus makes much of this point, embedding it in an attack on the "monstrosity" of the Ptolemaic universe. Howard Margolis offers an interesting interpretation of Copernicus's reasoning, arguing that Copernicus thought in terms of a particular pattern (see his 1987 chapters 10–12). In my terms, the process Margolis identifies would be *adequate,* but this is not to deny that adequate processes were also available to his opponents and critics.

proposes to explain the apparent motions in terms of the true motions and to provide an account of why the true motions are as they are. Decisions about the appropriate level for astronomy will be based on assessment of the past history of efforts to solve various kinds of astronomical problems and the likely possibilities for future solutions. Those who are relatively pessimistic will settle for level 1 and appraise astronomical schemes solely in terms of predictive accuracy and mathematical tractability. Considerations about the possibility of the moving earth then become quite irrelevant and Copernicanism can be defended along the lines of Osiander's famous preface. For those who adopt level 2, there will be complicated choices among Ptolemaic astronomy, Copernican astronomy, or some other system, based on evaluation of the competing advantages and problems of both, but, since the link between astronomy and Aristotelian cosmology has been broken, certain kinds of constraints from physics will no longer be in force. Finally, astronomers at level 3 will have an even more complex decision: they may try to rebuild the old Ptolemaic-Aristotelian universe, embed Ptolemaic astronomy within a new cosmology, or offer Copernicanism or some new system with cosmological and physical underpinnings.

In the second half of the sixteenth century many of these options are played out. Copernicanism is developed mathematically by astronomers at level 1.[34] It is largely rejected by astronomers at level 2, who either seek to defend versions of Ptolemy or else attempt to build new systems (Tycho Brahe is the most famous example).[35] Kepler is the most prominent figure at level 3, and he, of course, opts for Copernicanism. I claim that none of the initial decisions made by these men was irrational. In each case there is an adequate process that we can presume to have generated their modifications of individual practice—and, I suggest, there could have been adequate processes that led to some of the unexplored options as well.

The situation changed in the early seventeenth century. Kepler's labors issued in the production of a more accurate scheme of planetary astronomy than any hitherto available (although Kepler's mode of presentation initially prevented his ideas from securing a wide audience). At the same time, Galileo's telescopic researches and his incisive critiques of aspects of Aristotelian cosmology, physics, and methodology turned back the most important challenges to Copernicanism. The debate crystallizes in Galileo's famous *Dialogue Concerning the Two Chief World Systems,* where the principal arguments for rejecting Copernicanism are systematically evaluated.

The target of the *Dialogue* includes several of the options I have distinguished: Galileo attacks attempts to combine Ptolemaic astronomy with Aristotelian cosmology and negative evaluations of Copernicanism that are based on either Aristotelian physics or cosmology or commonsense ideas about

---

34. For one facet of this development, see (Westman 1975).

35. However, Tycho is by no means the only person who attempts to fashion a compromise system. For a fascinating account of the intricate history of geoheliocentric systems, see (Gingerich and Westman 1991).

motion. In a sequence of skirmishes, he adduces phenomena that his opponents cannot accommodate *without abandoning their conception of the structure of nature.* There is a trivial type of underdetermination that affects his reasoning.[36] With respect to the major points he presents—the sunspots, the phases of the moon and of Venus, the behavior of projectiles and of falling bodies, the existence of comets and novae—there are logically possible ways of squaring the pieces of Aristotelian doctrine under fire with Galileo's claims. It is to the credit of Galileo's Aristotelian spokesman, Simplicio, that he seeks to counter the arguments *while retaining a conception of the structure of nature.* Galileo's ingenuity is displayed in his ability to block off the specifiable possibilities. Time and again, Simplicio ventures an account of the phenomena that will enable him to preserve the Aristotelian explanatory schemata. Time and again, Galileo's spokesman (spokesmen?) reduce him to silence.

The thesis that scientific decisions are underdetermined by the evidence available rests on the idea that there are always possible ways of squaring bits of doctrine with reports of observation *without cognitive loss.* But there is cognitive loss if one is forced to abandon instantiations of an explanatory schema without any available alternative. This is precisely Simplicio's predicament. He consistently finds himself unable to fit Galileo's phenomena into his Aristotelian schemata, and, lacking other schemata to offer, he is compelled to concede that there must be some unknown explanation.

So, I suggest, Galileo's *Dialogue,* like the argument developed during the debate about Darwin's *Origin,* encapsulates a process of reasoning which is strikingly better designed for promoting cognitive progress than any available alternatives. As that process became accessible in the 1630s and 1640s, as criticisms were countered and as the arguments were strengthened, consensus on Copernicanism was finally reached.[37] It is, I believe, equally wrong to insist on the presence of decisive reasons for being a Copernican in 1543 and to deny that there were decisive reasons after 1632.

Examination of the requirements for applying the *compromise model* highlights some features of the low-budget history I have told, and also brings out interesting nuances. (C1) is satisfied because, in the 1630s and 1640s, major groups of people have been sufficiently convinced of Copernicanism so that heliocentric hypotheses can, relatively unproblematically, be taken as the starting point for discussions in cosmology and physics. But an important aspect of this result is a shift in the sites at which natural philosophy is done and in the kinds of individual who undertake astronomical, physical, and cosmological inquiries. Those who matter in the intellectual discussions of

36. The approach to questions about underdetermination adopted here will be articulated further in the next chapter.

37. One important vehicle for strengthening the astronomical arguments was Kepler's *Epitome of Copernican Astronomy.* Robert Westman has suggested to me that this work was more influential in securing the acceptance of Copernicanism than was Galileo's *Dialogue.* For my purposes here, it is not important to divide the laurels between Kepler and Galileo. The significant point is the emergence of a strong line of pro-Copernican reasoning in the 1620s and 1630s.

the midseventeenth century are no longer just the university men and (to a lesser extent) the church scholars. Court patronage of both individuals and groups has created networks of investigators who self-consciously deride the stuffiness of traditional universities, and there are men of independent means ("gentlemen") who dedicate their resources to "philosophy." Meanwhile, of course, the university curriculum has been under a broad attack, and, in some universities, there has been reform of the traditional curriculum. These changes are partly dependent on the course of the Copernican debate, partly independent, but they create a social milieu in which the arguments of Galileo secure a far more sympathetic hearing than they might have expected a half-century before. While the closure of the controversy is, I believe, to be understood in terms of the acquisition of epistemic authority by newly emerging groups on the basis of an appreciation of superior reasoning, it is important to be aware both of the extent to which the set of groups with epistemic power does not remain constant through the period and of the variety of factors that play a role in the waxing and waning fortunes of various constituencies.

The presence of nonepistemic interests, as claimed by (C2), is plain in the concerns about the religious implications of Copernicanism, voiced by some Lutherans in the midsixteenth century and by some Catholics in the early seventeenth century. Worries about credit within the community of scholars also surface. Tycho's bitter polemic with Ursus shows how deeply the former was concerned to enhance his reputation. Galileo's quest for patronage reveals both his interest in securing a livelihood that would give him the leisure for his scientific researches and his attachment to *la dolce vita*.

Cognitive variation, as demanded by (C3), is also easy to find. Copernicus himself made few observations but was steeped both in the technicalia of astronomical geometry and in the humanistic tradition. Tycho's observations were unparalleled among his contemporaries. Kepler's knowledge of optics, harmonics, and numerology contributed to making his astronomical writings virtually unreadable. In the century between Copernicus and Galileo, we find numerous men struggling with the problems of the solar system; applying different bodies of background information to different questions that they hail as significant, with very different cognitive temperaments (contrast the enthusiasm of Rheticus with the conservatism of Osiander); and reaching very different conclusions. The Copernican debate is not simply a matter of heliocentrism against geocentrism—with the Tychonic system thrown in for good measure. Almost every participant has unique ideas about which problems are important and how those problems should be solved.

Full defense of the applicability of (C4) and (C5) is impossible without more careful elaboration of the reasoning of the various participants and the kind of analysis that I shall undertake in the next chapter. However, the main lines of the defense are easily indicated. Given the astronomical advantages of Copernicus's scheme and its physical disadvantages, four different responses are readily comprehensible. Perhaps the physical difficulties of Copernicanism could be solved. Perhaps the Ptolemaic system could be revised to share the Copernican virtues. Perhaps there is a way to combine the attractive features of heliocentrism with a fundamentally geocentric scheme. Or perhaps the idea of realistic cosmology is just a mistake, and astronomy will best make progress by trying to solve, as accurately as possible, the problem of the apparent motions. There is enough to be said for the reasoning of the sixteenth century astronomers who pursued these options to make any dismissal of them as unreasonable or irrational seem fatuous.

However, by 1633, with Galileo's resolution of the physical problems for Copernicanism and his destruction of the methodology and cosmology of Aristotelianism, a clear line of pro-Copernican argument had emerged. With no bar to the earth's motion, the Copernican explanations of the differences in the behavior of the inferior and superior planets could be adopted without loss, and, in marked contrast to the Tychonic system, Copernicans could venture a tentative explanation of the motions predicated on the idea that smaller, lighter bodies revolve around larger, heavier ones. After Galileo, traditionalists were left struggling to find reasons for preserving geocentrism. For the first time in the debate, the reasoning available to one side was so markedly superior that the other side ought to have capitulated. Because of their cognitive blindness, the last traditionalists were excluded from astronomical consensus.

10. Rocks and Reasoning

My last example is less familiar than the well-studied examples of the last two sections. The "Great Devonian Controversy," which began with some uncertainties about how to date ancient strata in Devon and ended with the recognition of a new geological period, is especially useful for the present study because it was both finely documented by the participants and finely described by Martin Rudwick (1985) in one of the most complete accounts of any scientific controversy. I shall greatly simplify Rudwick's treatment and use my abridged version to show the success of the *compromise model*.[38]

The controversy began in 1834 with the discovery of some anomalous plant fossils in the apparently ancient Greywacke strata in North Devon. The fossils presented an anomaly because of their apparent kinship with plants found in Carboniferous strata, assumed to be much younger than the Greywacke. Henry De La Beche, employed by the government to produce a geological map of Devon, initially claimed that the strata bearing the fossils were deep in the Greywacke. He concluded that quite similar organisms can occur at quite different geological times, so that the attempt to use characteristic fossils to order and correlate strata is misguided. Roderick Murchison, an aspiring geologist of independent means and a defender of a fossil criterion

---

38. Obviously there are worries that the success of the model is an artefact of the precis that I make. I shall do what I can to forestall such concerns, by indicating how my story relates to Rudwick's far more complex narrative.

for correlating strata, countered by suggesting that De La Beche had wrongly placed the fossiliferous strata. Instead of being in the middle of the North Devon Greywacke, they were, he claimed, really at the top. The principle that strata deposited at the same times would include characteristic fossils could thus be sustained by supposing that the topmost layer of Greywacke was Carboniferous. However, the major British Carboniferous beds, in the Western Midlands, were smoothly underlain by a distinctive formation—the Old Red Sandstone—which was not found in the Devon Greywacke. Hence Murchison was forced to conclude that there must be an unconformity between the fossiliferous strata at the top of the Devon Greywacke and the more ancient strata below. (At least the "Old Red" had to be missing from the Devon formations.)

De La Beche was compelled, relatively quickly, to abandon his original interpretation of the Devon rocks, conceding that the plant fossils were indeed found at the top of the Greywacke.[39] But he was adamant that Murchison's hypothetical unconformity could not be found: the Greywacke strata graded smoothly into one another. Continuing to maintain the traditional claim that the Greywacke was very ancient, De La Beche concluded that the North Devon strata must be much older than Carboniferous and must be separated from the Carboniferous rocks bearing similar fossil plants by a period corresponding at least to the deposition of the Old Red Sandstone.[40] Correlating any part of the Devon column with the "Old Red" seemed quite impossible for both De La Beche and Murchison because of their great difference in appearance, because of the absence from the Sandstone of any of the distinctive shells and corals of the Devon Greywacke, and because of the absence from the Greywacke of the fish of the Old Red Sandstone.

For several years, De La Beche, Murchison, their various allies, and other interested parties explored variations on the two basic approaches to the Devon strata described: either the topmost strata were young (Carboniferous) and there was an unconformity, or the Greywacke was a continuous, ancient (pre-Carboniferous, pre–Old Red Sandstone) sequence. George Greenough

39. As Rudwick makes very vivid for the reader, this concession caused De La Beche considerable pain—not least because he worried that admission of an error, even an excusable error of the type he had committed, would be interpreted by his governmental superiors as a sign of incompetence, with dire consequences for the future of the publicly funded survey and for his own employment. De La Beche was thus constrained by factors external to the controversy, but, with the help of his friends, particularly Greenough, he was able to accept the correction without losing professional credibility.

40. Another aspect of Rudwick's story that is slighted in my treatment is the way in which the geologists most intimately involved in the controversy were forced to rely on the classifications offered by the fossil experts. The available "relevant evidence" was broadly distributed across the group of people involved, some of whom had at their fingertips detailed knowledge of the Devon strata but little background either in general paleontology or in taxonomy, others who had never seen the rocks in question but were considered authorities either on fossil classification or on the issues about whether highly similar fossil faunas and floras could be expected to be present at very different times.

proved to be De La Beche's most reliable ally, while Murchison found in Adam Sedgwick an energetic, although often ungovernable, collaborator. Robert Austen, a rising young man in the geological world, pursued a more independent course.[41] At a less prestigious level, both sides had ardent devotees with some local knowledge. De La Beche was supported by the Reverend David Williams, while Murchison had the backing of Thomas Weaver, once a consulting mining geologist in Ireland, who, like Williams, spent most of the controversy in Somerset. While in Ireland, Weaver had made a similar interpretative mistake to that error De La Beche originally committed in his interpretation of the Devon strata. With all the zeal of one who has seen the error of his former ways, Weaver committed his energies to the Murchisonian cause.

During the late 1830s, a number of the protagonists began to explore the possibility that the Devon rocks might not be as ancient as the Cambrian and Silurian strata that Murchison and Sedgwick had mapped (and annexed) in Wales. Murchison especially devoted his energies to attempting to relate the Devon Greywacke to various strata in Europe, first in the Rhineland, where his ability to interpret the rocks seemed to fluctuate erratically.[42] In early 1840, after previous partial successes and failures with the same basic theme, Murchison again returned to the idea of identifying the Devon Greywacke with the Old Red Sandstone, claiming that both were sandwiched conformably between the Silurian below and the Carboniferous above.

The publication of Murchison's book *The Silurian System* elicited a letter from a German geologist, Christian von Buch, who informed him that the scales of the Old Red Sandstone fish *Holoptychius* were found in Russian sandstone deposits in association with shells and corals. In the summer of 1840, Murchison set out for Russia, where he discovered unmistakable Silurian deposits, equally unproblematic Carboniferous deposits, and, between the two, beds of limestone with "Devonian" shells and beds of sandstone with Old Red Sandstone fish. He concluded, "The Q.E.D. is accomplished" (quoted in Rudwick 1985 360).

So it proved to be. All the elite geologists who had participated in the controversy quickly accepted Murchison's conclusion that the Devon rocks belonged to a distinct system, the Devonian, containing also the Old Red Sandstone, intermediate in age between the Silurian and the Carboniferous.

41. In an illuminating series of diagrams, Rudwick analyses both the variations on the basic themes (figures 15.2, 15.3, 15.4) and the trajectories followed by the participants (figure 15.5). Austen and William Buckland emerge as the earliest adherents of something like the eventual consensus.

42. Here I offer a ludicrous abbreviation of an extremely complicated story. Rudwick makes it very clear how Murchison sometimes experienced apparently insuperable difficulties both with the Continental strata and with his ally Sedgwick. Interestingly, many of the geologists involved either were moving toward the final consensus or had adopted Murchison's Devonian interpretation in 1839, on the basis of the analysis of rocks in the Rhineland. See Rudwick's figure 15.5 and chapters 11–13.

De La Beche had been right about the absence of an unconformity, whereas Murchison's original claim about the age of the plant fossils was also vindicated. So "the great Devonian controversy" was resolved.

Yet if the major parties to the debate formed a consensus, two less prestigious naturalists did not. Both Williams and Weaver held out for the interpretations they had previously espoused. Neither man had the chance to attend the metropolitan discussions of the scholarly gentlemen, as they reached harmony on the recognition of the Devonian. Weaver's last contribution to the controversy was to urge that the North Devon strata were younger than the Silurian but older than the Old Red Sandstone (Rudwick 1985 363), an interpretation that committed him to the existence of *two* elusive unconformities, one in Devon between the Carboniferous strata and the Greywacke and one in the Welsh Borderlands between the Old Red and the Silurian. This doubly problematic proposal consigned the (aging) Weaver to the margins of geological discussions. Similarly, although Williams continued to insist that Murchison's interpretation made no allowance for the differences between the Old Red strata and the North Devon Greywacke (Rudwick 1985 388), his reiterations of an approach that De La Beche himself had abandoned were ignored.

The Great Devonian Controversy was brought to a close between September 1840 and the beginning of 1842, as all the powerful figures in British geology, together with their principal continental collaborators, adopted Murchison's hypothesis[43] that there was a distinct system between the Silurian and the Carboniferous, the Devonian system, to which both the Devon Greywacke and the Old Red Sandstone belonged. As a direct result of the decisions made among the elite, consensus practice in paleontology was modified to embrace the new system. Thus (C1) is satisfied.

Rudwick's detailed narrative, rich with excerpts from letters, field notebooks, and diaries, reveals clearly how nonepistemic motives influenced the major and minor players in the controversy. Murchison's ambition to make a name for himself in geology is palpable both in his early determination to save "his" Silurian system and in his later efforts to be recognized as the discoverer of the new Devonian system. His counterpart, De La Beche, a once prosperous gentleman fallen upon hard times, worries throughout much of the controversy that the debate will undermine the case for a publicly funded geological survey (a cause to which De La Beche was committed both in the abstract and in terms of his own livelihood). Austen has the professional goal of earning a place among the geological elite, while Weaver seems driven by the desire to show that he has learned from his mistake in interpreting the Irish strata. Rudwick's story leaves us in no doubt about (C2).

Similarly, the cognitive variation within the geological community is evident throughout. De La Beche and Murchison have different methodological

43. As Rudwick makes very clear, the hypothesis is credited to Murchison not because he was the first to think along these lines, but because he pursued it more energetically and with more resources (money and time) than anyone else.

commitments, the former suspicious, the latter enthusiastic, about characteristic fossils. People with special expertise play a role in the debate: in London, Charles Lyell offers general advice on the large issues of deposition of strata, while William Lonsdale provides authoritative identification of the fossil specimens; in the country, local fossil hunters guide the visiting gentlemen to appropriate sites. Nor do all the participants see the same things. Murchison surveys far less of Devon than does De La Beche. Finally, there are differences in inferential propensities. Murchison's zeal for arriving at a general view of the strata exceeds that of his more cautious co-worker Sedgwick, who is, in this respect, closer to De La Beche. Cognitive variation is everywhere, in accordance with (C3).

As in the previous examples, the issues surrounding (C4) and (C5) are more murky, but even here, it is possible to see how the great Devonian controversy fits the *compromise model*. Consider the initial situation of the participants. There appear to be two available ways of interpreting the Devon strata: either the plant fossils are found in ancient rocks (and it is possible for very similar plants to have existed at widely separated times) or the fossiliferous strata are Carboniferous (and there is an unconformity between them and the ancient Greywacke below). The final solution seems impossible because of the radical differences between the Old Red Sandstone and the Devon Greywacke. However, both available schemes of interpretation face severe problems. De La Beche has to explain how, in this instance but apparently not in others, very similar fossils can be found in temporally distant strata. Murchison has to find the unconformity.

De La Beche's early reconstruction of the North Devon rocks, in which he placed the plant fossils deep in the Greywacke, was flawed. The strategy deployed in forming his belief about the relations of the pertinent strata was cognitively inferior to that underlying Murchison's interpretation, and De La Beche acknowledged the fact. Yet it is futile to suggest that De La Beche was "irrational": like the mathematics student who tackles a complicated problem and makes a mistake, De La Beche did not reason perfectly, but the processes underlying his conclusions do not involve any egregious blunders. Overlooking a possible rival interpretation of his findings, De La Beche did not extend his observations into areas that would have revealed problems with his favored reading of the stratigraphy. Murchison did. Moreover, he investigated parts of Devon that De La Beche had previously ignored precisely because he had his own axes to grind and was thus required to dissent from De La Beche's interpretation. The superior cognitive strategy emerges here partly from the social aspects of the situation.[44]

After acknowledging his mistake, De La Beche's cognitive strategies are no worse than that of Murchison—and no better. Both men make serious attempts to overcome the large problems facing their views. De La Beche canvasses other instances in which apparently Carboniferous fossils have been

44. I am greatly indebted to Martin Rudwick for helping me recognize this aspect of the case.

found in much older rocks. His efforts involve him in skirmishes with Murchisonians, who argue that the interpretations of strata are suspect in these cases. Murchison and his allies offer various hypotheses about where in Devon the unconformity is to be found, but De La Beche and his supporters win the resulting skirmishes. Neither side can be accused of undergoing inferior cognitive processes, but, by the same token, neither side makes any headway with its major problem.

Murchison's Continental ventures enlarge the scope of the controversy by bringing into contention a position that initially appears impossible. To defend the Devonian system it is necessary to show kinship between the North Devon Greywacke and the Old Red Sandstone. Murchison's attack on the Rhineland (with its many vicissitudes) goes some way in this direction, but, as Murchison himself recognized, the case was clinched by the co-presence of the Old Red fish with the Devon shells in the Russian strata. "It was the Russian evidence of Old Red Sandstone fish in close association with Devonshire shells, Murchison recalled, that had 'entirely dispelled any doubts' about the Devonian interpretation" (Rudwick 1985 395).

I claim that, to a first approximation, the acceptance of the Devonian system is based upon the following reasoning, articulated in different ways by participants who focus on different details:

1. There are three available interpretations of the Devon strata: (a) they are ancient, and the plant fossils at the top of the Greywacke antedate the Old Red Sandstone; (b) the topmost strata, including those bearing the plant fossils, are Carboniferous, while lower strata antedate the Old Red Sandstone; (c) they form a continuous sequence from lower strata contemporaneous with the Old Red Sandstone up to Carboniferous strata at the top.

2. If (a) were correct, then it would have to be possible for similar fossils to be found in rocks of radically different ages; other apparent examples in which this occurs have proved to be based on errors in interpreting the strata; hence, there is no independent reason to believe that the Devon strata contradict the generalization that highly similar fossil faunas or floras are not found in strata that are widely separated temporally.[45]

3. If (b) were true, there would have to be an unconformity between the Carboniferous strata at the top of the North Devon Greywacke and the ancient rocks below, but sustained efforts to find this unconformity have failed.[46]

45. As the controversy evolves, the kinship of the North Devon plants and Carboniferous fossils becomes ever clearer. By the same token, the disparity between the fossils at the top of the Devon Greywacke and the fossils of Murchison's Silurian system is more and more evident. In consequence, De La Beche's problem of defending the ancient age of the topmost North Devon strata is further aggravated.

46. Here, too, the problem becomes more recalcitrant through the controversy, as Murchison and his allies explore the most promising places for discovering an unconformity, with no success.

4. The initial implausibility of (c) rests upon the lithological differences between the Old Red Sandstone and the Devon Greywacke, and on the fact that the fossil assemblages are mutually exclusive; the sequence of strata in Baltic Russia shows both types of rocks occurring between Silurian and Carboniferous deposits, with the marine fossil faunas found in close association.

5. Thus the major problems with (a) and (b) remain unresolved, despite several years of effort, while the apparently critical difficulty with (c) has been overcome.[47]

Different members of the British geological elite would have found different aspects of this especially vivid. Murchison, and those who had talked with him at length about the Russian strata, felt the force of the geological evidence from the Baltic. De La Beche, and others who had detailed knowledge of Devon, were struck with the conformity of the strata. Yet, I suggest, something like (1)–(5) moved each of the major figures in the controversy to accept the Devonian system, and this reasoning is cognitively superior to the earlier processes that underlay their beliefs and actions.

As I have noted, not all those who had contributed to the controversy adopted the consensus. Williams and Weaver held out for diametrically opposed viewpoints. Rudwick's treatment of them has been criticized (Pinch 1986, Collins 1987) for its (relatively mild) departure from the requirement of symmetry.[48] I shall be bolder. There is no reason to think that Williams and Weaver should be seen as undergoing processes that are as cognitively virtuous as those that moved the elite geologists, no reason to bow before a shibboleth of symmetry. People's cognitive performances vary. Given the way the world is, some will undergo processes that are more likely to issue in progress-promoting transitions. Williams and Weaver, I claim, held out on the basis of processes that were cognitively inferior.

I offer a simple account of their resistance. Remote from the London scene, neither could appreciate the force of Murchison's Baltic evidence. Williams, lacking a broad perspective on the distributions of fossils in temporally separated strata and the differences among contemporaneous strata, failed to understand how anomalous it would be to suppose that the plant fossils could span so wide a stretch of the geological column (thus not appreciating (2)) and insisted on a demonstration of the similarities of the

47. As in the example of Murchison's attack on De La Beche's initial interpretation of the Devon strata (see text to footnote 43), the improved method of belief formation emerges in part from the social interactions. The nonepistemic motivations of the protagonists drive them to engage in encounters with nature from which results a far more convincing form of reasoning than any antecedently available. This point about the social promotion of epistemic goals will be developed in a more general and abstract way in Chapter 8.

48. Rudwick's analytic chapters provide a serious and resourceful attempt to find a *via media* between the traditional philosophical image of rational scientists, devoted to the ideal of truth above all, finding out how nature is constituted, and the proposals of relativistic sociology of science. While I share Rudwick's concern to find a position that captures the insights of both extremes, my own compromise is closer to the philosophical pole than his.

Greywacke and the Old Red *within Britain*. For the elite, the kinship of the strata *in Russia* was enough, because their wider knowledge of rock types and fossils generated no expectation that contemporaneous rock types need exhibit their similarities *within some arbitrarily designated area*. Thus Williams failed to appreciate both (2) and (4). Weaver, on the other hand, seems to have been so devoted to making up for his Irish mistake that he was determined to seek out the elusive unconformity no matter how long or fruitless the search might be.[49]

My view of Weaver and Williams can easily be illustrated with an analogy. Imagine that a group of people are judging the finish of a race. By the finish line, in prime position to register judgment, stands a cluster of individuals who have wide experience in calling the result. Up in the grandstands are two others, neither with an unoccluded view, who, if they stand on tiptoe and crane their necks, can just see the finish line. Neither has ever judged a race before. The runners cross, and everyone announces a verdict. All those by the finish line agree. The two in the grandstands disagree with the consensus and with each other. No devotion to "symmetrical explanation of belief" should lead us to overlook the circumstances that render one set of verdicts more likely to be correct and that cast doubt on the other. So it should be with cognitive appraisal generally, and with Williams and Weaver in particular.

There is no imputation of irrationality here. Weaver, perhaps, is the victim of an idée fixe, and Williams, through the circumstances of his vocation and his training, is in no position to recognize the full cogency of the elite reasoning. But dissolving the notion of rationality does not mean abandoning cognitive appraisal. I claim that the framework of this chapter enables us to develop Rudwick's account of the great Devonian controversy and to see how, in this instance too, the compromise model applies. Out of socially guided interactions with nature, there emerge superior ways of forming beliefs, and, because these were incorporated into the psychological lives of virtually all the participants, the great Devonian controversy was brought to a cognitively appropriate close. Society, nature, and sound individual reasoning combined to drive the social learning machine to a new success.

---

49. Speculation about Williams is easier because Rudwick provides more information about his ultimate reactions to the consensus. However, in neither instance would I want to claim that my psychological accounts are correct. They are simply intended as articulations of the idea that both men underwent processes that were less likely to promote cognitive progress than those that moved the geological elite. Explaining human belief requires us to take seriously the causal processes that generate and sustain belief, without assuming (on a priori grounds?) that all such processes will be equally likely to disclose the truth about nature.

7

The Experimental Philosophy

1. Models of Empirical Knowledge

At the heart of Legend is an epistemology articulating the simple idea that scientific knowledge rests ultimately on observation and experiment. Much twentieth-century philosophy, including the versions of logical empiricism that provide detailed articulations of Legend, adopts a *static* model of human knowledge. Abstracting from the complexities of human belief formation, one conceives of an idealized knower, in possession of a body of evidence statements that represent the contribution of experience, and the project is to identify the relations that must hold among statements if some are to justify others, and thereby show how the evidential corpus warrants claims of theoretical science that may both be universal in scope and also purport to describe entities remote from sensory experience.[1]

Aficionados of the static model of human knowledge regard history as irrelevant. Justification of an individual's beliefs is to be sought in the here and now. By contrast, *dynamic* models of human knowledge take the problem of justifiable *change of belief* to be central. From this perspective, one offers an account of the cognitive states of individuals and asks for the principles that govern rational transitions between states. The justificatory status of the claims and commitments of current science thus comes to rest on the possibility of tracing a sequence of justified transitions—not necessarily coincident with the actual course of history—that will link the states of our remote predecessors with contemporary acceptance of science.[2]

---

1. The approach I sketch here is plainly "foundationalist" rather than "coherentist," and in this I take it to represent the main currents of epistemology of science in the twentieth century. However, it is not hard to see how to formulate a rival static model of human knowledge that captures the main ideas of epistemologists who have pursued coherence theories of justification.

2. There are obvious questions which I shall sidestep for present purposes. Should a historicist epistemology take the final member of a sequence of beliefs to be justified only if it can be connected by justified transitions to an original *justified* state? Or is it sufficient that there be justifying transitions to some ur-state, whatever the epistemic status of that state? Does it matter that the justifying sequence diverges from the actual history so that the connecting states are never found in that history? Given the views about conceptions

Since the early 1960s, dynamic models of human knowledge have been popular among philosophers of science. Even before that, however, Popper had made the issue of the growth of knowledge central to his treatment of epistemological questions.[3] Moreover, many of the insights of theories of confirmation advanced within the static model can be reformulated within a dynamic model, for example that offered by contemporary Bayesians.[4] While these articulations of a dynamic model are less artificial than the static approaches, which presuppose the availability to the subject of a vast corpus of "evidence," they are still at an enormous remove from the practice of science.

Since the publication of (Kuhn 1962), several philosophers have broadened the conception of cognitive state to include other types of commitments. So, for example, both Laudan and Shapere think of the growth of scientific knowledge in terms of the evolution of language, methodology, aims, and instruments. This broadening of the notion of cognitive state is continued in my introduction of individual and consensus practices in Chapter 3. The aim, of course, is to add realism to philosophical accounts of science, accounts that have often been criticized for their apparent lack of connection with the phenomena.

Introducing the notion of an individual practice to capture the complexities of scientists' commitments is only one facet of my departure from previous models of scientific knowledge. The notion of experience as the source of evidence, celebrated in the static model, is transformed in prior versions of the dynamic model by supposing that experience is the motor of scientific change. In my account, there are *two* processes that spark the modification of individual practices: encounters with nature and conversations with peers.[5] The second of these is ignored in traditional models of knowledge: perhaps because it is taken for granted that the effects of social interactions are somehow reducible to interactions with asocial nature. By contrast, those who offer the sharpest challenges to philosophical accounts of scientific knowledge frequently contend that this neglected source of change in belief is paramount in science, and that the role of "encounters with nature" is insignificant.[6] By

---

of rationality that I proposed in the previous chapter, it will hardly come as a surprise if I claim that these issues dissolve under disambiguation of the kindred notion of justification.

3. Of course, in Popper's version the concept of justification gives way to a different notion of appraisal: transitions are never *justified,* although they might be *scientific* or (in Lakatos's formulation) *intellectually honest* (Lakatos 1970).

4. See, for example, (Jeffrey 1983, Howson and Urbach 1989, Earman 1992). Non-Bayesian approaches that share the commitment to thinking in terms of assignments of probabilities to statements can also articulate a dynamic model of knowledge. See (Levi 1982).

5. It is worth reemphasizing that encounters with nature include those occasions on which scientists reason by themselves. Thus, I want to allow for the creative process of reasoning to a new idea, which resolves some of the antecedent tensions in practice. As I noted in Chapter 3, such tensions are always present. The account of mathematical knowledge offered in my (1983a) saw them as sources of scientific change and as *dominant* sources of mathematical change.

6. Although they would not put the point in this way, I think that Harry Collins (1985) and Andrew Pickering (1984) suggest a dynamic model of human "knowledge" in which

contrast, I believe in recognizing both factors and in avoiding a priori judgments about which (if either) plays the dominant role in the modification of belief.

So much for situating my epistemological project with respect to other approaches. I have offered a picture of the modification of consensus knowledge through the alteration of individual practices. What kinds of changes occur? How ought they to occur?

Individual practices are multidimensional. They consist of language, explanatory schemata, questions, statements, techniques and instruments, assignments of authority to others, and methodological claims. The general form of the transitions that occur in the lives of scientists is thus

$$\langle L, E, Q, S, T, A, M\rangle \rightarrow \langle L', E', Q', S', T', A', M'\rangle$$

where some of the components of the later practice may be the same as the corresponding parts of the earlier one. When such changes occur, they result from processes that may involve interactions with other people, the reception of stimuli from the asocial environment, and the activation of inferential propensities. Our *descriptive* task is to understand the kinds of changes in practice that occur and the processes that stand behind them. The *normative* project is to consider whether making particular kinds of changes on the basis of particular kinds of processes is likely to promote cognitive progress.

I shall operate with a simple taxonomy of cases, one which is not meant to be complete but that will cover the episodes of principal interest. Distinguish first those occasions on which the reception of a stimulus modifies the set of accepted statements. Among these, I shall consider separately instances in which the change involves reliance on the trustworthiness of an instrument or technique, instances in which the modification of the set of accepted statements is dependent on the assessment of the authority of another person, and those simple instances in which there is no reflection either on the credibility of others or on the reliability of the instrumentation. Another major class of transitions is based on inferential processes in which some parts of existing practice are used as a basis for the emendation of others. The emendations can affect the set of accepted statements; the significant questions; the language; the explanatory schemata; the assignments of reliability to instruments, observers, and techniques; the explicit claims about methodology. Finally, there are numerous ways in which such processes and transitions can combine, and it will be important for us to understand how resolution is achieved when processes of different types lead in different directions.

---

modifications of the complex states of individual scientists are caused by social interactions among them. The model can probably be traced to (Bloor 1974). Perhaps the best way to contrast it with the usual philosophical approaches is to see that whereas philosophers typically take the social input from others to be ultimately reducible to the impact of nature, the sociologists whom I have mentioned regard encounters with nature as so thoroughly mediated by society that they do not provide any independent constraint on the cognitive states of scientists.

2. Observation

Start with a relatively simple situation. A behavioral biologist is observing a baboon troop. Over a period of several hours he records the episodes in which one of the animals grooms another, carefully noting the names of the animals (who groomed whom) and the time interval through which grooming occurred. Each entry in the notebook records the perceptual acquisition of a belief. Focus on any one. The observer is initially scanning the troop. He sees the male he calls "Caliban" approach the female he calls "Miranda." There is a sequence of facial expressions and gestures, at the end of which Caliban crouches behind Miranda and plucks at her fur. Our biologist presses a button on his stopwatch and quietly moves to a position from which he can gain a better angle on the interaction. After a few minutes, Miranda shrugs and moves away. Another button on the stopwatch is pressed, and the biologist writes in the notebook, "Caliban—Miranda, 6:43." That notation serves as an extension of declarative memory, something from which the biologist can later retrieve the belief that Caliban groomed Miranda for a period of six minutes and forty-three seconds.

The "observation reports" in the notebook are highly unlikely ever to figure in public scientific discussion. Assume that the biologist's principal interest is in differences in times spent grooming between relatives and non-relatives. The "data table" of the published report will divide up the interactions by applying background views about the relationships among the animals and then sum the time periods in various divisions. Caliban and Miranda may vanish from the scene; the individual episode of the last paragraph will almost certainly do so.

Assembling the published "data" typically relies on inference as well as observation, and almost always involves trusting the credibility of others. Simple observation rarely warrants, by itself, any modification of individual practice: justifications of the acceptance and rejection of statements, even at the level of "empirical data," are usually complex (see Bogen and Woodward 1988, Galison 1987). But even baboon-watching is more subtle than it might initially appear.

Consider the pitfalls that attend the performance of the imaginary observer. Perhaps the participants have been misidentified. Perhaps the interaction between them has been misclassified. Perhaps the stopwatch is not functioning properly. Perhaps the presence or the movement of the scientist disrupts the normal activity of the troop. (This last point is important, for the "data" eventually published will purport to characterize the behavior of baboons under undisturbed conditions. Indeed, the scientist may include a defense of the claim that his presence did not distort the pattern of behavior.) These are not instances of hyperbolic doubt, but live possibilities for an unskilled or unlucky observer to go astray. Typically, skilled observers do not proceed by running through a checklist, assuring themselves that the possible pitfalls are absent. They operate by habit.

The skilled observer has abilities to reidentify and classify that the neophyte lacks. While the newly arrived graduate student must painfully look for clues to a baboon's identity, long experience of watching the troop enables the biologist to pick out Caliban at once. Moreover, the ordinary, unselfconscious habits of observation can be inhibited if the environment signals the threat of a pitfall. When the stopwatch has recently fallen into the waterhole, the biologist will take pains to check its working before relying on it in further observations. And, as I intimated in describing the original situation, when he moves to improve his view of an interaction, the biologist will consciously try to avoid startling the baboons.

The trained biologist differs from the neophyte in two ways, using different concepts for reporting what is seen ("Caliban groomed Miranda" versus "A big baboon sat down next to a smaller one") and having different propensities for belief formation. The newly arrived research student is intermediate, possessing the concepts that the biologist employs but applying them with greater self-consciousness and less fluency. Even the apparently simple observational episode with which we began is shaped by the past, in the training of the skilled observer and in the history of the consensus that stands behind it. The approach to progress that I developed and defended in Chapters 4 and 5 is optimistic about what has been accomplished. The trained observer has learned to see *well:* using more adequate concepts in reporting what is seen, making discriminations that the untrained observer cannot make, acquiring propensities that work both more efficiently and more reliably than the explicitly clue-following processes of the research student.

This dependence on the past can be viewed more pessimistically.[7] Since the beliefs that people form depend not only on the stimuli they receive but on their background cognitive states—on the languages they are able to use and the propensities that have been stored through training—it is possible that two observers should receive the same stimuli and form quite different beliefs. This is unworrying if one (and only one) of the observers can be presented with stimuli that generate beliefs from which, by reasoning that accords with shared canons,[8] it is possible to arrive at the distinctive features of the cognitive state of the other. In this case there is an epistemic asymmetry, and the superiority of one of the ways of responding to the stimuli can be defended. But if habits of observation and observational reporting sometimes change without the possibility of justifying the modification, then there is

7. The pessimism is apparent in the writings of Kuhn and Feyerabend on the theory-ladenness of observation, especially (Kuhn 1962/70 chapter X and Feyerabend 1975). Both Kuhn and Feyerabend start from the insight that there can be variation in reporting of observations that cannot be traced to any obvious malfunctioning of one of the observers and conclude that there are possibilities of championing alternative doctrines that cannot be discriminated through appeal to observation. As I argue in the text, the conclusion depends on acceptance of a type of epistemic symmetry that goes beyond their initial insight.

8. I assume here that the shared canons of *good* reasoning, that is, that they are likely to lead to the inculcation of true beliefs. Even though both scientists might fail to employ *optimal* belief-modifying strategies, I shall also suppose that use of such strategies would break the symmetry in the same fashion.

room for concern that observation might be unreliable. Perhaps the "skills" inculcated by training are not skills at all.

Let us examine this argument from epistemic symmetry more carefully. Imagine that there are two observers with background cognitive states C_1 and C_2, and that there is a stimulus s that induces in the first a belief that p_1 and in the second a belief that p_2, where p_1 and p_2 are incompatible. There is a condition of epistemic symmetry only if (i) there are no stimuli that could induce a belief that p_2 in a subject with background state C_1, (ii) there are no stimuli that could induce a belief that p_1 in a subject with background state C_2, (iii) there are no processes of accepted good inference available to a subject with background state C_1 that would warrant belief that p_2, (iv) there are no processes of accepted good inference available to a subject with background state C_2 that would warrant belief that p_1, (v) there are no processes of accepted good inference available to a subject with background state C_1 that would modify the cognitive state in such a way that there would be stimuli inducing belief that p_2, and (vi) there are no processes of accepted good inference available to a subject with background state C_2 that would modify the cognitive state in such a way that there would be stimuli inducing belief that p_1.[9] It is now claimed that, in the history of science (specifically in the history of scientific controversy), there are rival observers with different background states who are led to incompatible beliefs by the same stimulus and with respect to whom a condition of epistemic symmetry prevails. One of the rival positions and the associated concepts and propensities for observation survive in our contemporary science. But we cannot claim that this survival marks an advance in ability to observe or that the "skills" are reliable generators of perceptual knowledge. For *at least one* of the systems of concepts and propensities must generate error (since they yield incompatible beliefs), and, by the hypothesis of epistemic symmetry, there is no way of telling which.

Making this skeptical argument work is far more difficult than appeals to the theory-ladenness of observation typically suggest. It is not enough simply to declare that rival scientists would have been inclined to respond to the same stimulus by asserting incompatible statements: Priestley announces that the stuff in the flask is dephlogisticated air, Lavoisier contends that it is oxygen, Galileo maintains that a point of light in the sky is a translunar comet, his Aristotelian opponent counters with the claim that it is a sublunary exhalation. What must be shown is that the conditions for epistemic symmetry are met, and, since the psychological states of Aristotelians, Galileo, Lavoisier, and Priestley are not easily accessible, the skeptic has to proceed indirectly by using information about the contemporary performances of

9. Intuitively, the symmetry must be unbreakable by further observation, unbreakable by inference, and unbreakable by a combination of observation and inference. In specifying the conditions of the text, I assume that the use of further observation and inference does not lead to a situation of further symmetry: so, for example, if (i) is satisfied (ii) is not.

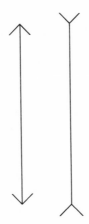

Figure 7.1 The Müller-Lyer Illusion. Although the line on the right looks longer than that on the left, both lines have the same length.

psychological subjects (hence the emphasis in Kuhn's discussion—1962/1970 chapter X—on the deliverances of "new look" psychology).

There are two general strategies for resisting concerns about our observational skills. One, articulated with vigor by Jerry Fodor (1984), is to argue that the plasticity of observation has been greatly overrated. The second is to show that in the allegedly problematic situations there are shared beliefs that allow for the resolution of the difference in favor of one of the parties (our precursors).

At the heart of Fodor's argument is a simple idea. It is a mistake to think that people will always be able to accommodate stimuli to their prior beliefs. The recalcitrance of the stimulus can be recognized by considering the familiar Müller-Lyer illusion (see Figure 7.1). Changing your beliefs from the naive state (in which you do not have any prior commitment about the relative length of the lines) to the sophisticated state (in which you believe that the lines are equal in length) makes no difference to the illusion: one line still looks longer than the other. Fodor draws the conclusion that the processing of perceptual information (the formation of perceptual belief) is often independent of features of the background cognitive state. Hence instances in which the same stimulus generates incompatible beliefs in observers with different background conceptual states ought to be rare.

The obvious reply to this (developed by Paul Churchland) is to deny that the differences in background cognitive state that are relevant here are merely differences in belief. Churchland pointedly asks, *"Who ever claimed that the character of a scientist's perception is changed simply and directly by his embracing a novel belief?"* (1988 175; italics in original). He goes on to note that Kuhn's defense of the theory-ladenness of observation began from the idea that there is far more to a scientist's knowledge than the disposition to assent to certain statements. People can be trained to respond differently to stimuli, and they come to acquire propensities that the untrained lack: music students, as Churchland rightly emphasizes, gain the abilities to discern properties and relationships that others lack; they can recognize chords, relative pitches, and, *sometimes,* absolute pitches. Perhaps there are also ways to train subjects to respond differently to the persistent illusions, the Müller-Lyer and its relatives.

The Fodor-Churchland debate throws valuable light on the skeptical argument with which I am concerned. Fodor's argument serves as a corrective to the overinterpretation of theory-ladenness. Not every change of belief issues in modification of observational propensities. Churchland's counter reminds us of the possibility of altering those propensities through training. Now there is no comfort for the skeptic if the differences in perceptual response that occur in the history of science are like those between the trained musician and the tyro. For the musician has ways of showing that she does indeed have superior discriminatory abilities—she can agree to be blindfolded and name the notes that the tyro plays on a keyboard (to cite one simple example)—to the extent that trained observers, primatologists, geneticists, neurophysiologists, spectral analysts, astronomers, particle physicists, paleontologists, can display their virtuosity in analogous ways there will be no room for concern that the skills they claim are not genuine.

We can now specify more clearly what the argument from epistemic symmetry must try to show. The rival cognitive states must include sets of different observational propensities. The subjects regard *their* propensities as skills and those of their rivals as bungling. Moreover, there must be no way of resolving this difference, no analogue of the demonstration of what a trained ear can discriminate.

At this point, we can take up the second strategy for combatting the skeptical argument. The mere fact that a scientist has been trained in a particular way, so that the presentation of a stimulus *s* standardly elicits the belief that *p,* does not mean that the scientist is doomed to disagree with others who have been trained differently and who are also presented with *s.* Recognition of the differences in training can itself modify the scientist's cognitive state so that propensities that would normally have been activated are inhibited in favor of more primitive—and widely shared—propensities.[10] The plasticity of perception may actually help. It is not hard to imagine how dialogue might go: after exchanges in which each protagonist uses his own preferred idiom, both retreat to a more neutral vocabulary—'dephlogisticated air' and 'oxygen' give way to 'vital air.'

Tempering one's observations does not yield reports that are entirely

10. This point has been appreciated in some recent, sensitive studies of observation and scientists' reporting of their observations. See, for example, (Shapere 1982, Pinch 1985), both of whom recognize that, at moments of controversy, scientists might respond to stimuli using more minimal—and unproblematic—categories. Shapere and I share the view that the use of more recondite concepts can be understood in terms of reliable processes of reasoning; Pinch prefers to explain such usage by appealing to social processes of negotiation.

innocent of commitment to doctrine, neutral descriptions of what is given to the senses.[11] The point of tempering is to achieve descriptions that only commit the subject to points of doctrine that are shared with her rivals and detractors. At this level, the opponents can agree on what they see, and they can use their agreement to attempt to decide whether either of their more ambitious ways of reporting observations is warranted.

Residual skeptical concerns must center on the idea that the corpus of tempered observations is too weak. Given the views about the properties of various chemical substances that Lavoisier and Priestley share, can Lavoisier articulate a chain of reasoning that will show the superiority of his preferred theoretical descriptions? The issues here are akin to those reviewed in the last chapter. At early stages of the debate, it may be correct to view the choice as underdetermined. Antirationalists will contend that the underdetermination is permanent, and that the ultimate endorsement of certain richer categories is to be explained by identifying how some of the participants were successful in processes of social negotiation.[12] Because problems of underdetermination will be addressed in far more detail later in this chapter, I shall postpone discussing them here.[13] Suffice it to note that the suggested resolution of the problems about acceptance of observation *may* exacerbate the larger epistemological issues.

The case of the debate between Copernicans and their opponents leads into more complex questions about observation. The apparent impasse between Kepler and Tycho, Galileo and the Aristotelians, about how to report the movements of the sun, the earth, and comets, can readily be resolved by talking about the apparent relative motions. Historically, supporting the Copernican propensities for reporting astronomical observations required the defense of a far more interesting and powerful skill, namely the ability to make telescopic observations. As Feyerabend emphasises (1975 126) Galileo operated with an imperfect instrument, one that he had to train himself and others to use.[14] How could he convince the skeptical that the surprising things he found in the heavens were really there? How could he show that he had really acquired an observational skill?

11. Thus there is no reversion to an observational "given" or to the idea of an *absolutely* neutral observation language. The important insights of (Hanson 1958, Kuhn 1962/70, Hesse 1970, Feyerabend 1965, 1975) that there are deep difficulties with the logical empiricist conception of observational evidence and observation language can thus be sustained. Even before the critiques of these writers, the fundamental epistemological points were made with great power in (Sellars 1953 chapter 5).

12. This appears to be the attitude of Pinch (1985). By contrast, Shapere (1982) proposes that the legitimacy of a richer observational vocabulary is to be understood in terms of "building on what we know." Shapere and I concur on this point, although the account of the pertinent reasoning that I offer later differs from his.

13. See Section 9, where Lavoisier's debate with the phlogistonians is treated in detail.

14. Thus, in my judgment, by far the most interesting and sophisticated skeptical argument about observational skills is that launched by Feyerabend in his (1975). Although I shall contend that Feyerabend's skepticism can be resisted, his discussion is valuable for raising the crucial issues.

Here we trespass beyond the bounds of the simple kinds of observations with which I began, but in the important respects the issues remain the same. How would a trained behavioral biologist persuade a doubtful beginner that her observational propensities are skills worth acquiring? By engaging in displays of discriminatory virtuosity. She draws the tyro's attention to traits of individual baboons that were previously overlooked and shows that she can recognize Caliban even at a considerable distance from subtle properties of his gait. She identifies features of various types of recurrent interactions and shows how they fall into regular complexes, associated with special names. What initially appeared as a jumble of monkeys engaging in meaningless movements begins to fall into a definite pattern, and the beginner finds himself able to predict what he is likely to see next.

How did Galileo convince his contemporaries that he had a reliable instrument and a special skill? In similar fashion. The power of the telescope could be detected *on earth* by using it to predict the detailed appearance of objects that are only just coming into view: you can easily astound an audience by inspecting an approaching ship and describing the captain's attire.[15] Moreover, Galileo could train others to perform the trick for themselves, and he could introduce technical improvements that made the telescope both more powerful and easier to use.[16] Aristotelians could still object. Perhaps the telescope, constructed from terrestrial materials, would only be accurate when turned on terrestrial objects. Perhaps it would be deceptive when used to look at the heavens.

Return, for a moment, to our behavioral biologist observing the baboons. Her display of virtuosity would be of no avail if her audience were not prepared to project from the context in which identifications can be checked—Caliban

15. Numerous reports in the writings of Galileo and his contemporaries attest to the success of *terrestrial* demonstrations of the veracity of the telescope: Galileo describes how the elder statesmen of Venice puffed up the stairs to turn the telescope on the sea, the members of the Lincean Academy tell of using the telescope to read the inscriptions on distant buildings, and even the critical Martin Horky concedes that *"below* it works wonderfully." See (Galileo/Favaro III, Drake 1978, Van Helden 1989). Feyerabend is very clear that there was no problem in checking the reliability of the telescope on earth (1975 107–108).

I should note that Ian Hacking's extended discussion of observation and the establishment of observational skills (1973 chapters 9–11) brings out in an exemplary way the kinds of strategies that historical figures—like Galileo—have employed. My own treatment is indebted to Hacking's, although I place far less emphasis on the role of manipulation. My account is also akin to Richard Miller's discussions of the extension of observation in his (1987), although I am concerned to emphasize the general structure of the ways in which skills are established and thus dissent from his suggestion that confirmation is merely local. For explicit contrast, see (Kitcher 1992).

16. For example, Galileo introduced the technique of stopping down a lens—possibly as the result of his own eye trouble, and his practical discovery that he could improve his vision by partially closing his eyes (Drake 1978 148). As we shall see, Galileo also made better telescopes than those who tried to emulate him, and an important part of his campaign for the telescope depended on his devoting large amounts of time, energy, and money to making and circulating instruments with lenses ground to his specifications.

and Miranda close at hand—to contexts in which they have not been independently vindicated. But *background practice* provides no grounds on which to distinguish (say) episodes of identification that have been carried out up to now (taken to be reliable) and episodes of identification in the future (assumed not to be reliable). By contrast, the practices of Galileo's contemporary Aristotelians generated principled reasons for thinking that Galileo's instrument might only work well on the earth.

Now the obvious and direct way of extending Galileo's demonstrations to undercut this objection would be to show some superlunary phenomenon that could not be appreciated with the naked eye and then put oneself in a position to check it directly through naked-eye observations. Lacking spaceships, Galileo was unable to do this.[17] Forced to proceed in a more roundabout way, Galileo responded to the Aristotelian objection by combining three strategies: first, he appealed to *tempered* naked-eye observations to argue that the celestial-terrestrial distinction is misconceived; second, by distributing his own telescopes and providing specifications of how to use them and pictures of the phenomena he claimed to have seen, Galileo could assemble a body of *tempered* reports of observations, produced by observers working in different places, that would testify to the regularity of his phenomena; third, he could connect his telescopic observations to phenomena that are barely visible with the naked eye.

The first strategy had already been implemented in the debate over the placing of the nova of 1604. Committed Aristotelians attempted to resist the claim that a new star had appeared in the allegedly changeless heavens. Faced with calculations of distance based on parallax which seemed to place the nova well beyond the sphere of the moon, they tried, in various ways, to explain away either the apparent distance or the apparent novelty.[18] Galileo continued to argue, from the early 1600s through into the 1630s, that the celestial/terrestrial distinction was untenable, not only because of the presence of comets and novae in the heavens but because of internal difficulties in the Aristotelian conception.[19]

17. Thus van Fraassen's claims about observability, while they capture the importance of checking putative observational skills against already accepted observational abilities, are quite detached from the ways in which scientists actually convince themselves and others that they have legitimately extended the senses (Van Fraassen 1980 16). In principle, however, Galileo might have been lucky enough to observe a comet approaching the Earth so that he could have used the telescope to predict its appearance. This would be analogous to his demonstrations with incoming ships.

18. One noteworthy effort was that of Ludovico delle Colombe in 1605, who suggested that the nova was an old, distant star that appears to us magnified by the crystalline spheres. In a devastating reply, allegedly "The Considerations of Alimberto Mauri," an unknown author (or authors) attacked Colombe by criticizing arguments for the incorruptibility of the heavens, by showing that Colombe's account gave a ludicrously bad estimate of the time of visibility of the nova and demonstrating how it was incompatible with everyone's astronomical theories. On the basis of both style and content, Drake speculates that Galileo may have had a hand in this work (1978 118).

19. The former types of argument culminate in Day Three of the *Dialogue,* the latter in Day One.

The third strategy, that of trying to show that telescopic observations are continuous with naked-eye observations, could be pursued in showing how the telescope apparently added greater detail to a picture whose outlines were known in advance. Thus Galileo's drawings of the moon and of two constellations (Orion and the Pleiades) reveal familiar features—the "darkish and rather large" spots on the surface of the moon that have been seen since antiquity, the previously identified members of the constellations in their well-known positions—but embed these familiar features in a richer context. Potentially available to Galileo was the tactic of showing how successively more powerful telescopes would disclose more and more of the detail. Since his own construction of telescopes began with a low-magnification instrument, it is quite possible that his own conviction that he was seeing more and more of the fine structure of the heavens rested on a sequence of inspections of the same astronomical phenomenon using progressively more powerful telescopes, but I do not know whether he ever appealed to any such sequence in defending the reliability of the telescope.

By far the most prominent strategy in Galileo's attempt to establish his observational virtuosity centered on his campaign to help others to see the four "Medicean planets" (the moons of Jupiter). The reasons for this are not hard to discern: Galileo's primary goal in 1610 was not simply to establish the fact that the telescope could give accurate information about the heavens or to employ the telescope to vindicate Copernicanism and debunk Aristotelianism[20]; his aim was to show that he had made a specific discovery, worthy of being rewarded with a position at the Florentine court.[21] Because he made the existence of the Medicean planets central to his own claims, his opponents, particularly Horky and Sizzi, joined debate with him on precisely this issue.

How can anyone simultaneously campaign for a new phenomenon and for the reliability of the instrument that is used to uncover it? We have already looked briefly at ancillary arguments that would undercut various kinds of Aristotelian challenge. The centerpiece of Galileo's case, however, is the possibility of obtaining, from astronomical observers working in different parts of Europe, a systematic collection of sightings that show both (a) agreement with one another on locations assigned on the same night and (b) a sequence of assigned locations that concur with those represented in the *Sidereus Nuncius* in showing apparent regular rotation about Jupiter. By 1612 Galileo achieved just this, as observers from around Europe jointly contributed to a corpus of claims satisfying (a) and (b). In consequence, many of

20. Here Feyerabend goes astray in suggesting that Galileo's Copernicanism played a central role in his defense of the telescope. (Westfall 1985) shows how Galileo only turns to a Copernican program of astronomical observation after he has successfully defended the telescope.

21. This is evident not only from the dedicatory strategy of *Sidereus Nuncius*, but from the actions Galileo took to ensure that the Medici in particular would have ample confidence in his discovery of the moons of Jupiter. See the penetrating analyses by R. S. Westfall (1985) and Mario Biagioli (1990).

Galileo's eminent contemporaries accepted the reality of the Medicean planets and the reliability of the telescope to show at least some new things in the heavens.[22] By 1613, Galileo's initial critics either had converted (Sizzi) or had been discredited (Horky), and, as a contemporary writer put it, those who first jested at the alleged discoveries came to have no doubts about them (Drake 1978 211).

Plainly, Galileo's demonstration of his virtuosity cost him a great deal of work. Success was neither immediate (as perhaps the historical and philosophical expectations of Legend would take it to be) nor indefinitely postponed (as some of Legend's most vocal opponents would lead us to expect). My preferred explanation of the episode supposes that Galileo pursued a complex strategy, needing considerable work to implement, that induced in his contemporaries processes that led them to modify their practices, countenancing both the moons of Jupiter and the reliability of the telescope; these processes were epistemically virtuous when judged by the criteria of the last chapter. This interpretation is at odds with Feyerabend's suggestion that Galileo was victorious because his contemporaries did not think very hard and because of his effective propaganda. I shall simultaneously try to defend my reading and oppose Feyerabend's.

There is no doubt that the moons of Jupiter were initially hard to see. In the wake of publication of *Sidereus Nuncius,* readers attempted to construct telescopes for themselves, and, although these devices enabled them to observe some of the phenomena Galileo claimed, they were too weak to reveal the Medicean planets. Nor was Galileo always successful in showing the moons through his own telescope. On one famous occasion, of which Feyerabend makes much, Galileo took a telescope to the house of the astronomer Magini and, according to Horky,

> Galileo Galilei, the mathematician of Padua, came to us in Bologna and he brought with him that spyglass through which he sees four fictitious planets. On the twenty-fourth and twenty-fifth of April I never slept, day and night, but tested that instrument of Galileo's in innumerable ways, in these lower [earthly] as well as the higher [realms]. On Earth it works miracles; in the heavens it deceives, for other fixed stars appear double. Thus, the following evening I observed with Galileo's spyglass the little star that is seen above the middle one of the three in the tail of the Great Bear, and I saw four very small stars nearby, just as Galileo observed about Jupiter. I have as witnesses most excellent men and most noble doctors, Antonio Roffeni, the most learned mathematician of the University of Bologna, and many others, who

22. One particularly important group whom Galileo convinced were the Roman Jesuits. Late in 1610, Jesuit observers in Rome were able to report that they had seen the Medicean planets, and, in 1611, at the instigation of Cardinal Bellarmine, they produced an official report on Galileo's findings. As I shall indicate, agreement with Galileo often exists at the level of *tempered* observations, and Jesuits such as Clavius explore alternative explanations of these tempered observations (see Drake 1978 165). But this is already to recognize that the telescope is reliable about some aspects of the heavens (even if, from a traditionalist perspective, Galileo is inclined to overinterpret its deliverances).

with me in a house observed the heavens on the same night of 25 April. But all acknowledged that the instrument deceived. And Galileo became silent, and on the twenty-sixth, a Monday, dejected, he took his leave from Mr. Magini very early in the morning. (Van Helden 1989 92–93)

Magini's own comments (also expressed to Kepler) were more guarded. He reported that "nobody has seen the new planets distinctly" and doubted whether Galileo's discoveries would stand up.

Galileo's dejection seems only to have been temporary. Throughout the second half of 1610 he distributed high-quality telescopes to prominent centers of culture in Europe, providing detailed instructions for the observation of the Medicean planets. In the light of the growing number of reports that the combination of Galileo's instruments, his instructions, and his diagrams for showing the various arrangements of the moons (essentially indicating what the phenomena were supposed to look like!), even those who had initially had difficulty in making observations with the telescope changed their minds. After Horky published his attack on Galileo, Kepler broke relations with him, Magini chased him out of the house, and Roffeni, on behalf of the Bolognese faculty, Magini, and himself, published an apologetic reply to Horky's tract.

By late 1610, those who were still undecided about the reliability of the telescope had two options: either they could view those who reported the positions of the Medicean planets as duped *in some systematic way* or they could suppose that failure to see the planets resulted from the absence of observational skill. The latter, of course, conceded to Galileo everything he claimed. The former required acceptance of minimal reports of observation about the appearance of points of light in specific places at specific times and shifted the issue to find an explanation of those appearances in terms of the "deceptions of the telescope." With no such explanations available, it is hardly surprising that *thoughtful, conscientious* people came to adopt Galileo's interpretation and accept the deliverances of the telescope at face value.

Feyerabend's skeptical argument fails to take into account both the coordination of the observations reported by different people and the orderliness of the successive observations of single observers. Both are expressed in discussions of the period of the satellites. He also neglects Galileo's technological improvements: by the end of 1610 his telescopes were better than they were at the beginning, and it was thus easier for him to obtain agreement from others. However, at no stage did Galileo have an adequate theoretical account of his telescope, and in this Feyerabend finds a crucial flaw.

According to Galileo's own account, his original construction of the telescope was based on his understanding of "the science of refraction" (Van Helden 1989 37). Yet no satisfactory explanation of the properties of lenses had been published by 1610, and, as Feyerabend points out, Galileo's "knowledge of optics was inferior by far to that of Kepler" (1975 105). But Galileo, as nobody else in the period, combined a knowledge of optics with a rich repertoire of "craft knowledge." Trial and error, guided by both his practical experience with lenses and his acquaintance with mathematical optics, led him to a series of telescopic improvements (see Van Helden 1977 18, 26–27). No doubt a full theory of the telescope would have been very helpful to him, but it was not needed for devising his instrument, for improving it, or for vindicating its accuracy. Ignoring the multiplication of systematic observations of the heavens and focusing on the most celebrated incident in which demonstration of the telescope failed, Feyerabend concludes that only the provision of a theory of the telescope would have underwritten Galileo's claims. As I hope to have shown, Galileo used quite different means to display his observational virtuosity, means that cannot be written off as propaganda.

I conclude that we can recognize the dependence of observation and of observational reporting on background cognitive states without abandoning the thesis that the propensities acquired by specialists are observational skills. Trained professionals can demonstrate their virtuosity in ways that are appreciable by the laity, and so turn back the challenge that their alleged skill is merely masquerade.[23]

3. Inductive Generalization

Scientific research, like everyday life, would be impossible without the activity of generalization. The evaluation of observational techniques, of others' credibility, requires using the past record as a guide to future performance.[24] Even more obviously, to anticipate the future course of experience, to guard against danger, and to exploit opportunity, people have to use the information available—and what is available is, of course, information about the past, about what has been observed. The issue is not *whether* the things we have checked should guide our beliefs about what remains unchecked, but *how* we should be guided.

Let us begin with a common inductive situation. Imagine that scientists (or people pursuing practical projects) have isolated a set of entities, the A's,

23. This is not to say that there will never be difficult cases. For it is quite possible that the kinds of strategies I have described for checking a new instrument (or those that Hacking discusses in his (1983)) would be unavailable because of the absence of any overlap with the deliverances of other, unproblematic, modes of observation. If an instrument is designed to do a very specific job which is discontinuous with the functions performed by other pieces of established technology, there may simply be no basis from which to judge its reliability. Perhaps this is the situation with Weber's apparatus for detecting gravity waves, described with great force by Collins in chapter 3 of his (1985). However, for a significant class of complex instruments, it is possible to develop defenses of their reliability along the lines indicated here. See the lucid account offered by Peter Galison in his (1987).

24. Thus a full account of the establishment of an observational skill—such as Galileo's defense of the use of the telescope in astronomical observation—depends on recognizing the effective strategies of generalization. My discussion in the last section thus proceeds at an intuitive level, relying on everyday views about how to generalize without identifying them precisely.

and that the question "How do the A's exemplify $D(B)$?" is significant for them—where B is some determinate property (for example, being blue) and $D(B)$ is the corresponding determinable (in this instance, color). For problems of this form the possible answers can be represented as n-tuples $\langle p_1, \ldots, p_n \rangle$ such that the p_i sum to one.[25] The number n is the number of determinate forms of the determinable $D(B)$. Cases in which one of the p_i is one and the rest zero are *pure* generalizations, in which the answer is "All A's are C" (where C is a determinate form of $D(B)$). *Mixed* generalizations are answers that assign nonzero frequencies to two or more of the determinate forms— as, for example, when half of the A's are declared to be B and half C. An inductive problem situation is constituted by the initial demarcation of the set over which generalization is sought (the A's), the determinable with respect to which generalization is sought ($D(B)$)—these, as we have just seen, fix the set of possible answers—and the *sample knowledge*. The latter consists in the specification of the frequencies with which the determinate forms of $D(B)$ have been found in those A's that have been observed—the members of $O(A)$—which can be represented as an n-tuple $\langle p_{01}, \ldots, p_{0n} \rangle$, together with statements about some of the characteristics of $O(A)$. Since $O(A)$ is a finite proper subset of A there are always two or more possible answers that are consistent with the observed frequencies. Adding an answer to the set of accepted statements always involves the risk of error, even if the specification of the observed frequencies is correct.

I suggested earlier that the important question about inductive generalization is *how* to let past experience guide our anticipations of the future. We can now see that this question divides into two parts: *When* should we accept some one of the possible answers? *Which* answer should we accept? Plainly, the need for relief from ignorance may be more or less acute. In everyday life, we sometimes have no choice but to select some one of the answers. Similar situations arise within the practice of science, particularly when scientists have to assess the reliability of techniques, instruments, and potential informants. In other instances there is less urgency, and the choice for a scientist is initially whether to pick some answer now or whether to seek more information (typically by expanding $O(A)$). If the first option is chosen, there is then a decision about which answer to pick; if the second, the next step is to decide how to obtain more information.

Inductive propensities are far more complex than is usually supposed. Even for the simple type of situation with which we are presently concerned, I shall conceive of the inductive propensity as involving (i) the framing of the problem through the specification of A and $D(B)$, (ii) a disposition to accept some answer to the question "How do the A's exemplify $D(B)$?" if and only

25. Strictly speaking, this holds for discrete cases in which there are a finite number of determinate properties corresponding to the determinable $D(B)$. Some problems involving the investigation of magnitudes can be conceived as either discrete or continuous. Thus a scientist might inquire how the entities in a set are assigned values of the magnitude in question (continuous form) or which of a *finite* set of functional dependencies is realized within this set (discrete form).

if the subject believes that $O(A)$ meets a condition R, and (iii) a function f that maps the frequencies of the determinate properties within $O(A)$, $\langle p_{01}, \ldots, p_{0n} \rangle$, onto the answer, $\langle p_1, \ldots, p_n \rangle$, that is accepted when the subject believes that $O(A)$ meets R. The inductive propensities in which I am interested have to satisfy a condition of psychological realizability: people have to be disposed to identify the A's, disposed to pick out $D(B)$ and its determinate forms, able to judge whether a set meets R, and able to give a general specification of f. Methodologists should try to explain what realizable dispositions to pick out A, $D(B)$, R, and f will work well in combination.

The most obvious question about inductive generalization seems to be how to choose the function f. But, I suggest, we can narrow our focus by concentrating on the special case in which f is identity. By doing so, we generate a *family* of inductive propensities that correspond to the straight rule ("Infer that the frequency with which a trait is found in a sample is that with which it is found in the entire population"). Members of the family differ with respect to the ways in which they frame inductive problems (choices about combinations of A and $D(B)$) and their standards for judging that a sample is representative (selection of R).

An inductive propensity, then, consists in a disposition to generalize (follow the straight rule) with respect to certain ways of classifying entities and picking out their properties and given certain beliefs about the observed instances. Since Nelson Goodman's seminal (1955), it has been obvious that inductive generalization cannot be completely understood without taking into account the types of classification involved. For any set and determinable, we can consider the subsets observation of whose members would suffice to prompt us to generalize: intuitively, these subsets comprise what we take to be representative samples. If there are no such *proper* subsets, then the pair A, $D(B)$ is (in Goodman's term) *unprojectible*. We cannot expect to learn the answer to the question, How do the A's exemplify $D(B)$? by looking at instances. Conversely, well-posed inductive problems are those that do allow for representative samples that are proper subsets of the set over which we intend to generalize.

The normal form of an inductive generalization is as follows:

[1] All members of A have some form of $D(B)$.

[2] The distribution of the determinate forms of $D(B)$ among the observed members of A, $O(A)$, is $\langle p_1, \ldots, p_n \rangle$.

[3] $O(A)$ satisfies R.

Therefore

[4] The distribution of the determinate forms of $D(B)$ in A is $\langle p_1, \ldots, p_n \rangle$.

An inductive propensity can be viewed as a disposition to engage in *certain kinds* of inferences of this form. So it can be regarded as a disposition to frame inductive problems by selecting certain pairs of sets and determinables and to apply the straight rule when the sample is judged to meet a certain condition. Let us represent this as $\langle F, R \rangle$, where F selects the appropriate $\langle A, D(B) \rangle$ pairs and R is the condition on the sample for the activation of the propensity for straight generalization.[26]

The methodologist's goal is to discover which pairs $\langle F, R \rangle$ are good ones.[27] Our next task is to articulate an appropriate standard of goodness. There are various ways to proceed. The simplest is to consider a range of situations in which an inductive propensity might be applied. In some of these, the conditions for the activation of straight generalization will not be met, and a subject who employs the inductive propensity will not perform any inductive inference. Call these the *undecided cases,* and the rest the *decided cases.* Within the decided cases, the propensity sometimes yields true conclusions and sometimes false conclusions. The former are the *positive cases* and the latter the *negative cases.* Let the ratio of the decided cases to the undecided cases be the *relief index,* and the ratio of the positive cases to the negative cases be the *truth ratio.* Ideally we would like propensities with high relief indices and high truth ratios. However, it is not hard to see that the two desiderata pull in opposite directions: the more likely a propensity is to provide relief from agnosticism, the less cautious it is likely to be, and the abandonment of caution typically lowers the truth ratio. Thus one problem that has to be faced is how to make an appropriate trade-off between the two virtues. Another difficulty in articulating a standard is to decide what the total range of situations we envisage ought to be.

Just as the set of inductive propensities we consider is restricted to those that are psychologically realizable, so too the range of situations over which those propensities are evaluated should encompass only situations that arise in the actual world. The methodologist hopes to advise human beings investigating our world. However, to specify appropriate types of situations we have to turn to history and to science. History tells us what kinds of problems scientists and others have had to face in the past; science informs us about the solutions.[28] Drawing a representative sample, we evaluate inductive propensities by computing their relief indices and their truth ratios. So our present perspective on the world and the investigation of it is employed to judge the success of various ways in which our predecessors might have done their generalizing.

My talk of a representative sample is a fudge. There is no obvious way to select from the past some set of problem situations that will encompass

26. By the condition on psychological ascertainability, R cannot simply be presented as the property of being representative. There must be some conditions that the subject can apply.

27. Or, failing that, which strategies are best for modifying our current propensities $\langle F, R \rangle$ so as to improve them.

28. Thus, the standard for evaluating inductive propensities is a specification of the external standard, (ES), of the last chapter. At any given stage of inquiry, we employ what we think we have learned about the world to evaluate methods of learning more about the world.

the puzzles that typical people have to overcome. Human experience is sufficiently diverse that it might be better to assess inductive propensities by looking at their performance with respect to many sets, each of the *normal sets of problem situations*. Intuitively, we eliminate as pathological certain kinds of "inductive experiences"—for example, those in which a subject always encounters problems whose correct solution is indicated by the first object sampled and those in which long runs of instances of one kind are eventually followed by even longer reversals. The remaining inductive experiences, those consisting of sequences of problems that we think a subject might encounter, count as normal sets of problem situations. Different inductive propensities might, of course, fare better or worse on different normal sets. What we want to discover is which propensities do well across the board, which do reasonably well in all cases and extremely well in some, and so forth. I do not pretend that there is any single correct way to articulate the standard.

Even the simplest approach to setting standards for inductive propensities encounters two sources of indeterminacy, one surrounding the trade-off between relief from agnosticism and deliverance of truth, and one concerned with delimiting the sets of problem situations with respect to which the assessment is to be conducted. Matters are further complicated if we explore various refinements: (a) taking into account not only whether or not the conclusion coincides with the correct answer but also its *distance* from the correct answer, (b) considering the costs of various kinds of errors (so that incautious inductive propensities might be heavily penalized for delivering incorrect conclusions about certain kinds of problems), (c) recognizing the practical or cognitive significance of certain types of questions (so that highly cautious inductive propensities might be heavily penalized for not delivering answers). Taking these factors into account allows yet more indeterminacy, so that the range of appropriate standards for evaluating inductive propensities is further increased. In accordance with the discussions of Chapters 3 and 6, I draw the moral that the notion of a single "correct" or "rational" inductive propensity is a myth. That is not to declare that all inductive propensities are equally good. Some may be classed as definitely inferior whatever standard we choose.

4. Eliminative Induction

I want to defend an old-fashioned idea about human inductive propensities— or, more exactly, about the human inductive propensities that are activated in scientific inferences. According to this idea, induction proceeds through the elimination of alternatives. The main task of this section will be to show how to characterize an inductive propensity that works eliminatively and how to understand certain types of scientific inference in terms of the activation of this propensity. In later sections, I shall try to show that this propensity can be defended against some powerful objections.

The motivating idea behind the eliminative propensity is that the prior state of individual practice is used to select appropriate candidates for inductive generalization and to construct a space of alternative possibilities. Thus, for the question, How do the A's exemplify $D(B)$?, to present an inductive problem for subjects with the eliminative propensity, those subjects must accept the statement that each A has some determinate form of $D(B)$ and they must use a language in which the set A is recognized as projectible with respect to the determinate forms of $D(B)$.[29] The space of possibilities that they construct depends on the accepted explanatory schemata, specifically on views about how the determinate forms of $D(B)$ depend on other properties for objects of a kind $K(A)$ to which the members of A all belong.

If there are several explanatory schemata that provide views about the dependencies of the determinate forms of $D(B)$ on other properties for different kinds that subsume A, then the space is constructed by considering the narrowest of these kinds. The procedure for construction is as follows: Call the properties on which the form of $D(B)$ is explanatorily dependent across $K(A)$, the *background conditions*. Partition A into cells such that the members of each cell exemplify a conjunctive condition each of whose conjuncts is a background condition or its negation and which contains, for each background condition, either that condition or its negation. (So, for example, if there are two background conditions, X and Y, the partition of A has the four cells, XY, $X-Y$, $-XY$, $-X-Y$). The eliminative propensity relies on the idea that the background conditions specify *all* the properties on which variation in determinate forms of $D(B)$ across $K(A)$ depends. In cases in which the subject accepts the statement that the presence of a form of $D(B)$ in an A is deterministic, the possible hypotheses are reduced to a finite number, all of which claim that each cell is homogeneous for the form of $D(B)$. With respect to such situations, subjects with the eliminative propensity generate the only hypothesis compatible with the findings in $O(A)$ just in case $O(A)$ is judged to contain at least one member of each cell.

Consider the simplest type of situation in which the eliminative propensity works. The problem is to explore the way in which A's exemplify $D(B)$, and we can imagine that m A's have been examined each of which possesses the property B. The individual practice of the subject counts A as projectible with respect to $D(B)$ and recognizes k background conditions, $C_1, \ldots, C_k$. Moreover, it is part of individual practice that the presence of determinate forms of $D(B)$ in members of A is deterministic. A is partitioned into 2^k cells, each of which is taken to be homogeneous with respect to the determinate form of $D(B)$. The subject generates the conclusion that all A's are B's just in case she believes that $O(A)$ contains at least one member from each of the

29. This condition requires that there be accepted generalizations linking sets similar to A to determinate forms of $D(B)$. The similarity is fixed by background views about which sets of objects form kinds and which are subkinds of the same kind. The general line I advocate is thus akin to that endorsed by Goodman in his own treatment of the "new riddle of induction": projectibility stems from entrenchment and entrenchment comes from the past history of successful projection (see chapter 4 of Goodman 1955).

cells. Otherwise she seeks representatives of those cells that are not yet represented in $O(A)$. (Since there are 2^k cells it seems that the cardinality of $O(A)$, m, has to be at least 2^k if a conclusion is to be generated. I shall explore this point later.)

The inference performed by the subject is deductive. Among the premises is a claim that there are only a finite number of possibilities. Other premises jointly imply that all but one of these possibilities is incompatible with the premise that reports the properties of those A's that have already been examined. The conclusion, in the words of Sherlock Holmes, is that the remaining hypothesis, however improbable, must be the truth.[30]

So far I have only been attempting to characterize the eliminative propensity.[31] That attempt raises some obvious questions: Where do the strong premises—the claims about explanatory dependence and so forth—come from? Isn't inductive generalization needed to obtain them? How did the whole process ever get going? Should an eliminative strategy allow for the possibility that the currently specified background conditions are not complete? These questions deserve answers, and I shall take them up later. First, however, I want to show how the eliminative strategy functions in scientific generalization.

Consider, first, the behavior of a chemist investigating reactions involving a newly synthesized compound. The chemist hopes to arrive at generalizations about the rates at which the reactions proceed, the energies released or absorbed, and so forth. The investigation is framed by drawing on the explanatory schemata of chemistry and singling out those properties (pressure, ambient temperature, and so forth) on which variation in rates of reaction and absorbed energies depend. Once the appropriate variables have been "controlled for," the chemist is ready to announce whatever generalization emerges unscathed.

Turn now to the philosopher's favorite, generalizations about the colors of bird plumage. These are poor examples for strict inductive generalization, because background biological theory suggests that there will be exceptions: mutant alleles that direct the formation of abnormal pigments can easily arise. However, there is a related form of generalization, prevalent throughout biology, in which judgments about the overwhelming majority of members of a species (or a higher taxon) are made on the basis of sampling ("*Dictyostelium discoideum* aggregates in the presence of cAMP," "*Impatiens* grows

---

30. In his (1992), John Earman defends eliminative induction and explicitly makes the link with Sherlock Holmes. My own treatment has been influenced by Earman's ideas about scientific inference, although I believe that I am far more a partisan of eliminative induction than he currently is.

31. Moreover, given the informal nature of some of my characterizations, I have only succeeded in picking out a cluster of eliminative propensities. The exact details of how background practice determines the set of pairs $\langle F, R \rangle$ have not been specified, for I have not stated precise conditions on background practice for the set A to count as projectible with respect to $D(B)$, nor have I done more than indicate the way in which the spaces of possibilities are constructed. (In accord with the pluralistic approach of the last section, we might well allow for variation here.)

well in the shade," "The gestation period in elephants is over eighteen months"). To make such generalizations, biologists rely on received ideas about explanatory dependencies to specify relevant alternatives.[32] Eliminating all but one alternative, they generate their conclusion about what almost all members of the target group do.

The hoary example of the ravens will illustrate how this works. Background practice assures the biologist from the beginning that it is unlikely to be strictly true that all ravens are black. Albinism is always a possibility. However, this type of counterexample to the generalization is not common (again the voice of prior practice!) so that if only such kinds of counterinstances are found it would be fair to conclude that almost all ravens are black. What major sources of color variation might there be? Background practice recognizes that there can be differences between sexes, differences at various stages in ontogeny, differences due to variation in habitat (local adaptations to ecological conditions), possibly differences produced by *past* adaptations to ecological conditions that no longer obtain. The sample has to include instances that rule out these potential sources of variation, and so the biologist will look at old birds and young birds, male birds and female birds, birds from many geographically separated populations. Once the sample has included the potential sources of variation, the generalization will be accepted if it survives.

How many instances have to be examined? The initial thought (discussed earlier) is that $O(A)$ has to be of cardinality at least 2^k (where k is the number of background conditions). However, the search can be truncated if prior practice supplies claims about the absence of interactive effects. In general, if I believe that the C_iC_j condition will produce a deviation from the generalization if and only if the $C_i - C_j$ condition does, then I can simplify my search by looking at what happens under C_i without worrying about the presence or absence of C_j. In our ornithological fantasy, $O(A)$ need only contain male ravens, female ravens, juvenile ravens, mature ravens, ravens in each of the relevant climatic and vegetational zones, and not all the possible combinations (a juvenile female inhabiting scattered copses in a temperate climate, and so on). Prior practice tells us that certain kinds of interactions among background characters make no difference: if we discover that ravens are not sexually dimorphic in one habitat, then we can ignore sexual dimorphism in other habitats and simply select any raven (male or female) in the appropriate habitat. Depending on how much information about absence of interactions is furnished by prior practice, the size of $O(A)$ can be substantially reduced—

32. Hence biological predicates picking out large sets of organisms are often taken to be "almost-projectible" with respect to physiological, anatomical, behavioral and developmental properties, and this status shows up in virtue of the presence within the accepted statements of practices of generalizations about the overwhelming majority of members of the set. As Wesley Salmon reminded me, strict generalizations and probabilistic generalizations (including "almost all" generalizations) have different logical properties: specifically, the latter do not admit of contraposition. This point can be used to evade the paradox of the ravens, although, as I shall argue in the next section, my approach to induction allows for a general solution of that paradox.

perhaps to k, perhaps even to 1. (There are generalizations for which prior practice selects an ideal instance with which to check a putative generalization, one that concentrates in itself all the possible background conditions that might cause deviation. The strategy of seeking that instance, the putative "worst case," has been noted by Richard Boyd (1973). The strategy misfires if there are possibilities of compensating interactions among the background conditions.[33])

The philosophical literature on induction contains periodic efforts at turning inductive inferences into types of deduction, and a reasonable response to my account is to complain that it passes the buck. Plainly, revealing a deductive structure beneath an apparently ampliative argument requires one to strengthen the apparent premises, and this seems to transfer the original puzzle—"What are the rules for good ampliative argument?"—to a question about the status of the claims that are added to the premises. Aren't ampliative arguments required to accept these? If such ampliative arguments are then viewed as covertly deductive, then the burden will be thrown back onto yet further premises. So it will go, apparently, until we finally reach a stage at which it is necessary to countenance ampliative arguments that cannot be given the deductive treatment.

In my version, contemporary scientists acquire their languages and conceptions of explanatory dependencies from their predecessors. If there was some initial stage at which the eliminative propensity could first be put to work, then how did people arrive at the views about explanatory dependencies that were embodied in the practice of this stage?

My answer is that the eliminative propensity is overlaid on a more primitive propensity to generalize. As a consequence of our genotypes and our early developmental environments, human beings come initially to categorize the world in a particular way, to view certain kinds of things as dependent on others, to generalize from single instances of especially salient types. Moreover, just as there is a propensity to form certain generalizations, so too there is a propensity to restrict those generalizations in particular ways when matters go awry. I suggest that this *primitive apparatus* works tolerably well in confronting the problems that our hominid ancestors encountered: it is relatively well designed for enabling primates with certain capacities and limitations to cope with a savannah environment and with the complexities of a primate society.[34] Whether it is well designed for advancing scientific investigations,

---

33. Here there are also affinities to the ideas of Popper and Lakatos about "rigorous testing." Andrew Wayne has pointed out to me that there are many similarities between the arguments of this chapter and Lakatos's approach to scientific methodology. This seems to me most evident in my approach to problems of underdetermination and in my view of the shortcomings of Bayesianism.

34. To say that our primitive apparatus is relatively well designed is emphatically not to assume that it delivers truths about regularities in the behavior of middle-sized physical objects. See, for example, (Stich 1990) and the discussion in §10.3. The account of inductive propensities I give here is analogous to that advanced for our primitive mathematical knowledge in my (1983a).

the primitive apparatus stands behind our primitive scientific practices. With those practices in place, the eliminative propensity can be activated, and the use of that propensity (together with other types of inference to be considered in Section 7) allows for the modification of practice, the revision of the primitive categorizations and views of dependence.

5. Positive Instances

I want to approach other concerns about eliminative induction indirectly, by considering some familiar issues about confirmation of hypotheses. I shall start by using the well-known paradox of the ravens to introduce issues about the value of positive instances.

In a landmark article (Hempel 1945), C. G. Hempel pointed out that the statement that all ravens are black is logically equivalent to the statement that all nonblack things are nonravens. Suppose we hold that generalizations are confirmed by their instances: the statement that an object a that is A is also B lends further support to the claim that all A's are B's.[35] Examining a raven and discovering it to be black therefore support the hypothesis that all ravens are black. However, if we add the plausible claim that a statement e that confirms a statement h confirms any h' logically equivalent to h, then we reach apparently counterintuitive conclusions. Suppose that we come to accept the statement that a certain object that is a shoe is white. By the principle about instances, this confirms the hypothesis that all nonblack things are nonravens. Now, from the principle about confirmation of logical equivalents, we can conclude that it supports the hypothesis that all ravens are black. But surely that is dubious. Observing white shoes is not a cheap way of increasing reasonable confidence in the generalization about ravens.[36]

Viewing induction as eliminative enables us to solve the puzzle. The problematic idea is that *any* positive instance of a generalization lends it support. From the eliminativist perspective, instances support generalizations by eliminating potential rivals. An instance that does nothing to eliminate a potential rival does nothing for the generalization. *No* report of the whiteness of a shoe does anything to eliminate rivals of "All ravens are black"; *some* (but not all) reports of the blackness of ravens eliminate rivals and thus lend support to the generalization.

Consider the puzzle in more detail. Suppose that the background conditions derived from the explanatory schemata of practice are $C_1, \ldots, C_k$, that

35. Here I avoid specifying the order in which the properties of a are recognized. Some commentators on the raven paradox (e.g., Mackie 1963) have held this to be especially significant.

36. The paradox has spawned an enormous secondary literature, containing a large number of ingenious ideas. For a review of *some* prior attempts at solution, see the discussion in (Horwich 1982). There are some affinities between my own approach and that adopted by Horwich, which ultimately derives from an insight of Janina Hosiasson-Lindenbaum (1940).

prior practice yields conclusions about the absence of interactions that permit them to be treated separately, and that the instances already sampled reveal that ravens meeting conditions $C_{m+1}, \ldots, C_k$ are black. The generalization under focus is

$$(x)\ (Rx \supset Bx)$$

and its rivals are statements of the form

$$(x)\ (Rx \supset (-Dx \equiv Bx))$$

where D is one of the C_i.[37] Some of these rivals have already been eliminated. Those that remain are of the preceding form with D as one of $C_1, \ldots, C_m$. To obtain further support for the generalization we need reports of objects that satisfy R, B, D, where D is one of $C_1, \ldots, C_m$. White shoes *and lots of black ravens* don't do the trick.

The original puzzle was generated by using the logical equivalence principle, and, specifically, by considering the hypothesis

$$(x)\ (-Bx \supset -Rx).$$

Does a report of a white shoe confirm this hypothesis? That depends on whether it excludes any rivals. If the rivals take the form

$$(x)\ (((-Dx\ \&\ -Bx) \lor (Dx\ \&\ Bx)) \supset -Rx)$$

where D is among the C_i, then they are logical equivalents of our original set of rivals, and the exercise in which we are engaged is one of finding inconvenient logical equivalents for the statements under focus in the last paragraph. A report of an object can eliminate one of the rivals (while remaining compatible with the generalization) provided that that object satisfies D, B, R. This, of course, is our old condition, and, since white shoes do not measure up, they confer no support. Providing white shoes with confirmatory power requires consideration of a different set of rivals. Thus, if prior practice generated background conditions $E_1, \ldots, E_n$ so that there were rival hypotheses of the forms

$$(x)\ (-Bx \supset (-Rx \equiv -Fx))$$

where F is among the E_j, then a report of an object would eliminate a rival (while remaining compatible with the generalization) if that object satisfied $-B$, $-R$, F. *In principle* it is possible for a white shoe to accord with this demand. We can draw a conditional conclusion: if prior practice generates rivals of the most recent form, then it is possible for observation of nonblack nonravens (maybe even white shoes) to lend support to the generalization.

However, for prior practice to generate such rivals, there would have to

---

37. The rivals are contrary generalizations about the color of ravens. They can take the form presented in the text or the more specific form of specifying the colors of variously characterized subsets of the ravens. The argument of the text goes through on the second interpretation as with the first.

be views about the dependencies of the presence of properties like *being a nonraven* in things that are *not of such-and-such a color*. Our ideas about explanatory dependence countenance no such dependencies. Hence, activating the eliminative propensity with respect to contemporary practice, there is a firm judgment that the white shoe does nothing to lend support to the generalization. The paradox of the ravens is resolved.

This resolution has a cost. I have had to concede that not all—perhaps even not many—positive instances of a generalization lend it any support. Moreover, the same discovery that could have supported a generalization had it been made earlier might be preempted and its evidential force undercut. Can these suggestions be reconciled with the apparent fact that scientists sometimes repeat experiments, sometimes look for instances of types that have already appeared, and sometimes continue to check predictions made by hypotheses all of whose rivals have been eliminated?

Eliminative induction is obvious in neo-Darwinian research. Yet there are recurrent complaints by antiadaptationists that important rival hypotheses are not explored, that the space of rival hypotheses *actually considered* is incomplete. More significantly, perhaps, there are evolutionary problems in which there are grounds for doubt that the rival hypotheses that can be generated from prior practice are complete. In considering the coloration of organisms, evolutionary biologists naturally think about camouflage, sexual displays, and thermoregulation, but they are rarely confident that these exhaust the possibilities. Perhaps there is something else going on, something that nobody has yet articulated.

Positive instances can sometimes play a role in eliminating what I shall call *residual doubt*. There are two different (but related) kinds of residual doubt. One is that there are instances that fall outside the scope of accepted schemata because those schemata are incomplete, so that reliance on the space of hypotheses generated from those schemata would omit crucial alternatives. A second is that the accepted schemata lump cases that ought to be separated, so that the cells of the partition to which they give rise are not really homogeneous after all. The former kind of concern lies at the bottom of the example about the selective value of coloration. Reflection on the history of evolutionary biology provides sobering reminders that rival hypotheses about selective value (or absence of selective value) are frequently not formulated, so that it is hard to be confident that the adaptive value of a style of coloration must stem from increased ability to evade predators, improved thermoregulation, enhanced chances for capturing prey or for attracting mates. The second worry attends generalizations whenever we have doubts that our inventory of potential perturbing conditions is complete. Under such circumstances, there are currently unstatable potential rival hypotheses which are eliminated in practice by supposing that the cells of our partition are homogeneous. Perhaps we have taken for granted the absence of an interaction.

Accumulating extra positive instances can help to resolve such residual doubts. Such instances do not eliminate any of the formulated rivals, but they do guard us against overlooking *unknown* rivals. Effectively, we recognize that prior practice may be incorrect or incomplete in certain respects, so that even a process of elimination which it countenances as complete may lead to an incorrect conclusion. Since we do not know how to state the unknown rivals, it is impossible to proceed systematically to eliminate them. We simply cast our net as widely as we can, and, by accumulating instances, attempt to remove the forms of residual doubt that stem from worries about prior practice. In many instances, past success of prior practice makes residual doubts about the completeness of our schemata small (for example, it is, I think, entirely reasonable for contemporary transmission geneticists to think that the schemata of contemporary genetics cover the vast majority of instances). In other cases (as with respect to many evolutionary problems) there is room for residual doubt and scientists reasonably multiply instances. Even here, however, there are some kinds of properties that can be dismissed as irrelevant since our background views about explanatory dependence give no credence to the notion that they could affect the presence or absence of the properties with which we are concerned. If we are concerned that some unknown property might have been neglected in an apparently successful eliminative induction to the conclusion that the molecular structure of common salt is that of a rectangular lattice, it is unlikely to be assuaged by trying to repeat some already performed experiment subject only to variation in the hair color of the experimenter!

By expanding $O(A)$ to include instances that do not eliminate any of the rivals generated from prior practice, extra support can accrue to a generalization. Does this affect the solution to the paradox of the ravens? No. A report of the presence of a white shoe would only support the hypothesis if there were rivals of a particular form, and, even though we might wonder whether our list of the factors relevant to the presence of bird coloration is complete, there is no basis for believing that some hypothesis of the pertinent form might be true. Our ideas about explanatory dependencies would have to be *radically* misguided for that to be the case. More formally, the legitimate worry about an apparently successful eliminative induction to the conclusion

$$(x) \ (Rx \supset Bx)$$

is that there should be some unknown property U which is not represented in our inventory of conditions. As a result we have not formulated the appropriate rival hypothesis

$$(x) \ (Rx \supset (Bx \equiv -Ux))$$

and, because the sample we have examined includes only objects that are $-U$, we have managed only a pseudoelimination. The remedy would be to discover an object that is R, B, U. Of course, since we have no idea what U is, we cannot do this systematically. We simply amass objects that are R, B, in the hope that *if* there is an unknown relevant property it will show up. Finding white shoes can play no part in this process.

Let me close this section by addressing a different question. Methodologists sometimes debate the relative virtues of prediction and accommodation. According to one tradition, hypotheses only gain support from the novel predictions they make; their ability to accommodate statements that were already accepted is irrelevant to their status. The rival view contends that prediction and accommodation should be equally valuable; the order in which information is acquired is epistemically irrelevant. Which of these is right?

Those who take a dim view of accommodation suggest that it is too easy to make up a hypothesis that will fit antecedently known statements. Because this game is so straightforward, no credit can accrue to those who achieve success at it. By contrast, presenting a hypothesis that is then able to predict something unexpected is a risky business, and victory here redounds to the credit of the hypothesis.

Opposing these ideas is the contention that evidence is evidence, irrespective of the order in which it is acquired. Why should it make a difference to the credentials of h if its consequences $e_1, \ldots, e_m$ were all known in advance or if some of them—say $e_{k+1}, \ldots, e_m$—were first discovered after h was proposed? (Practitioners of historical sciences—for example, evolutionary biologists—are likely to be sympathetic to this question.) After all, in both instances, the total set of statements available for assessing h at the end of the evaluation period is the same.

From the eliminativist perspective I have been developing, we can sustain this positive view of accommodation while explaining why the worries about accommodation might sometimes be well founded. Suppose that prior practice generates a set of rivals to h, $\{h_1, \ldots, h_n\}$. The critical issue is whether, in combination, the statements $e_1, \ldots, e_m$ eliminate all the h_i. If they do, then *in neither case* is there any basis for doubt about h other than the residual doubts about the completeness of prior practice. If they do not, then *in neither case* should h be accepted, for there is an outstanding uneliminated rival. Surprising novel predictions may be striking, but the touchstone of their evidential force is the ability to reduce the space of rivals sanctioned by prior practice. Beyond that, as I have noted, they help to allay residual doubts about the completeness of the space of recognized rival possibilities.

Accommodating the evidence is often *not* an easy game, because the constraints from prior practice are so powerful that they make difficult the genesis of even one hypothesis that will fit accepted findings (more on this later). However, when the constraints are lax or when confidence in the completeness of prior practice is (quite reasonably) low, there is room for doubt about hypotheses that accommodate accepted results. The problem in this case is not *accommodation* itself, but the state of prior practice. I conjecture that opponents of accommodation have been moved by examples in which the constraints from prior practice are lax, and there is no serious attempt to explore a space of rival hypotheses. That type of reasoning should be frowned upon, but the troubles should be traced to failure to activate the eliminative propensity, not to some supposed defect in the strategy of accommodation.[38] But, of course, surprising predictions are always welcome

---

38. The approach indicated here is, I believe, consonant with the views about prediction

when they can be had, for, in these cases, we discover that something that might so easily have subverted the constraints—leading to a position in which there was *no* available solution—is, in fact, in accord with the constraints. The result is that our confidence in the legitimacy of the constraints is increased, and, derivatively, residual doubts about the hypothesis (the single solution) diminish.

6. Underdetermination

For all my pleading on its behalf, the notion that individual scientists can profitably modify their practices by using eliminative induction is likely to seem (at best) quaint. Surely the idea that there are crucial experiments (or observations) in science was disposed of long ago, specifically by Pierre Duhem (1906). Thanks to Quine's revival and development of the Duhemian theme (1951, 1960, 1970), the notion that theories are inevitably underdetermined by experience has become a philosophical commonplace. Scientists, however, sometimes greet this allegedly mundane point with incredulity. "It's hard enough," they complain, "to find *one* way of accommodating experience, let alone many. And these supposed ways of modifying the network of beliefs are changes that no reasonable—sane?—person would make. There may be a *logical* point here, but it has little to do with science."[39] I think that the complaining scientists are, at least roughly, right. The underdetermination thesis, in its usual guise, is a product of the underrepresentation of scientific practice.

Duhem offered two arguments for the impossibility of using eliminative induction to justify a hypothesis (Duhem 1906/1953 183–187, 189–190, 211–212, 216–218). His first claim, less prominent than the underdetermination thesis but no less a staple of philosophical orthodoxy, is that any hypothesis has an infinite set of rivals. My response to this point is already prefigured in previous discussions. Against the background of prior practice there may be only a finite set of serious possibilities.[40] Consider the extinction of the dinosaurs. There are any number of *logically* possible hypotheses that could account for their demise: perhaps little green aliens exterminated them or

---

and accommodation advanced by John Worrall (1985). For Worrall and Elie Zahar, what matters is the evidence that is used in generating the hypothesis, and their concern is that, in my terms, all the constraints on the hypothesis should come from the evidence. Hence, I see the general idea that accommodation is unobjectionable when there are severe constraints imposed by background practice (and when background practice is, itself, not problematic) as articulating their insight.

39. Concerns along these lines were voiced by Stephen Jay Gould. In several conversations in the early 1980s, I tried to convince Gould that Duhemian holism spells doom for any idea of conclusive falsification. He resisted my arguments. I now think that he was correct.

40. Similar approaches to problems of underdetermination have been pursued by Laudan and Shapere, both of whom take reliance on background knowledge to be crucial (Laudan 1984, Shapere 1984). Obviously, much depends on the notion of a "serious possibility." Much of the rest of this chapter is devoted to explaining how such possibilities are generated and to illustrating my claims with detailed historical examples.

there were spontaneous combustions at all the right places to eliminate all the females. However, the *serious* hypotheses are relatively few in number, and paleontologists spend their efforts in trying to discriminate accounts in terms of meteoric impact from those that appeal to volcanic eruptions, atmospheric changes (increased oxygen tension), climatic changes, and a few other factors.[41] Why do these deserve greater attention? Because our prior practice recognizes certain kinds of processes as occurring in nature and not others. Of course, by abandoning that practice and adopting something different we could change the assessment of the serious possibilities. There is a potential modification of practice that would bring into consideration hypotheses that are presently outside the space of possibilities. As things stand, however, investigation of the demise of the dinosaurs goes forward under some stringent constraints.[42]

Even if the constraints do not manage to reduce the serious rivals to a finite number, it is still possible for the eliminative strategy to succeed. For the infinite set of hypotheses may be divisible into a finite number of subsets, subject to wholesale elimination. (The subsets may, for example, correspond to ranges of values for continuously varying parameters, and one may be able to show that the parameters do not fall into some of the ranges. See (Earman 1992) for a convincing illustration). To oppose eliminative induction on Duhem's first grounds, it is necessary to suppose that attempts to delineate a space of rivals encounter systematic failure. This can only be done by emphasizing the revisability of the constraints imposed by prior practice. The Duhemian point must eventually come down to the thesis that we have a choice of infinitely many rival hypotheses because it is always possible to introduce new—hitherto "nonserious" —rivals by giving up parts of our practice. This, of course, is simply a version of the underdetermination thesis. Duhem's first challenge to eliminative induction presupposes the success of the second.

That second challenge is often seen as deadly. Hypotheses, Duhem claimed, are never tested in isolation (1906/1953 187). Any attempt to confront a hypothesis with empirical results takes for granted auxiliary hypotheses that may be revised if things go wrong. So the testing situation is seen as involving three statements H (the hypothesis taken to be under test), $-O$ (a report accepted on the basis of observation), and A (a conjunction of auxiliary statements that are required for the derivation of observationally ascertainable

41. To see how the practitioners view the debate, see *Scientific American* (1990), no. 9 (September).

42. This point about scientific practice is readily accommodated in some accounts of science. Thus, Kuhn's conception of "normal science" allows for determinate resolution of issues because of the constraining role of background practice (the "paradigm"). Making this comparison raises the worry that my critique of underdetermination will be flawed because of the possibility of breaking free from the constraints in a scientific revolution. This is a serious challenge, and I shall try to address it later by showing how, even in such large transitions in the history of science as the chemical revolution of the eighteenth century and the acceptance of minimal Darwinism in the nineteenth, underdetermination can be resisted.

statements from *H*). Since *H,A⊢O*, the set {*H, A, −O*} is inconsistent. Something has to give. But what is to be amended is, allegedly, not fixed. Scientists have the options of (i) abandoning *H*, (ii) abandoning or modifying *A*, (iii) refusing to accept −*O*. Because any of the options can be exercised, there are any number of systems of beliefs that scientists can square with their total experiences. In particular, since *H* can always be retained *no matter what H may be* there is an infinite number of total theories each compatible with our total experience. (To construct the infinite set of theories, simply form a denumerable infinity of distinct statements and embed each of them in a total theory.)

The natural response to this point is to contend that not all of this panorama of possible theories, and not all the responses to experience that underlie them, are equally good.[43] Consider the strategy, explicitly canvassed by Quine in a famous passage (1951 43), of "pleading hallucination" in the face of disconfirming observations. Sometimes pleas about the unreliability of observation are in order—considerations drawn from prior practice counsel that the current observational situation should be viewed with suspicion. But imagine that, in the case at hand, the observational rules of prior practice insist that the current situation is excellent. If those parts of practice are given no power to constrain the current options, then the core of the underdetermination thesis is the following logical truism: any consistent statement can be embedded in a consistent set of statements. The thesis adds to this the simple idea that some of the statements in this set express "rules" (carefully chosen) for evaluating observational situations.

Nobody ought to be astonished by the suggestion that there are numerous ways of forming sets of beliefs if there are no constraints imposed by observation, history, or anything else (save logic—and, of course, Quine urges the possibility of revising logic as well [1951 43]). Descartes's program for epistemology failed, and the basic point made by the underdetermination thesis records that failure. However, there is a more interesting version of the doctrine, at which Quine sometimes hints,[44] according to which there will be alternative ways of revising beliefs even when all the constraints imposed by methodology are given their due.

To evaluate this doctrine, we first need to formulate it. Let a *Duhemian situation* (or *predicament*) be an occasion on which scientists accept a hypothesis *H*, accept certain auxiliary assumptions *A*, where *H,A⊢O*, and where, following their normal observational procedures, they would accept the statement −*O* on the presentation of a stimulus to which they are actually exposed. I shall suppose that the methodological maxims of accepted scientific practice

43. Again, my approach to this issue is akin to those of Laudan and Shapere. All three of us recognize grounds for distinguishing systems of belief beyond mere consistency with "the evidence"—although we formulate the distinctions rather differently.

44. The hints are scattered throughout (Quine 1960) in those passages in which he insists that scientific theories are underdetermined, even when we imagine "an ideal organon of scientific method." For discussion of Quine's oscillations on the question of whether any such organon is to be considered in evaluating underdetermination, see (Laudan 1990).

include rules that direct scientists to suspend their observational judgments when certain conditions are found. Otherwise, they are to endorse the statements that the presented stimuli incline them to believe. Part of prior practice consists in rules that may be summarized as follows:

> (OR) Endorse the PI belief that p unless one of the defeating conditions $D_1, \ldots, D_q$ is found to be present. Otherwise, suspend judgment unless there are independent causal routes to belief that p.

Here, I shall suppose that the PI belief may be triggered by interactions with some complex piece of apparatus, so that the defeating conditions embody views about what would subvert the workings of the apparatus. Of course, doubts about the correctness of these views can always be expressed by retreating to a more primitive observational idiom (see §2) and assigning the claims about the functioning of the apparatus to the set of auxiliary assumptions.

With respect to each of the auxiliary assumptions conjoined in A, there are (i) chains of reasoning that appear to support that assumption and (ii) instantiations of schemata in which that assumption figures as an essential part of a prediction or an explanation. The *costs* of a proposed way of amending the Duhemian situation by dropping a component of the auxiliary assumptions (*possibly* replacing it with something else) are measured by considering (i) the number of supporting chains of reasoning for the abandoned assumption for which no error can be diagnosed and (ii) the number of apparently acceptable explanations and predictions that have to be given up by dropping the assumption. I shall not suppose that there is any unique way of counting the costs, but, in accordance with the ideas of Chapter 6, I shall assume that there is a class of *admissible cost functions*. A Duhemian situation counts as a crucial falsification of H just in case (OR) requires the acceptance of $-O$ and, for any admissible cost function, the costs of any amendment to the auxiliary assumptions A are greater than the costs of abandoning H.

Interesting underdetermination can arise if, for the same admissible cost functions, there are equally good alternative ways of amending $\{H, A, -O\}$ to make it consistent or if there are different admissible cost functions that single out rival emendations as optimal from the point of view of minimizing costs. Quine's version of the underdetermination thesis seems to be grounded in the former kind of consideration.[45] The possibility that underdetermination

45. The obvious motivation for believing that rival modifications are equivalent with respect to the same cost functions comes from studies of the conventionality of geometry (Poincaré, Reichenbach) and of the conventionality of simultaneity (Reichenbach). In such examples, one is able to demonstrate the existence of rivals which are deductively equivalent with respect to what is singled out as the class of observational statements. Provided that (i) these observational reports exhaust the total evidence and (ii) the only admissible cost functions count theories (practices) as equivalent when they are deductively equivalent, then the cases of conventionality in space-time theories demonstrate the possibility of Duhemian underdetermination.

might arise from alternative ways of measuring costs of changes recalls the ideas about shifting standards that have been emphasized by Kuhn and some of his followers (see Kuhn 1962/1970, 1977 essay 13, Doppelt 1977). I shall consider the possibilities in turn.

The simple reason for believing that there are many ways of accommodating recalcitrant observations with minimum mutilations is that troublesome cases can be treated as local exceptions. If I want to escape the $\{H, A, -O\}$ situation, why cannot I make a small amendment in (OR) to add a defeating condition that will apply just to the special circumstances that generate the belief that $-O$? Or why can I not circumscribe some generalization in A so that its new range is just like its old except for not including the particular instance to which it was applied on this occasion? Such solutions by gerrymandering only appear cost-free so long as one overlooks the fact that observational rules and auxiliary generalizations are constrained by the explanatory dependencies and the views of projectibility present in prior practice. A claim that there is a local error in investigating things of a very special type needs to be integrated with our background ideas about the potential sources of error. That integration cannot succeed if our explanatory categories provide no way of distinguishing the special type from a much broader class in which the mode of investigation is to be endorsed as reliable. Faced with the contention that we simply cannot trust our observation to record the properties of the shadows of rapidly rotating small circular objects, the natural riposte is, What's so special about this observational situation? By contrast, the insistence of some Aristotelians that the telescope could not reveal the character of the immutable heavens was grounded in distinctions that were employed in their explanatory schemata.[46]

Once the constraining character of prior practice is understood, there is no quick argument for the widespread presence of underdetermination on Quinean grounds.[47] The harder task is to defend against the possibility of underdetermination that results from rival systems of counting costs. Kuhn and others have rightly insisted that champions of rival points of view might assign different significance to different questions. *In principle,* this can generate occasions on which the relative merits of strategies for escaping Duhemian situations are oppositely assessed because one preserves solutions valued by one side while the other continues to endorse the solutions valued by the other. Moreover, I agree that such occasions sometimes arise in the

46. Thus, in the dispute over the veridicality of telescopic observations of the heavens, the legitimacy of the celestial/terrestrial distinction became a matter for debate. As I noted earlier, Galileo's defense of the telescope rested *in part* on undermining the distinction through appeal to such "naked-eye" phenomena as the novae.

47. In the end, everything will have to turn on close examination of individual cases. Hence I regard the brief discussion the triumph of Copernicanism in this section, and, even more, the extended review of the transitions to Darwinism and the "new chemistry" of Lavoisier, as crucial parts of a *philosophical* argument. Although each of these examples has been discussed by many commentators, it seems to me important to combine a look at the details with the epistemological perspective that I favor.

history of the sciences and that debate is sometimes *temporarily* inconclusive. The compromise model of the last chapter explicitly allowed for this, while maintaining that such controversies are ultimately resolved through the provision by the victor of successes in the terms demanded by the rival.

Faced with claims about rival admissible cost functions, we should look at the costs on both proposals for emendation and ask whether there is a coherent conception of scientific significance that underwrites them. For many writers influenced by Kuhn (for example, Doppelt 1977) it seems that the recognition of the possibility of counting costs in various ways leaves no objective grounds for appraisal of rival responses to a Duhemian predicament. Rash judgments can be tempered by reminding ourselves that there are instances in which preferring one response requires dismissing the significance of a large number of predictions and explanations and taking debate to revolve solely around one difficulty. Making that evaluation presupposes a conception of scientific significance that must itself be defended.

Let me hammer the point home with an analogy. Most of us, at some stage of our lives, have bought a second-hand (used—or, as the upmarket advertising has it, "preowned") car. Buying decisions typically involve traipsing around various lots and comparing the virtues of a number of possible vehicles. Few people would be confident that they have ended up making the best decision. Most would surely recognize that there are competing advantages and disadvantages, and that the final decision is made by weighing some factors more heavily than others. Rival systems of counting pluses and minuses might well have generated different decisions, and some of those decisions might have brought the agents closer to satisfying their preferences. I shall suppose that those preferences involve wanting a car that will transport people from place to place, not wanting a car that demands frequent repairs, wanting a car that will last for a while, and so forth. While acknowledging the existence of rival schemes of accounting that would yield different *sensible* decisions, we should not overlook the fact that some such systems are ludicrous. Faced with two cars, one with a functioning engine, transmission, good tires, and the other without engine, transmission, or wheels, we tend to prefer the former. That tendency would not be overridden for decision makers who were reasoning well, if the sole virtue of the second car were the fact that its front right bumper had less rust than the front right bumper of its rival. There is, of course, a possible scheme of valuation in which presence of rust on the front right bumper is accorded overwhelming priority. But that scheme cannot be backed with any coherent conception of significant factors.[48]

Let us now turn to a problematic example, the debates about Copernicanism between 1543 and 1632. As I pointed out in the last chapter, there

---

48. As the referee insightfully pointed out to me, there is a way of bringing my approach within the Bayesian framework. Instead of thinking in terms of probabilistic conditionalization, one can regard scientists as facing Bayesian decision problems in which the values of various alternatives are set by just the types of considerations to which I would appeal in "counting the costs." I shall leave it to Bayesians to decide exactly how much of what I claim can be accommodated within a natural extension of their views. But see §10.1.

were a number of "grades of involvement" with the doctrines of *De Revolutionibus* that could be—and were—adopted by Copernicus's successors: one could use heliocentrism for the purpose of astronomical calculations, espouse heliocentrism as a true account of the motions, seek to embed heliocentrism in an explanation of the motions. For present purposes, I shall focus primarily on the second of these attitudes and ask what kinds of estimates of relative significance would underlie acceptance of or opposition to the basic Copernican thesis that the planets (including the Earth) move around the sun.

Throughout the midsixteenth century, there was no system of astronomical motions that was as accurate as Copernicus's. This was acknowledged from 1543 on, and, in consequence, the mathematical representations of *De Revolutionibus* were bound to be used. Nonetheless, it was not difficult for the geometrically sophisticated to see that any heliocentric system could be transformed into an equivalent geocentric system. For post-Copernican astronomers, therefore, whether or not they *used* the new system for making calculations, there was a serious question as to whether a heliocentric or a geocentric account should be preferred *as the truth about the motions.*

I shall not try to provide an exhaustive list of the costs and benefits which were assigned to different lines of response to the astronomical observations. Three main kinds of consideration stand out for the sixteenth century astronomers:

(1) The incompatibility of the thesis that the Earth moves with accepted physical theory and with arguments from commonsense observations (worries about birds, cannonshots, clouds, and falling bodies that Oresme had addressed in some detail in the fourteenth century; that Copernicus sought to downplay by offering a precis of Oresme; and that Galileo would eventually tackle with unprecedented thoroughness)

(2) The failure to observe stellar parallax, which constituted a prima facie objection to heliocentrism

(3) The possibility of dispensing with some apparently arbitrary assumptions in geocentric systems: for example, Ptolemaic astronomy had been forced to differentiate those planets which reveal only limited elongation from the sun (Mercury, Venus) from those whose elongation from the sun is unlimited (Mars, Jupiter, Saturn) by supposing that the motions of the deferents on which the former planets revolve are tied to the motion of the sun.[49]

Those (few) sixteenth century astronomers who believed that heliocentrism is true gave far greater weight to (3) than to (1) and (2).

On what grounds might one defend relative estimates of the importance of (1), (2), and (3)? Those who emphasized (1) did so by defending the available (Aristotelian) account of local motion and criticizing the counter-

---

49. This is one example—although the most striking—of the "mathematical harmonies" on which Copernicus set such store. For others, see (Margolis 1987).

arguments of Oresme (and, derivatively, Copernicus). Others, acknowledging that incompatibility with a *well established* body of results would diminish the credibility of a novel doctrine, contended that, to the extent that the alleged anti-Copernican findings were well-established they were not incompatible with heliocentrism (the commonplace observations of birds and clouds overhead could be reconciled with the idea of a moving Earth along lines explored by Oresme or by a rethinking of ideas about local motion), while, to the extent that they were genuinely incompatible with heliocentrism, the results were not well established. Thus, different emphases on (1) resulted from more fundamental differences in evaluating the successes of Aristotelian accounts of local motion and the prospects for producing a novel treatment that would be consistent with heliocentrism. The few who advocated heliocentrism saw Aristotelian physics as beset with difficulties and pinned their hopes on the possibility of replacing it with something better.

With respect to (2), the issues are far more sharply defined. Initially, it appears that from the hypothesis that the earth revolves around the sun (H) and some parts of elementary trigonometry one may conclude that observations of fixed stars should display stellar parallax (O). Since stellar parallax was not observed—and was not observable using sixteenth (or seventeenth or eighteenth) century instruments—the rules for forming PI beliefs dictate acceptance of $-O$, so that there is an apparently conclusive refutation of H. Copernicans, however, could—and did—reply that the auxiliary hypotheses needed for deriving O from H must include not only the elementary trigonometry (which is hard to deny) but also a much more controversial claim about the distance to the fixed stars. Could this claim be jettisoned without cost? Apparently so, for there were no antecedent arguments for thinking that the dimensions of the universe must be confined to any particular size and there were no explanations and predictions in which the size of the universe played a crucial role.[50] Hence, in this instance, the falsification of heliocentrism could be—and was—dismissed as inconclusive, so that (2) lent no weight to the case against Copernicus.

Just as Copernicans placed little weight on (1), so they stressed (3). From the central assumptions of Ptolemaic astronomy—for example, the claim that heavenly bodies describe circular orbits around the Earth, ideas about planetary ordering—one cannot derive the conclusion that Mercury and Venus will display limited elongation from the Sun while Mars, Jupiter, and Saturn will not. Ptolemaic systems have to add an assumption to the effect that the deferents of Mercury and Venus are linked to that of the Sun, but there is no deeper explanation of why this should be. Within a heliocentric system, by contrast, the central assumptions about planetary motions and planetary ordering immediately generate the conclusion that Mercury and Venus, alone

50. Despite contemporary unease at the vastness of the Copernican universe, there was no explanation of why the distance between the outermost planet and the sphere of fixed stars should take on some specific value, large or small. The collapse of cosmological hypotheses about crystalline shells made any Ptolemaic claim about the proximity of the farthest planet to the fixed stars a brute fact.

among the planets, should display limited elongation from the Sun. Copernican astronomy does not have to invoke any curious brute fact about the relations among deferents.

The principle of unification, briefly introduced in Chapter 4, plays an important methodological role here. Heliocentric systems reduce the number of facts one has to accept as brute. In principle, geocentric astronomy could make up the loss by embedding the geometrical account of planetary motions within a physical system that would show why the deferents of certain heavenly bodies are linked together. But, in the sixteenth century, this type of physical explanation was nowhere in sight, and as the search for the causes of the motions of the heavenly bodies was undertaken in the seventeenth century, heliocentrism rather than geocentrism received the epistemic dividends.[51]

Seventeenth century astronomy would enlarge the differences of costs and benefits by adding further phenomena—notably the phases of Venus—which could be easily derived through central Copernican assumptions but which were problematic for Ptolemaic astronomers. These findings, together with the increased accuracy of Kepler's elliptical orbits, Galileo's rebuttals of physical objections to Copernicanism, and the provision of embryonic dynamic accounts of planetary motions, shifted the balance of costs and benefits, so that on any choice of admissible accounting functions Copernicanism would emerge as superior. Yet, for the sixteenth century and the opening decades of the seventeenth century, it is easy to see how reasonable people could make alternative assessments of costs and benefits, some emphasizing (1) and siding with Ptolemy, others stressing (3) and following Copernicus. This judgment is, of course, in line with the compromise model of scientific change, offered in the last chapter.

Underdetermination does occur in science, but, as I suggested in Chapter 6, major controversies are only resolved when the sources of underdetermination have been addressed, when one practice has shown itself able to succeed in the other's terms. Individual scientists may have to make choices in the meantime. They do so by evaluating the costs of the formulated ways of escaping from Duhemian situations, thinking hard about what options have

---

51. I would suggest that one of the failures of the Tychonic system to appeal to some mathematicians and astronomers lay in its inability to furnish any clues to a physical explanation of the motions of the planets. By the turn of the seventeenth century, there are clear indications that two ideas about celestial influence are gaining currency: large heavenly bodies have more influence than smaller ones, and bodies have greater influence on closer neighbors. Versions of these ideas are evident in William Gilbert, in Kepler, and in Galileo. Against the background of these dynamic hypotheses, the Tychonic system appears hopeless. For it is a mystery how the Sun can both be subject to the influence of the Earth and able to influence bodies that are sometimes closer to the Earth than to it. Given both the vague dynamic ideas, the rotation of the Sun around the Earth could be explained if the Earth were larger than the Sun, but this would yield trouble for the idea that Mars, Jupiter, and Saturn (and, on some versions, Venus), which are all sometimes closer to the Earth than the Sun, rotate around the Sun rather than around the larger Earth. So far as I know, detailed historical investigation of the role of embryonic dynamical hypotheses in the rejection of the Tychonic system has not yet been undertaken.

been explored and what possibilities remain open. They make their delib-
erations by relying on parts of prior practice. After all, what else do they
have to go on?

Yet practice can itself be modified. The constraints that favor some lines
of solution and eliminate others can themselves be changed. Our next task
is to investigate how this is possible.

7. Adjusting Constraints

For any individual practice and any set of inconsistent statements that a
proponent of that practice is inclined to accept, we can recognize what I shall
call the *escape tree*. A path through the tree is determined in the following
way: Start with the set of inconsistent statements. Pick any one of them.
Consider the consequences of deleting this statement from the set and represent all the losses that would be incurred on the branch. A *response* to these
losses consists in the addition of further statements that (i) are consistent with
the remaining members of the original set, (ii) compensate for the losses, and
(iii) conform to the constraints derived from practice. A response is *threatened*
if there is some set of statements that proponents of the practice are inclined
to accept that are inconsistent with the newly added statements. In such cases,
the branch can be continued by attempting to remove the inconsistency from
the new set. If the scientist(s) can think of no way of continuing then the
path is *blocked*. On the other hand, if at any stage of the branch a response
is unthreatened, then the branch offers an escape from the original predicament and the problem of removing the inconsistency has been solved.

Escape trees presuppose constraints. When do scientists begin to consider
modifying the constraints? Such situations start with an inconsistency whose
escape tree has the following characteristics:

(a) every path through it is blocked.
(b) the scientist believes that it represents the range of possibilities allowed
    by the constraints.
(c) the scientist assesses the costs of the least costly path as severe.

Under these circumstances it is reasonable for a scientist to explore ways of
modifying the constraints and to pursue practices whose languages, schemata,
and derivative conceptions of significance are different, even when those
practices have not yet achieved the full range of solutions of the displaced
practice. The judgments that are made in these cases depend on weighing
the possibility that some available line of solution has been overlooked against
the chances of abandoning some of the successes of traditional practice.

The eventual standard for evaluating new systems of constraints is the
ability to find substitutes for explanations and predictions generated within
the old practice or to explain why apparently satisfactory explanations and
predictions were not genuinely adequate, to avoid the inconsistencies with

blocked escape trees, and to furnish a unified system of explanatory schemata. The principle of unification plays an important role here.

As I remarked earlier, apropos of the suggestion that recalcitrant observations can be ignored by pleading hallucination, the strategy of claiming local exceptions has epistemic costs. In similar fashion, there are cognitive losses in contending that explanations that have been abandoned in responding to a Duhemian predicament can be made up by lifting the requirement that explanations must fit the schemata of the practice. Breaking with old views of explanatory dependencies, codified in the schemata of prior practice, requires offering a new set of schemata (possibly only in tentative outline) whose unifying power is alleged to be superior to that of the set replaced. Recall that unifying power is to be assessed in terms of the paucity of the patterns employed, their stringency, and the breadth of the set of consequences they generate. Merely tacking on a new, unrelated, pattern to cope with a single recalcitrant phenomenon would produce a decrease in unifying power, and should thus be rejected. By contrast, shifts in views of explanatory dependencies often come about by finding a perspective from which previously successful schemata are viewed as special subpatterns of a more generally applicable pattern.[52]

Consider, in more detail, how a Duhemian predicament might lead to the reform of constraints. Suppose we have a Duhemian predicament that is insoluble against the background of the accepted language and schemata. Divide the possible lines of escape into two main types: rejecting some statement (or set of statements) that one is inclined to adopt on the basis of observation or abandoning a more general statement that has been used successfully elsewhere in explanation and/or prediction. The costs of the former strategy will be determined by the available views about the processes of observation pertinent to the acceptance of the statement. In some instances, the problem can be shifted by identifying a hitherto unrecognized condition for accurate observation *which can be introduced in accordance with prevailing views about explanatory dependencies*. A successful escape along this branch thus involves introducing a condition that background explanatory dependencies sanction as subversive of the process of observation involved in acceptance of the recalcitrant statement(s) and showing that the condition is present on the occasions that give rise to the predicament. When potential lines of escape that shift the problem to background ideas about observational processes have been explored without success, scientists regard the escape tree as blocked, at least in this direction. Sometimes the process of canvassing options is a relatively simple matter, as for example in naked-eye observations of objects in good view and in good lighting. Quite frequently, however, with the use of various pieces of apparatus, there are various options that can be

52. The weak realism described in §8 of Chapter 5 provides a ready explanation of the status of the unification principle, but, as admitted there, weak realism has a much more complex task of reconstructing talk of kinds and causes. Here, as elsewhere in philosophy, epistemological advantages are purchased at metaphysical cost.

checked—familiar interfering conditions that have to be eliminated and more recondite possibilities that cautious scientists consider—and the process of exploration is far more lengthy.

The alternative strategy is to eliminate some claim that does apparent explanatory or predictive work in other instances and attempt to make good the losses involved. So, for example, one might try to abandon a hypothesis *H* that had been used in providing an explanation of some phenomenon *P,* an explanation in accordance with the accepted schemata. Thus the problem is shifted to explaining *P* in some way concordant with accepted schemata. But it may be relatively easy to show that, given beliefs that are invoked in a vast range of apparently successful explanations, *H* is the only possibility for explaining *P* in accordance with the schemata. Or there may be some possibility of undermining the argument for the uniqueness of *H* as a component of an explanation of *P,* by identifying some relatively unemployed premise in that argument, *H'*. Abandoning *H'* would only cost us the explanations of *P'* and *P''*, and perhaps we could make these up in accordance with the prevalent explanatory schemata. So the escape tree ramifies further as the explanation of *P'* and *P''* now becomes a problem with increased significance. Yet exploration of this possibility too may fail, so that the tree remains blocked.

The elements of scientific decision making are, I contend, quite simple. A scientist struggles to eradicate inconsistencies, maintain a unified account of the phenomena—conceived in terms of the background repertoire of concepts and explanatory schemata—and project regularities in accordance with the views about representative samples and projectible predicates embodied in his practice (views which are, as I have suggested, dependent on ideas about relevant causal factors and, thus, on the accepted explanatory schemata of the practice). These elements are far simpler than those typically introduced in theories of confirmation. I have invoked no intricate logical machinery for computing degrees of confirmation, no precise assignments of probability to hypotheses or processes of conditionalization on experiential input.[53] Yet, as should be clear from the discussion of the preceding paragraphs, the activity of scientific decision making is anything but simple. For the elementary processes are combined in elaborate ways as a scientist seeks a path through an escape tree. A central problem of scientific inference consists in showing how

---

53. For thorough expositions of approaches to confirmation which involve logically or mathematically more recondite principles, see (Carnap 1951), (Hintikka 1965), (Rosenkrantz 1977), (Horwich 1982), (Howson and Urbach 1989), (Earman 1992), and (Glymour 1980). As I shall suggest in the final section of this chapter, there are ways of relating such more sophisticated views of confirmation to the simpler (naive?) picture of scientific decision making which I offer here. For the moment, it suffices to note that Glymour's account, with its emphasis on the use of background beliefs, is far closer to my own suggestions, and, in particular, I would endorse many of the points that Glymour makes in using the apparatus of bootstrap testing to reconstruct the inferences made by scientists in historical episodes (see chapter V of Glymour 1980).

endeavors to find paths through escape trees yield various types of modifications of practice.

Consider some of the kinds of changes that may be produced by efforts to follow a path through an escape tree. First, there may be modifications of the significance of questions. Suppose, for example, that the attempt to evade a Duhemian predicament shifts the initial problem to that of finding an alternative explanation for the phenomenon P. This may not only dramatically increase the significance of the question, Why $P?$, but also bestow derivative significance on other questions, answers to which would contribute to the provision of an explanation of P. So, for example, if a potential explanation of P is ruled out because it would seem to require values of a particular quantity that are too high, and if the scientist suspects that the apparatus used in the past to measure values of the quantity is not as reliable as has been assumed, then she may come to take as significant the question of how that quantity can be determined more reliably. Alterations of the perceived significance of a question can cascade throughout the appraisal of other problems and can affect *actions* taken with respect to instruments and apparatus.

Revised estimates of the significance of a question can also be the direct product of conceptual reform. The problems scientists set for themselves depend on their conceptions of the boundaries of kinds. Exploration of the idea that a single predicate covers two quite different types of entity will typically lead to the rejection of general questions involving the predicate in favor of more narrowly specified problems. But how is conceptual reform itself initiated?

Distinguish three important ways in which scientists may be led either to introduce new language or to adjust the reference potentials of old terms. (I do not claim that these are the only such ways or that all others are less important.) First, one may try to respond to a Duhemian predicament by finding a generalization about a set of entities. The available information about these entities may suggest that their instantiation of determinate forms of a particular determinable property is sufficiently variable and confused to suggest the possibility that distinctions should be made among them, even though the scientist does not yet see how to draw such distinctions in terms allowed by the available explanatory schemata. This kind of conceptual revision may precede modification or further articulation of the explanatory schemata to underwrite new distinctions, but, in accordance with my earlier remarks about the loss of unification produced by simply declaring exceptional cases, it will ultimately prove important to integrate the distinctions with a set of explanatory schemata. A second style of conceptual reform is generated by drawing distinctions that are motivated by prevailing explanatory schemata: one distinguishes those A's that are C from those that are not, adducing some previously unrecognized condition C that exemplifies a type which accepted schemata view as relevant to the behavior of entities like A. Third, and perhaps most straightforwardly, conceptual change can be fuelled by modifications of explanatory schemata. As we change our views about which kinds of characteristics are dependent on which, we simultaneously alter our visions of the boundaries of natural kinds.

Changes in the accepted set of explanatory schemata can themselves be generated in several ways. Sometimes, the view that particular phenomena are dependent on others results from a straightforward extension of other beliefs about explanatory dependencies (I shall offer an illustration of this in the discussion of Lavoisier in §9). However, the most obvious way in which Duhemian predicaments lead to the modification of explanatory schemata is that to which I have alluded several times in earlier discussions. When efforts to find a path through an escape tree consistently founder because there are, apparently, no available explanations that will meet the conditions required by prevailing practice, then, after more or less effort to force the phenomena into the schemata, a scientist will eventually explore the possibility of amending the schemata.[54] As I have emphasized, the challenge here is to do so without forfeiting unifying power.

My aim in canvassing these possibilities is a relatively modest one.[55] I am concerned to show how it is possible to adjust the constraints that are brought to a scientist's inferential practice, not to offer a complete inventory of the possibilities nor to identify precisely the forms of the modifications in practice that occur in particular instances.

Nevertheless, the account I have offered is incomplete in another important respect. Because I have introduced the process of constraint adjustment by beginning with the problem of underdetermination, an important type of predicament has been overlooked. Duhemian predicaments have escape trees, and searches through those trees often generate what may reasonably be called *open problems*. Attempting to abandon some hypothesis that seems to be involved in an inconsistency, one needs a substitute that will do the lost explanatory work. Just as the escape tree extends beyond the point at which one contemplates possible surrogate explanations, so too, *starting from* the open problem of how to find an explanation of *P,* where *P* is a phenomenon that is judged akin to those covered by accepted schemata, it is possible to

54. My language here is deliberately reminiscent of that employed by Kuhn. As I shall suggest later in the text, my discussion of escape trees, of strategies for finding paths through such trees, and of blocked escape trees is intended to capture some of his insights about the transformation of puzzles into anomalies. However, the idiom that I have chosen enables me to do so without any commitment to separating episodes of normal science from episodes of revolutionary science. The modifications induced by searches through escape trees can be large or small along several different, independent dimensions (the dimensions indicated in my account of individual practices). Thus there seems to me no need to draw the hard-and-fast lines that have bedeviled Kuhnian historiography.

55. In a future study I hope to explore the possibility of mapping various strategies of changing individual practice. For the present I am concerned simply to show the possibility of objectivity by identifying clear methodological exemplars. It is possible that this is as far as the methodologist can go: demonstrating clearly in concrete instances that certain modifications would prove too costly (see, in particular, §§8–9). But it is surely worth investigating to see if the instances can be subsumed under general principles that constrain the class of admissible cost functions.

generate a search through a tree structure (perhaps the label "escape tree" is no longer apposite here, but I shall preserve it because of the idea that, in such instances, what is sought is escape from explanatory ignorance). Duhemian predicaments signal failure to achieve *one* of our scientific goals, the attainment of true belief. The predicaments corresponding to open problems record difficulties with the other dimension of cognitive value, showing that our account of the structure of nature is, as yet, incomplete.

Open problems frequently arise in our efforts to escape Duhemian predicaments. By the same token, responses to open problems often spawn Duhemian predicaments. But the notion of a Duhemian predicament is simply an instance of a more general type of predicament, inconsistency among things one is inclined to believe, and it is this more general type of predicament that is the true counterpart of the open problem. Efforts at avoiding inconsistency often yield open problems, just as efforts to solve open problems often bring inconsistency (which the aspiring solver hopes to be temporary). Given the bipolar nature of the account of cognitive goals developed in Chapter 4 (significant truth, where significance derives from the provision of a comprehensive explanatory structure), this relationship is hardly surprising.

I shall close this section with some brief remarks intended to link my discussion to the proposals of others and to themes that have been prominent in recent philosophy of science. First, a point that emerges directly from the symmetry between open problems and Duhemian predicaments (or, better, general predicaments in which individual scientists are threatened by inconsistency): Philosophers since Reichenbach have often wanted to separate the context of discovery from the context of justification. They are surely correct in recognizing that the processes which initially engender belief (or some other form of cognitive commitment) may not warrant any such attitude. Moreover, there is a legitimate enterprise of elucidating the structure of the reasoning that is available to support modifications of practice that have already been made, an enterprise that can abstract from the idiosyncrasies that led an individual initially to make those changes. On the other side, it is equally important to recognize that, on some occasions at least, the processes that engender shifts in practice warrant the individual in making these shifts. Moreover, in elucidating the structure of that individual's reasoning *or in retrospectively assessing the epistemic status of the shift,* it may be necessary to take into account considerations about the possibilities that particular types of discoveries will ensue.

So, for example, a modified practice may allegedly be valuable because it responds to a prior inconsistency. Among the costs of modification may be the loss of statements that were previously employed in giving apparently successful explanations of various phenomena, so that, from the point of view of the new practice, there are significant questions which its predecessor could apparently tackle but which are now viewed as open problems. In assessing the credentials of the shift, it is thus necessary to appraise the possibilities that solutions will be found. Constructing an argument for the new practice requires exploring possibilities for *discovering* answers to open problems. By the same token, scientists actively involved in trying to discover solutions to open problems frequently generate what they hope will be temporary inconsistencies, doing so by outlining possible lines of escape that they intend to explore in more detail later. So, from the perspective I have offered, it is easy to understand the senses in which Reichenbach's celebrated distinction is important and correct and those which vindicate the criticisms that it has been superseded.

My outline account of constraint-adjustment reflects in obvious ways points that have been made by Kuhn and Laudan. Aficionados of Kuhnian language might insist that puzzles are those predicaments (Duhemian predicaments, more generalized inconsistencies, open problems) which are resolvable by finding paths through escape trees that do not call for modification of constraints; puzzles turn into anomalies as a scientist recognizes that all paths through the tree are blocked, and so begins to tinker with the constraints. There is obviously a relationship between my language and Kuhn's, but I would prefer to avoid the puzzle/anomaly distinction (and its kindred dichotomy of normal science/revolutionary science). Identifying a path through an escape tree can involve small modifications of accepted beliefs or large changes. Recognizing a blocked tree and adjusting constraints may produce small changes in language, assignments of significance, or schemata or quite sweeping ones. The concept of practice allows for the magnitudes of changes to vary along several independent dimensions, and I am reluctant to see any one of the dimensions as privileged in that modifications along it produce some special form of scientific change. Within my framework we can appreciate two polar types of case. Open problems or inconsistency predicaments are sometimes resolved by introducing or abandoning some small number of statements, leaving one's language, schemata, questions, and the body of accepted statements relatively unaltered. Occasionally, changes of practice are quite sweeping, and the response to blocked escape trees entails modifications of practice along all dimensions, including a large-scale replacement of previously accepted statements by new statements that are formulated in an apparently more adequate idiom. If we attach the label "normal science" to one pole, and "revolution" to the other, little harm will be done—provided that we do not forget that there are all kinds of other modifications of individual practice for which we should allow.[56]

Similarly, there are obvious affinities between my account and Laudan's talk of problem solving. Like Laudan, I believe that it is important to recognize not only *empirical* problems, corresponding to Duhemian predicaments and some open problems, but also *conceptual* problems, corresponding to other kinds of inconsistency predicaments and other types of open problems. Finding a path through an escape tree is my counterpart to Laudan's general

56. Strictly speaking, one ought to reserve the labels for similar types of modifications of *consensus practice*. Nonetheless, it is an excusable extension of the concepts of normal and revolutionary science (and one which Kuhn himself employs in discussing the particular changes in belief of great figures, such as Copernicus and Lavoisier) to attach those concepts to the modifications of individual practice that spark the shift in consensus practice.

notion of solving a problem, and here, I believe, my language has representational virtues. For, in Laudan's discussions (Laudan 1977), very little is said about what counts as an apparent solution to a problem (presumably a *genuine* answer to a problem is a true sentence that answers the associated question), and, in consequence, it is not easy to see how the problem-solving powers of various scientific alternatives are assessed. Can one claim to have solved a problem (in Laudan's sense) by saying anything at all, no matter what its implications may be for other beliefs that one holds? Presumably not, but nothing in Laudan's treatment makes clear the constraints on admissible problem solutions. Recognizing the need to find open paths through escape trees, and to do so in ways that minimize epistemic costs, takes us some way toward seeing how to fashion an account of problem-solution—even though I would not claim to have provided precise criteria for judging when predicaments have been escaped or problems solved.[57]

I have been attempting to sketch a picture of reputable scientific reasoning, the kind of reasoning involved not only in everyday research but also in assessment of the merits of proposals that would change practice in large ways. It is high time to make this sketch more exact and more vivid by examining some illustrative examples.

8. Darwin's Appeal to Biogeography

I hope to show that the kinds of inferences I have been discussing play a role within scientific decision making even at times of large modification of scientific practice and that they can be employed without risk of underdetermination. Previous chapters have offered both a characterization of the overall strategy of Darwin's argument for descent with modification (minimal Darwinism) and some suggestions about how the argument was transmitted, refined and accepted. (These topics were considered in Chapters 2 and 6, respectively.) The present aim is to look at Darwin's reasoning at a much finer grain—as if, in our survey of his decision making, we were turning up the power of our microscope.

My discussion will be based on the most fully developed version of Darwin's reasoning, that presented in the *Origin*.[58] Darwin contends that the phenomena of biogeographical distribution offer strong support for the thesis

---

57. In line with the emphasis on cognitive variation in Chapter 3 and the suggestion that social interactions among proponents of different individual practices craft cognitive strategies that are superior to anything that underwrites the beliefs of any single individual (see Chapter 6), I would see the task of mapping the reasonable ways of articulating escape trees as typically allowing alternatives. As noted in footnote 55, I defer this project to a future study, recognizing that the provision of methodological exemplars may ultimately be all that methodology can offer.

58. This differs only in level of detail from that offered in the notebooks, so, in this instance, we need not worry whether we are scrutinizing his initial reasoning or a retrospective justification.

that organisms of different species are linked through networks of descent with modification. We can start with the general problem of biogeographical distribution: Why are G's found in R (where G is some taxon and R is a geographical region)? Darwin and his contemporaries agree that answers to questions of this form should trace the histories of current ranges, explaining why organisms are where they are now in terms of their dispersal from where they used to be. There are two ways of conceiving such historical narratives: either they begin from the point at which the species was created, specifying the locus of creation, or they extend further back into the past, claiming that members of the species under discussion are modified descendants of an ancestral species. The latter option is, of course, Darwin's. The former can be attributed to characters whom we may reasonably call "creationists"— although it should not be assumed that such people share the ideas and misconceptions of contemporary "scientific" creationists.

Since the Darwinian account links species (or, more generally, taxa) to one another, it is hardly surprising that the crucial questions in biogeography, the ones that most sharply divide Darwin from the creationists, are those that turn on *comparisons*. Thus the relative merits of the two types of account are to be considered by asking questions of the form, Why are G's found in R and G^*'s in R^*? In Darwin's discussion there are several important versions of this general form of question.

(A) *The Neighborhood Problem*—Suppose that G and G^* are anatomically and physiologically very similar, and that R and R^* are relatively close (where proximity is defined by the dispersal powers of the organisms under discussion). Less pedantically: neighborhood problems ask why we find similar organisms in the same neighborhood.

(B) *The Barrier Problem*—Suppose that G and G^* are anatomically and physiologically very similar, and that all paths between R and R^* would lead through a region which no organism of the kinds G, G^* could traverse (a *barrier* for short).

(C) *The Disconnected Range Problem*—Suppose that $G = G^*$ and that R and R^* are disconnected. Less pedantically: a single species inhabits separate regions and does not live in the places between them.

(D) *The Displaceable Organisms Problem*—Suppose that R and R^* are disjoint, and when G^*'s are introduced to R they displace G's. Less pedantically: organisms not found in a region do better in that region than the native inhabitants.

Before we see how problems of these four forms play a role in Darwin's reasoning, it is important to identify the commitments of creationism. *Minimal* creationists would see their task, in issues of biogeography, as one of simply tracing the histories of ranges back to the initial creation of the species concerned. Minimal Creationist accounts start from claims to the effect that a species S was created at a place P. They immediately confront the question, Why was S created at P?, and this question can be exacerbated by posing the neighborhood problem (A). Given the wealth of instances in which similar species are found in geographically adjacent regions, there is pressure on creationists to go beyond the minimal version of their doctrine and to replace the host of disconnected facts about similar species in neighboring places with a more unified treatment.[59] Appealing to the principle of unification, we can argue cogently that minimal creationism is epistemically unsatisfactory, that we need a coherent account of why organisms are created where they are.

So it should hardly be surprising that Darwin's contemporaries moved beyond minimal creationism and offered an obvious explanation of the sites of initial creation. They suggested that there is a link between the character of the place and the traits of the organisms created in it. Assuming that the Creator acts through secondary causes, they could even think of the qualities of the site as calling forth, in some unspecified, mysterious, way, the types of organisms that were created there. In any case, to whatever extent creationists involved or distanced the Creator in the epoch-to-epoch business of creation, they maintained a principle of well-adaptedness:

(WA) The organisms created at *P* are well adapted for living in *P*.

(WA) admits of a strong and a weak reading. On the strong construal, well-adaptedness conforms to the idea that the Creator arranges for an optimal distribution of organisms to sites of Creation: among their contemporaries of the same type, the organisms created at *P* are the *best adapted* for living in *P*. The weak reading supposes merely that they do well in *P*, allowing for the possibility that some of their contemporaries might do even better.

Armed with the *strong* version of (WA), creationists can tackle (A)–(C). The treatment of the neighborhood problem begins from the idea that adjacent places are physically similar, and hence call for organisms with similar traits: the optimally adapted organism in *P* is likely to be akin to the optimally adapted organism in *P** if *P* and *P** are much alike. Both (B) and (C) can be handled by recognizing independent creations of the same, or of similar, species. However, the commitment to the strong form of (WA) introduces severe problems for the creationist, difficulties that will be highlighted by (D). Darwin's biogeographical argument begins with an acute diagnosis of why the strong form of (WA) is too strong.

Chapter XI of the *Origin* opens with a statement of the anticreationist thesis: "In considering the distribution of organic beings over the face of the globe, the first great fact which strikes us is, that neither the similarity nor the dissimilarity of the inhabitants of various regions can be accounted for by their climatal and other physical conditions" (346). Darwin follows up with an array of examples: Old World environments find their counterparts in the New World, and yet the faunas and floras are quite different; South Africa, western South America, and Australia have three "utterly dissimilar"

---

59. Here, of course, we see the force of Darwin's frequent appeals to the goal of unifying our account of geographical distribution, and of Huxley's desire to reduce the "fundamental incomprehensibilities" to the smallest possible number.

faunas, despite having extremely similar physical conditions. Much later, in discussing the distributions on oceanic islands, the attack is focused sharply as presenting conclusive objections to the strong version of (WA).

> In St. Helena there is reason to believe that the naturalised plants and animals have nearly or quite exterminated many native productions. He who admits the doctrine of the creation of each separate species, will have to admit, that a sufficient number of the best adapted plants and animals have not been created on oceanic islands; for man has unintentionally stocked them from various sources far more fully and perfectly than has nature. (390)

The displaceable organisms problem spells doom for the strong version of (WA). Schematically, the creationist is faced with an inconsistent set of commitments: $\{G$'s were created in P, G^*'s were created in P^*, P and P^* are distinct, G's and G^*'s are contemporaries of the same type, organisms created at a place are better adapted for living there than any of their contemporaries of the same type, G^*'s are better adapted for living in P than are G's$\}$. Faced with the example of the plants and animals of St. Helena, creationists can consider lines of escape from this predicament. It is hardly attractive to suppose that the inhabitants of St. Helena were created somewhere else: How did they get there? What has happened to them at their original place of creation? Nor is there joy in supposing that the introduced aliens were, in fact, originally created on St. Helena: What happened to them after their original creation there? The facts of competition, including near or complete extermination, make it impossible to deny that the aliens are better adapted for life on St. Helena than are the natives. Revisionist geography, that would question the thesis of disjointness, has nothing to recommend it, and it is similarly impossible to question the fact that the organisms that have competed on St. Helena are contemporaries of the same type. Under these circumstances, the escape tree is blocked in every direction but one. The strong version of (WA) has to go.

Now, however, we have to consider what happens to the general line of solution to comparative problems in biogeography. Initially, it seemed that the creationist would answer questions of the form, Why are G's and not G^*'s created at P?, by instantiating the following schema:

(1) G's are better adapted to life in P than contemporary organisms of the same type.

(2) G^*'s are not better adapted to life in P than contemporary organisms of the same type.

(3)   The organisms created at P are those which, among contemporaries of a particular type, are best adapted to life in P.

G's and not G^*'s are created at P.

Unfortunately, [3] is equivalent to the strong version of (WA). In abandoning the strong version of (WA), creationists reopen a large class of explanatory problems. The challenge for them is to reformulate the preceding schema in a way that will deploy only a weak version of (WA).

The force of this challenge is best appreciated by noting that versions of (WA) sufficiently weak to circumvent the problems posed by (D) have to allow that there is a relatively broad class of "well-adapted" organisms that are candidates for creation at a particular place. The organisms actually created need not be the best adapted of the candidates. But now it is plain that none of the problems (A)–(C) can be solved by simply substituting a weak version of (WA). Consider, for example, the neighborhood problem, (A). The weak form of (WA) licenses us in concluding that there is a pool of candidate organisms for creation at one place and another pool available for creation at nearby places. If organisms are simply drawn at random from these pools, then there is only a very small probability that they will be similar to one another. Each member of either pool has a counterpart in the other pool, but each is dissimilar to most of the members of the other pool. Thus the creationist fails to account for the similarity of organisms created in adjacent physical reasons.

Darwin has an alternative way of attacking the creationist attempt to solve the neighborhood problem. For similarities among neighboring organisms arise even when the environments in which they live are radically different.

> On these same plains of La Plata, we see the agouti and bizcacha, animals having nearly the same habits as our hares and rabbits and belonging to the same order of Rodents, but they plainly display an American type of structure. We ascend the lofty peaks of the Cordillera and we find an alpine species of bizcacha; we look to the waters, and we do not find the beaver or musk-rat, but the coypu and capybara, rodents of the American type. Innumerable other instances could be given. (349)

With the observation that the similarities among neighboring organisms outrun the similarities of their environments, Darwin exposes the inadequacy of creationist solutions that trace similarities of form to similarities of environmental demand.

But elimination of the creationist program will only work as an argument for Darwin's own solution if he can show that there are no promising available alternatives and that his own proposal is not subject to kindred severe criticisms. We have already examined the case for the first point. For any given species, the options of descent from a prior species or creation de novo appear to be exhaustive. Hence, we must choose between some version of descent with modification and some form of creationism. As I have already suggested, minimal creationism is untenable, except possibly as a position of last resort, a confession of ignorance in reaction to the failure of all positive efforts to explain the distributions of plants and animals. So the creationist must find some substitute for the idea of creation in response to environmental demand and the strong principle (WA) to which it gives rise. Lacking any clues about how to develop the needed account, it is hard to retain creationism as a live possibility—unless, of course, its rival is in trouble, and we find ourselves in the dismal situation of having eliminated *all* the initially promising candidates. The crucial issue, then, is the viability of Darwin's own account, and the bulk of chapters XI and XII is devoted to responding to problems and inconsistency predicaments. He faces apparently serious difficulties generated by problems (B) and (C).

We can begin to represent the inconsistency predicament generated by the barrier problem, (B), by noting that Darwin is committed to instances of statements of the following forms:

G's and G^*'s show repeated similarities (homologies).

[H]   Organisms with repeated similarities (homologies) are derived from a common ancestor.

G's are found in R and G^*'s are found in R^*.

R and R^* are now mutually inaccessible for G's and G^*'s.

If Darwin were to be forced to abandon [H], he would lose his successful solutions to the neighborhood problem (A), and much more besides. However, we do not yet have an inconsistency. To produce an inconsistent set of schematic sentences, we have to add the idea that present inaccessibility betokens past inaccessibility. More exactly, we would have to assume:

[PA]   If R and R^* are presently mutually inaccessible for G's and G^*'s then it would have been impossible for the common ancestor of G's and G^*'s, G_0, to radiate into both R and R^*.

In cases where he is committed to instances of the first four schematic sentences, Darwin can save his cherished principle [H] by denying the appropriate instance of [PA]. In other cases, he can resolve the threat of inconsistency by denying the claim of *present* inaccessibility. Pursuing both strategies in tandem, he turns back potentially damaging criticisms of his position based on the barrier problem.

After noting the difficulties posed by the presence of apparently impassable barriers, Darwin calls attention to the possibility that [PA] may be false.

> But the geographical and climatal changes, which have certainly occurred within recent geological times, must have interrupted or rendered discontinuous the formerly continuous range of many species. So that we are reduced to consider whether the exceptions to continuity of range are so numerous and of so grave a nature, that we ought to give up the belief, rendered probable by general considerations, that each species has been produced within one area, and has migrated thence as far as it could. (353–354)[60]

60. It is worth noting that Darwin is here pursuing both problems (B) and (C) by effectively supposing that barrier problems can (often? always?) be tackled by recognizing a continuous range of the ancestral species on both sides of the region in which a barrier emerges. His remarks also contrast his own line of solution with creationist approaches to (C).

Investigation of the hard cases will require Darwin to look both at the possibilities of geographical change and at the dispersal powers of organisms. On the former score, he canvasses various ways in which islands may become linked to continents, bodies of water joined, changes of climate have allowed a "high road for migration" (356). Yet, contrary to the enthusiasm of some of his contemporaries, Darwin demands that hypotheses about past changes in geographical position must be firmly founded in hypotheses about presently acting geological processes (357).

The second part of Darwin's strategy, to which he devotes greater length, explores the ways in which *current* organisms can be transported. Focusing on the apparently difficult example of the dispersal of plants across sea water, he reports a series of experiments to test the ability of seeds to germinate after floating on sea water for various periods. The argument deserves quoting at some length.

> It is well known what a difference there is in the buoyancy of green and seasoned timber; and it occurred to me that floods might wash down plants or branches, and that these might be dried on the banks, and then by a fresh rise in the stream be washed into the sea. Hence I was led to dry stems and branches of 94 plants with ripe fruit, and to place them on sea water. The majority sank quickly, but some which whilst green floated for a very short time, when dried floated much longer; for instance, ripe hazel-nuts sank immediately, but when dried, they floated for 90 days and afterwards when planted they germinated; an asparagus plant with ripe berries floated for 23 days, when dried it floated for 85 days, and the seeds afterwards germinated: the ripe seeds of Heliosciadium sank in two days, when dried they floated for above 90 days and afterwards germinated. Altogether out of the 94 dried plants, 18 floated for above 28 days, and some of the 18 floated for a much longer period. So that as 64/87 seeds germinated after an immersion of 28 days; and as 18/94 plants with ripe fruit (but not all the same species as in the foregoing experiment) floated, after being dried, for above 28 days, as far as we can infer anything from these scanty facts, we may conclude that the seeds of 14/100 plants of any country might be floated by sea-currents during 28 days and would retain their power of germination. In Johnston's Physical Atlas, the average rate of the several Atlantic currents is 33 miles per diem (some currents running at the rate of 60 miles per diem); on this average, the seeds of 14/100 plants belonging to one country might be floated across 924 miles of sea to another country; and when stranded, if blown to a favorable spot by an inland gale, they would germinate. (359–360)

Darwin apologizes for his "scanty facts," and indeed, if his task were to establish precise rates of dispersal of plant species, the trials he performed are obviously too crude. But the goal is the far more modest one of warding off the threatened inconsistency obtained by instantiating the schematic sentences presented earlier by taking the G's as species of plants and R and R^* as continent and oceanic island. To block the inconsistency it is enough to show that one of the contributing claims can be challenged, and Darwin's explorations permit him to cast doubts on the efficacy of the alleged barrier.

This is only the opening wedge of the argument. Sensitive to the point that the means of transport he has canvassed may be too tenuous—or that they may not allow for transport of the right species—Darwin indicates a number of other ways in which the threatened inconsistency could be removed. Stones embedded in the roots of trees may enclose portions of seed-bearing dirt and offer complete protection against being washed away; drift timber can thus serve as a vehicle for seeds. Birds can swallow seeds, which are excreted whole and which will germinate. Birds also carry dirt particles on their beaks or their claws. Icebergs "are known to be sometimes loaded with earth and stones" (363). When we recognize the variety of ways in which seeds can be dispersed, Darwin concludes, we should think it to "be a marvellous fact if many plants had not thus become widely transported" (364).

As I read this section of the *Origin,* Darwin has responded to a threat of inconsistency by sketching an escape tree with several open branches. He has not demonstrated that each of the problematic instances can be resolved. But he has provided grounds for thinking that the problems may not be insuperable, that there are resources within his recommended practice to find relief from inconsistency. By assembling his catalogue of potential modes of transport, Darwin indicates the lines along which solutions can be found and so provides *corrigible* grounds for thinking that solutions are available. Given his successes with other problems (for example with (A)), the existence of apparent barriers offers no block to the acceptance of his proposal for biogeographical explanation.

I shall close by considering Darwin's response to (C), the disconnected ranges problem, where his discussion focuses on a very particular example and thus supplements the general outlining of possible solutions with a concrete achievement. He begins by suggesting that the case he will treat poses an exceptionally severe challenge to his approach to biogeography.

> The identity of many plants and animals, on mountain-summits, separated from each other by hundred of miles of lowlands, where the Alpine species could not possibly exist, is one of the most striking cases known of the same species living at distant points, without the apparent possibility of their having migrated from one to the other. (365)

Darwin's predicament can be identified by the following set of schematic sentences:

G's are found both in R and R^*.

R and R^* are mutually inaccessible for G's.

G's descend from a single ancestral population.

[PA*] If R and R^* are presently mutually inaccessible for G's then it would have been impossible for the ancestral population of G's, from which all current G's descend, to radiate into both R and R^*.

When we take the G's to be alpine species of plants and animals, there are numerous instantiations of these schematic sentences which commit Darwin to the first three claims. If he is to escape inconsistency, he must therefore challenge the relevant instances of [PA*].

This is carried out in the section "Dispersal during the Glacial Period" by adducing geological evidence for large changes in the earth's climate.

> The ruins of a house burnt by fire do not tell their tale more plainly, than do the mountains of Scotland and Wales, with their scored flanks, polished surfaces, and perched boulders, of the icy streams with which their valleys were lately filled. So greatly has the climate of Europe changed, that in Northern Italy, gigantic moraines, left by old glaciers, are now clothed by the vine and maize. Throughout a large part of the United States, erratic boulders and rocks scored by drifted icebergs and coast-ice, plainly reveal a former cold period. (366)

During this previous cold period, Darwin claims, arctic species of plants and animals would have been able to extend their ranges throughout much of Europe and North America. But, as the climate grew warmer again, they would have been displaced from the new additions to their ranges by the return of temperate organisms, previously driven south by the advancing cold.

> As the warmth returned, the arctic forms would retreat northward, closely followed up in their retreat by the productions of the more temperate regions. And as the snow melted from the bases of the mountains, the arctic forms would seize on the cleared and thawed ground, always ascending higher and higher, as the warmth increased, whilst their brethren were pursuing their northward journey. Hence, when the warmth had fully returned, the same arctic species, which had lately lived in a body together on the lowlands of the Old and New Worlds, would be left isolated on distant mountain-summits (having been exterminated on all lesser heights) and in the arctic regions of both hemispheres. (367)

So Darwin provides evidence against [PA*], while simultaneously giving an explanation of the apparently anomalous distribution.

Darwin follows up this basic story with some detailed discussions of particular instances of distribution, designed to show that the glacial dispersal story not only resolves the initial difficulty but also accounts for particularities of distribution that have appeared peculiar on extant accounts of biogeography (see, for example, 376–379 on the distribution of temperate organisms in mountainous regions of the tropics). Even without probing his further arguments, I think that enough has been said to show the prevalence of those modes of reasoning discussed in previous sections. Darwin's overall strategy is to proceed by eliminating rival hypotheses. In doing so, he formulates inconsistency predicaments for his opponents and shows how the escape trees from those predicaments are blocked. He also responds to the inconsistency predicaments that threaten his own account, either showing in some detail how there is a line of escape (the discussion of the alpine fauna and flora) or at least indicating the possibilities of amending the inconsistent set without epistemic loss (the treatment of methods of dispersal). Notice finally that there is no serious threat of underdetermination. Although there are logically possible rival hypotheses at any number of points in Darwin's argument, none of them has any plausibility, given the state of practice in natural history from which Darwin begins. Of course, one *could* suppose that the cases in which new arrivals have displaced native inhabitants involve some hushed-up conspiracy on the part of colonizing humans, or that the Welsh mountains were scored by the hand of God (or the devil), or that birds used to secrete digestive juices that would destroy any seeds they might swallow. One could suppose that an evil demon has contrived the distribution of plants and animals to deceive us. Part of Darwin's achievement is to set forth an argument that leaves extravagant hypotheses akin to extreme forms of skepticism as the only refuge for his opponents.

9. Lavoisier against the Phlogistonians

If there is a single historical episode which has sustained the notion that traditional formal accounts of scientific reasoning are inadequate to reveal the epistemic worth of the decisions made by scientists at times of great change, the "chemical revolution" of the late eighteenth century is an excellent candidate. According to a once popular historiographical tradition, this revolution consisted in the replacement of the phlogiston chemistry, begun by Stahl and continued in the period by Priestley and Cavendish, with the "new chemistry" or "antiphlogistic system" of Lavoisier and his French colleagues (notably Berthollet, Fourcroy, and Guyton de Morveau).[61] Once, it appeared that the chemical revolution, conceived within this tradition, offered an illustration of Legend, for, as Conant described it in his eminently accessible narrative, the phlogiston theory crumbled under the cumulative force of Lavoisier's evidence. Yet, ever since the publication of (Kuhn 1962), the same conception of the chemical revolution as the overthrow of the phlogiston theory has sustained a radically different interpretation, according to which no objective evidence forced the abandonment of the phlogiston theory. Instead, we have been offered the picture of two different ways of doing chemistry, both with certain successes to their credit and both facing unsolved problems. Allegedly there was no cognitively superior reasoning available to the participants, which would have decided the issue in favor of Lavoisier.

I shall argue that this fashionable picture is a myth. Although it is correct that the older story, which Conant tells so well, is inadequate as an account of the transition in chemistry between 1770 and 1795, at least it is recognizable as a simplification of the actual decisions of the actual participants. I shall use the framework constructed in this and in previous chapters to offer what I hope is a more realistic version.

The first question prompted by recent historical scholarship is whether there was a "revolution" in chemistry in the late eighteenth century. I believe

---

61. Notable representatives of this tradition are Conant (1957), Partington (1965), and Partington and McKie (1937–38).

that the framework offered in earlier chapters provides a way of focusing this vexed issue. The work of Lavoisier and his associates certainly brought about major changes in a number of components of chemical practice: the language of chemistry was reformed, questions about the composition of substances and the weight relations among constituents were given far greater significance than hitherto, new schemata for answering these questions were adopted, instruments and apparatus for performing chemical experiments were enormously refined, and Lavoisier's famous "principle of the balance" furnished a new standard for assessing experiments. Appreciating the magnitude of these changes, we can readily understand why some of the protagonists saw themselves as living through a revolutionary period in chemistry (although, for obvious historical reasons, the vocabulary of "revolution" was more likely to occur to them than to workers in other periods). In line with my approach in earlier chapters, I suggest that nothing much hangs on whether we apply or withhold the label.

Part of the story does involve the repudiation of accounts of combustion, reduction, and other chemical reactions based on appeals to phlogiston. So, while we cannot identify the changes wrought by Lavoisier with the overthrow of the phlogiston theory, one facet of his work consists in replacing phlogistonian explanations with different analyses. Mindful of the motivations for the compromise model, adduced in the last chapter, we must be careful to allow that the availability of superior cognitive processes supporting Lavoisier and opposing the phlogistonians may vary with time. Moreover, we do not have to assume that repudiation of phlogistonian analyses necessarily involved acceptance of Lavoisier's accounts of *all* reactions: there are some parts of Lavoisier's chemistry about which he and his colleagues admit the tentative character of their conclusions. The thesis that I shall try to illustrate is that, by the early 1790s, Lavoisier and his co-workers had made available cognitively superior forms of reasoning supporting the thesis that there was no coherent phlogistonian analysis of central chemical reactions and that there was a coherent rival conception of those reactions based on the entities (oxygen, hydrogen, and so forth) which the French chemists recognized.

Both the opposing stories of the chemical revolution have a clear virtue and a clear defect. Conant enables us to understand why there was a pronounced trend to Lavoisier in the 1780s, so that, by the mid–1790s, virtually every chemist had abandoned the phlogiston theory.[62] However, the explanation of easy consensus is purchased at the price of making those who resisted Lavoisier at all seem unintelligent. By contrast, Kuhn, and, following him, Doppelt, are determined to understand how extremely astute people could

---

62. As is well known, Priestley remained unconvinced until his death in 1804. The parade of "converts" to Lavoisier is, however, impressive: in 1785, Berthollet; in 1786, Fourcroy; in 1787, Guyton de Morveau; in 1791, Richard Kirwan. See (Perrin 1988b) and, for the situation in Germany, (Partington and McKie 1937–1938). Priestley's refusal to join the band should be put into context by recognizing the breadth of his intellectual interests and the important changes that took place in his personal life in his last two decades. See (Schofield 1967).

engage in debate over a period of two decades and still not reach consesnsus. Their answer is that the available evidence could not force a decision, but this answer carries with it an inability to understand the overwhelming consensus of the 1790s. Doppelt, who is more explicit than Kuhn, suggests that decisions are ultimately made between "equally reasonable" points of view, because of "various sociological, psychological, biographical, and historical factors" (1978 54), but he does not identify which such factors were operative in the consensual repudiation of phlogiston.

According to Doppelt, the chemical revolution serves as an outstanding illustration of Kuhn's epistemological relativism.[63] The inability of reason and evidence to resolve major scientific disputes is supposed to rest on shifts in standards, whereby scientists make alternative, equally reasonable, decisions about what counts as science, what problems are important to solve and what counts as a solution, as well as on losses of data, instances in which experimental or observational results that were accounted for in earlier discussions are simply discarded because they cannot be explained from the new point of view. So, in Doppelt's story, the phlogiston theory

> explained the common properties of the metals as due to their possession of phlogiston, lacking in their ores. . . . In effect, the new "quantitative" chemistry of Lavoisier and Dalton abandoned any concern for these questions and these observational data—whose treatment constituted the main achievement of the earlier model of chemistry. (1978 43)

He goes on to suggest that the problem of understanding weight relations assumed far greater importance within Lavoisier's new chemistry than within the phlogiston theory, and that, in consequence, although phlogistonians could recognize the anomaly that metallic calxes (supposedly resulting from metals through the loss of phlogiston) weigh more than the metals, they could downplay this problem as insignificant (1978 44).[64]

The story of what occurred among chemists between 1770 and 1790 is far richer and more interesting than either the Conant or the Kuhn/Doppelt accounts recognize. It is, I believe, possible to honor both desiderata, showing how highly intelligent people could have disagreed for so long while understanding how the experimental evidence and the construction of reasoning based on that evidence could ultimately decide the credentials of phlogiston.

63. Of course, Kuhn has repeatedly denied that he espouses relativism, and it is therefore unclear whether or not he would endorse the arguments that Doppelt ascribes to him. Nevertheless, as Doppelt's extensive quotes show (1978 40, 48), it is hardly inexplicable that Kuhn's (1962/1970) has been read as a defense of relativism. I shall leave it to others to decide whether Kuhn is committed to the ideas that Doppelt derives from him. My aim is to show that those ideas make little contact with the reasonings of the chemists of the late eighteenth century.

64. As John McEvoy notes (1988 199), Doppelt's exposition presents a philosophical thesis about the history of science with remarkably little historical evidence. As I shall argue shortly, a closer look at Doppelt's primary example shows how far his account is from the phenomena it is supposed to analyze.

The solution to the historiographical puzzle lies in deploying the approach to scientific reasoning that I have taken in this chapter and showing how using the eliminative strategy in a situation where there are numerous constraints, some of which need careful scrutiny, involves cognitive tasks that tax even the most powerful intellect. Very roughly, Lavoisier, Fourcroy, Priestley, Cavendish, Kirwan, and Berthollet were all engaged in the project of offering a coherent analysis of phenomena of calcination, reduction, and acid-metal reactions.[65] That project involved enormous difficulties because apparently plausible solutions generated inconsistencies elsewhere. As Frederic L. Holmes (1985, 1988) has made abundantly clear, Lavoisier's own route to a position that satisfied the vast majority of constraints was long and tortuous.[66] By the mid–1780s he was, however, in a position to make available a perspective on the multitude of experimental findings that could demonstrate to his contemporaries the explanatory power of appeals to oxygen and the explanatory difficulties of appeals to phlogiston. In the end, I suggest, there was no "shift of standards," no "loss of data," nor was there a simple confirmation or falsification of the type celebrated in Legend. Instead, there was a highly complicated process of construction and exploration of escape trees, resulting in the provision of a superior cognitive strategy that led almost all of Lavoisier's contemporaries to modify chemical practice.

Let us begin by clearing out of the way some prevalent misconceptions about the debate between Lavoisier and Priestley. I argued earlier (Chapter 5) that the Kuhnian thesis of "inevitable partial communication across the revolutionary divide" is incorrect. There are semantic resources available so that claims made by phlogistonians can be formulated in Lavoisier's language, and conversely. Nor is this an abstract philosophical possibility. Lavoisier, Priestley, and Cavendish communicate quite effectively, deploying the kinds of resources mentioned in my account. Although, on occasion, the protagonists maintain that they have not been understood—witness Lavoisier's comment that Kirwan has not "formed perfect ideas of the doctrine which he distinguishes by the name of Antiphlogistic" (Lavoisier's note in Kirwan 1789)—these are failures of comprehension of the kind that attend any attempt to grasp a complex doctrine formulated in a language that one understands. Kirwan's misunderstandings are not semantic; he has simply misidentified some of Lavoisier's explanatory claims, and Lavoisier thinks that further enunciation of these claims will promote mutual understanding—as apparently it did. In general, the *interpersonal* difficulties of holding the whole of a complex proposal together are no different from the *intrapersonal* troubles

65. Here I oversimplify. Lavoisier, Priestley, and their contemporaries wanted to give chemical analyses of other reactions, and to integrate their analyses with explanations of vital processes. For Lavoisier, see Holmes (1985) and the middle part of Lavoisier (1789). For Priestley, see his (1775).

66. I shall return to the issue of Lavoisier's own difficulties in articulating his "antiphlogistic system" at the end of this section. My account will be heavily dependent on Holmes's pioneering excavations of Lavoisier's laboratory notebooks.

of keeping everything in mind. Lavoisier's own struggles toward a coherent perspective help him to clarify his views in response to the similar misunderstandings of others.

Second, contrary to what Kuhn (and, following him, Doppelt) suggests, the debate between Lavoisier and his opponents does not involve phlogistonian claims to the effect that the new chemistry fails to explain what all the metals have in common. This is a good thing, for the "explanation" offered by phlogiston chemistry is, to put it charitably, rudimentary and an analogous "explanation" can easily be furnished within the framework of Lavoisier's chemistry. Phlogistonians such as Priestley, Kirwan, Cavendish, and Gren do offer many arguments, appealing to a diverse set of empirical findings, but the invocation of explanatory successes of the kinds that Kuhn and Doppelt see as critical to their case is absent. Interestingly, *Lavoisier* does mention the suggestion that phlogiston can help supply qualitative explanations of colors and tastes (1783 638–639), and, with a crisp survey of substances supposed to contain or lack phlogiston, is able to show that there is no available coherent account.

Nor is there any plausibility in the idea that Lavoisier's introduction of the "balance sheet"—his weighing and measuring of reactants and products— was resisted by phlogistonians on methodological grounds. Much of Lavoisier's ingenuity as an experimenter consists in his design of apparatus and instruments that enable him to deal with closed systems. Thus, in many instances, he is able to demonstrate that the total weight of the reactants is equal to the total weight of the products, so that there are no grounds for thinking that anything "ponderable" has entered or escaped from the reaction. Attacking the use of the "balance sheet" would require his opponents to put forward some hypothesis that explains why deviations from equality of prior and posterior weights are to be expected in the case at hand, and Lavoisier's experimental designs make such hypotheses very hard to defend. As a result, it is hardly surprising that his interlocutors—Priestley, Cavendish, Kirwan, and others—not only accede to his principle but also employ it in designing and reporting their own experiments. Instead of a methodological dispute about the legitimacy of the practice of measuring and balancing, the documents reveal claims and counterclaims about the weights of substances collected in various experiments.[67]

Another popular myth about the debate between Lavoisier and his opponents centers on the idea that it was relatively easy for phlogistonians to evade difficulties about the weight gain in calcination of metals by supposing that phlogiston has negative weight. As I emphasized earlier, in discussing underdetermination, preservation of a favored hypothesis by adjusting aux-

67. For two examples among many, see Kirwan's discussion of reactions with nitrous acid (Kirwan 1782/1789 76–115) and (Priestley 1775 vol. III 82). McEvoy (1988 203–204) makes similar points about the thesis (defended in Schofield 1967) that Priestley had a methodological difference with Lavoisier about the use of the principle of the balance sheet, and he argues cogently that theses about Kuhnian incommensurability cannot be illustrated by reference to this example.

iliary assumptions is typically not free, and, in the case at hand, both physics and chemistry provide constraints that close off the escape trees generated by the notion of the levity of phlogiston. Lavoisier's (1783) points out that there are experiments which demand positive weight for phlogiston (more exactly, he shows that phlogistonian accounts of some experiments are committed to the ponderability of phlogiston). Moreover, efforts to integrate the levity of phlogiston with gravitational theory broke down in accounting for phenomena of falling bodies and pendulum motions.[68] The difficulties for those who endeavored to satisfy the principle of the balance without conceding that *something* is absorbed in calcination were increased by Lavoisier's experiments showing the proportionality of the weight gain to the volume of air lost. Thus it is hardly surprising that, for many phlogistonians, including the most prominent and influential champions of the "old doctrine," the idea of phlogiston with negative weight has no appeal whatsoever: as early as 1775, Priestley announces that he "never had any faith at all in that doctrine of the principle of levity" (1775 vol. II, 312).

Armed with these points, we can begin to analyze the debate over the credentials of phlogiston. In the first phase, which occurs in the 1770s, Lavoisier shows that there is a weight gain in the calcination of metals, and he demonstrates that the weight gain equals the weight loss of the air.[69] At this point, phlogistonians do not announce that they have qualitative explanations that Lavoisier cannot emulate, they do not question his principle of the "balance sheet," they do not (with few exceptions, chemists who quickly find themselves in a thicket of difficulties) campaign for the levity of phlogiston. Instead, they do something that is far more reasonable: to wit, accept Lavoisier's claim that something from the air is absorbed and try to combine this concession with the traditional idea that phlogiston is emitted.

I shall consider the structure of this transition in phlogistonian accounts shortly. Before doing so, I want to note that, at this stage of the debate, in the mid-1770s, Lavoisier has not yet arrived at a coherent account of what occurs in the various reactions that he and his interlocutors have investigated: calcination, reduction of calxes, acid-metal reactions. *There is room for genuine disagreement at this point of the controversy because nobody knows how to offer a consistent treatment of all the reactions on which the parties have data.* In particular, nobody knows whether or not phlogiston emission can be combined with absorption of a constituent of the air to yield a general analysis

68. See the discussion of Gren's attempt to assign negative weight to phlogiston and Mayer's critique of it in (Partington and McKie 1937–1938 II 33–38).

69. Thus the 1774 memoir "On the Calcination of Tin in Closed Vessels" proceeds by showing first that the weight of the whole system remains constant through the experiment, second that the weight of the excess air introduced to fill the vessel after calcination has occurred is equal to the weight gain of the calx. Lavoisier writes: "Thus, in this operation, more air is found in the vessel than before the calcination, and it is evidently this excess of air that yields the gain in weight; if therefore this same increase of weight is found in the metal, it will be shown [*prouvée*] that the excess of air which came in has served to replace the portion that was combined with the metal during calcination, and which increased its weight:" (1774 112). Lavoisier goes on to report the equality.

of combustion (and whether this can be extended to treatments of reduction and acid-metal reactions).

By the mid-1780s, however, Lavoisier has fashioned a general account which deals, in a unified and consistent way, with a far greater range of the experimental results than any extant version of the phlogiston theory. Moreover, he is able to expose systematic problems with the particular phlogistonian accounts developed by Priestley, Cavendish, and Kirwan. This is not to say that his own analysis is free of problems. Notoriously, Lavoisier had some trouble in integrating his treatment of chemical composition with his views about heat, and his discussions of some acids were, as he conceded, tentative.[70] There was still work to be done, after 1790, in showing that Lavoisier's preferred schemata could be instantiated with respect to some phenomena, but, well before this date, he and his colleagues had shown that they could apply their schemata to a class of instances well beyond the range of the rival phlogistonian approach.

Let us now look more closely at the kinds of reasoning involved both in the resolution of the initial phase of the controversy in the 1770s and in the abandonment of phlogiston in the late 1780s. We begin from the experiments on calcination in sealed vessels, which seem to show the following:

[1] The sum of the weights of the reactants equals the sum of the weights of the products.

[2] Weight of calx is greater than weight of metal.

[3] The air in the vessel decreases in volume during the reaction, and the weight loss of the air equals the weight gain of the calx.

Phlogistonians will find these conclusions troubling, for they seem to support the idea that combustion is a process of absorption, not one of emission. How can that consequence be avoided?

One suggestion might be that the inference from [1] to the closure of the system is faulty. But this is hard to sustain. If the proposal is that ponderable

---

70. I should note that, while Lavoisier's official doctrine is that heat is a subtle fluid, caloric, that can insinuate itself among the particles of matter (thereby causing transitions from the liquid to the gaseous state), there are passages in which he expresses a more ecumenical view. In a remarkable passage in the *Traité* of 1789 he suggests that caloric doesn't have to be regarded as a real substance: "It suffices, as will be better understood on reading what follows, that it should be some kind of repulsive cause that scatters the moelcules of matter, and one can thus regard the effects in an abstract and mathematical fashion" (1789 19). We should thus beware of assuming that Lavoisier's treatments of chemical reactions encountered problems caused by his commitment to the caloric theory of heat.

Similarly, although Lavoisier claims that oxygen is found in all acids, he explicitly bases this on experiments on carbonic, sulfuric, and phosphoric acids. "We have not yet succeeded in decomposing and recomposing all the acids; but we are assured at least that oxygen is a principle common and necessary to the formation of all those whose composition we are acquainted with.... But when from these particular facts [the analyses of carbonic, phosphoric, and sulfuric acids], the general induction is made that oxygen is a principle common to all acids, the consequence is founded on analogy, and here it is that the theory commences" (Lavoisier's note in Kirwan 1789, 19–20).

matter escapes from or comes into the system, then it must be explained how this transaction always occurs so as to give the appearance of equality within. On the other hand, if the suggestion is that imponderable matter enters or exits, then it is far from clear how this will solve the problem of the observed weight relations.

Asserting that phlogiston has negative weight allows the phlogistonian to accommodate [2], but this assertion now has to be brought into conformity with the phenomena of mechanics. Moreover, given that Lavoisier is able to show first that the weight gain of the calx varies with the diminution of the volume of the air in the vessel, and then that the weight gain of the calx is equal to the weight loss of the air, resistance to the idea that there has been an absorption has to account for the loss in weight of the air and the con-comitant loss in volume. If it is proposed that the weight loss of the air results from the fact that the air has absorbed phlogiston (with negative weight) then the diminution of volume must be accounted for by hypothesizing that the absorption of phlogiston by the air produces a gas which occupies a smaller volume. Now phlogistonians must explain why the diminution of volume is such that the weight of that volume of air agrees with the weight gain of the calx. The hypothesis that phlogiston has negative weight thus faces difficulties caused by constraints from physics and simultaneously supposes a mysterious coincidence of values.[71]

Thus, given the background assumptions shared by members of the phlo-giston theory (and by their opponents), assumptions that cannot be replaced without serious epistemic losses, the conclusion to be drawn from [1]–[3] is

[4]   When metals are calcinated in air, something is absorbed from the air.

71. These difficulties are concisely noted by Lavoisier in his retrospective account of the analytic situation in 1772, when he views himself as having established the absorption thesis (my [4]). Writing in 1792, he comments on Guyton de Morveau's proposal that phlogiston has "less weight than the atmospheric air" so that there would be the appearance of weight gain through loss of phlogiston when the substances were weighed in air.

This explanation would have been tenable, if the increase in weight of the metallic oxides had not been equal to that of the air displaced, or, which comes to the same thing, if it had disappeared when they were weighed *in vacuo*. But this increase is much too large to attribute to that cause, since it amounts, in some metals, to about one third of their weight. One must thus either abandon the explanation given by Guyton de Morveau, or go so far as to suppose that phlogiston has negative weight, a tendency to distance itself from the centre of the earth, a supposition that will be found in contradiction with all the facts maintained and recognized by the disciples of Stahl. (1792 102)

Although all the main points of the defense of (4) are assembled here, Lavoisier's reasoning seems to make an interesting conflation: the point about the weight of the displaced air has broader scope than the weighing experiment he suggests for discrediting Guyton de Morveau's proposal and can be deployed to attack the doctrine of negative weight (a doctrine that Lavoisier dismisses by pointing to its peculiar physical consequences).

However, it is possible to combine [4] with the fundamental principle of the phlogiston theory, namely

> [5]   When metals (and other substances) are calcinated, phlogiston is emitted from the metal (the calcinated substance).

[4] and [5] can be rendered consistent, simply by supposing

> [6]   Calx = metal − phlogiston + X.

This idea can be developed in one of two ways. Either the phlogiston is lost to the residue of the air used in the calcination, or the escaping phlogiston combines with the entering constituent of the air, which, joining to the dephlogisticated metal, produces the calx. We can represent these articulations as follows:

> [6a]   Calx = (metal − phlogiston) + X
> Initial air = $X + Y$
> Residual air = Y + phlogiston
>
> [6b]   Calx = (metal − phlogiston) + (X + phlogiston)
> Initial air = $X + Y$
> Residual air = Y

On the basis of the line of reasoning we have rehearsed so far, the principal versions of the phlogiston theory after the late 1770s accept [6] and, in consequence, either [6a] or [6b]. The debate of the 1780s is thus between phlogistonians who espouse some version of [6a] or [6b] and antiphlogistonians (initially Lavoisier, later Lavoisier and other French chemists) who try to show that neither [6a] nor [6b] is tenable.

Quite plainly, the experiments to which Lavoisier appealed in supporting [1]–[3] do not suffice to discredit versions of [6]: [6] is explicitly designed to accommodate them. By 1777, Lavoisier had arrived at an alternative story, one familiar to us. The air, he suggested, consists of two parts. One, "eminently respirable air" (or, sometimes, "vital air") is absorbed by metals during calcination; Mr. Priestley, Lavoisier notes, has "very improperly called [this] *dephlogisticated air*" (1777 184). The other part, "mephitic air" is a "mophette," which "contributes nothing to phenomena of combustion" (1777 192). Lavoisier indicates how this approach will explain the established results of calcination, and he announces that he is engaged on a series of experiments that will oppose the phlogistonian accounts of Stahl and of Priestley (1777 190). Six years later, at the beginning of *Reflections on Phlogiston*, Lavoisier alludes to the state of his argument in 1777, by viewing it as dependent on the use of Occam's razor. As of 1777, he suggests, he could show that the phenomena of combustion and other types of calcination could be understood without invoking phlogiston, so that this entity was purely hypothetical, unnecessary, and, according to the principles of "good logic," should be dismissed (1783 623). But the new attack is to go further and to cut deeper. Lavoisier tries to show that, in accounting for the wealth of experimental results that are now available, defenders of phlogiston become enmeshed in contradictions.

As we might expect, Lavoisier's assessment of the epistemic state in chemistry in the late 1770s reflects his partisan perspective, but, I shall suggest, he is correct in recognizing that there is an important shift in the 1780s. Having established [4] in the way that we have considered, Lavoisier could offer, in 1777, a general explanation of processes of calcination in terms of absorption of part of the atmosphere. *But this did not eliminate the possibility of explaining those processes by adopting* [5] *and* [6]. Moreover, the apparent simplicity of Lavoisier's account of combustion might readily be offset by difficulties with other experimental findings. Calcination was not the only process of interest to or recognized by the late eighteenth century chemists. Hence, the Occamite suggestion to dismiss phlogiston could easily seem premature, when the accounting for so few chemical reactions was in hand.[72]

The task facing both phlogistonians and antiphlogistonians in the late 1770s was to provide analyses of a variety of chemical reactions, including calcination, reduction of calxes, and the acid-metal reactions. Both sides were committed to the following schema:

Chemical Composition

[C1]   In reaction R the reactants are $R_1 \ldots R_m$ in relative proportions by weight $r_1, \ldots, r_m$.

[C2]   The constituents of R_i are $S_{i1}, \ldots, S_{iki}$ with relative proportions by weight $s_{i1}, \ldots, s_{iki}$.

[C3]   Under the temperature conditions of reaction R, the affinities of the constituent substances $S_{11}, \ldots, S_{mkm}$ for one another are.....

Therefore

[C4]   The products of R have the chemical constituents $T_{j1}, \ldots, T_{jkj}$ with relative proportions by weight $t_{j1}, \ldots, t_{jkj}$, and the products are in the relative proportions $p_1, \ldots, p_n$.

[C5] P_j is the substance which has the constituents $T_{j1}, \ldots, T_{jkj}$ with relative proportions by weight $t_{j1}, \ldots, t_{jkj}$.

Therefore

[C6]   The products of the reaction are $P_1, \ldots, P_n$ in relative proportions by weight $p_1, \ldots, p_n$.

---

72. As we shall discover, knowledge of the apparent reactants and products of other chemical reactions could make phlogistonian analyses seem attractive. Enormous work would be needed to lessen these attractions—and, as we follow the twists, turns, and complexities of the debate, it will hardly seem surprising that resolution of the controversy took so long.

In this schema, all parties to the dispute agree to the principle of the balance sheet—in other words, that the amounts of the constituent substances must agree in both reactants and products. Phlogistonians maintain that all assignments must be consistent with [5] and [6]. Antiphlogistonians contend that these assignments must be compatible with

> [7]   Atmospheric air = vital air + mephitic air
>   Calx = metal + vital air[73]

The task, then, is to instantiate [C1]–[C6] subject to the shared constraint of the principle of the balance sheet and to the *preferred* constraint imposed by views about calcination, for instances of R covering calcination, reduction, and acid-metal reactions. What Lavoisier tries to show is that efforts to provide systematic solutions of these open problems within the constraints adopted by phlogistonians inevitably fail.

The first step in giving a version of the phlogiston theory must be to arrive at a view about the substances involved in calcination. According to [6], calx = metal − phlogiston + X. The task is to specify X. Lavoisier's experiments showing that calxes can be produced by heating metals in vital air, and that the weight increase of the calx equals the weight loss of the vital air, combine to suggest that X = vital air. Attempts to defend [6a] now run into trouble, because, when calcination proceeds in vital air ("pure air," "eminently respirable air"), the residue seems to contain nothing but vital air. The absence of any detectable candidate for phlogiston in such experiments is discouraging. Furthermore, as Kirwan points out in his elaboration of the phlogistonian approach,

> That metallic calces are immediately united to pure air, is admitted by many, who yet are of opinion that metals contain phlogiston: yet this admission seems to me inconsistent with the latter opinion; for they allow that metals during their calcination give out phlogiston, and that they are incapable of calcination in any other than pure air; this air therefore meets the phlogiston, and must, with it, form either fixed air or water, one or both of which are absorbed by the calx, and augments its weight. (1782; 1789 183–184)

The analysis in this passage depends, in part, on Kirwan's substantive views, in particular his identification of phlogiston with "inflammable air" (hydrogen), but the central difficulty that he raises is independent of that assumption. On anyone's account, there has to be an answer to the question, What results from the mixing of phlogiston and vital air under conditions of heat?, and, given the problems with tracing the phlogiston that is allegedly emitted from

---

73. Here I ignore complications stemming from Lavoisier's ideas about the combinations of bases of substances with caloric. On his official view of heat, vital air consists in the union of the "basis" of vital air with caloric. Not only does the simplification not affect the epistemological issues concerning the dispute between Lavoisier and the phlogistonians, but, as mentioned in note 70, Lavoisier himself saw that the assumption of heat as a substance could be omitted from his system.

combustion, it is hard to resist the conclusion that they combine to form a new substance that is absorbed by the calx.

Kirwan thus uses the eliminative strategy we have canvassed in earlier sections, within the constraints of the general phlogistonian program, to argue, in effect, for the repudiation of [6a] in favor of [6b]. Now to adopt [6b] is to commit oneself to some instance of the following pair of schematic sentences:

[8] Phlogiston + vital air = Z
     Calx = (metal − phlogiston) + Z

The versions of [8] most prominent in the 1780s and 1790s are those canvassed by Kirwan in 1782: Cavendish supposes that Z = water; Kirwan himself believes that Z is often "fixed air" (carbon dioxide) but sometimes water.[74] To understand why these choices were made, we need to consider some other reactions which the chemists participating in the dispute hoped to analyze.

Three important types of reduction experiments were performed and discussed by the main protagonists. None would have disputed the existence of all the types, but the assignment of particular experiments to the categories involved important controversial elements. From the mid–1770s on, Priestley's classic experiment on the red calx of mercury served as an exemplar of one type of reduction:

(Ri) Metal calx → metal + vital air
              heat

However, calxes that could not be decomposed through the gentle heating employed in this first type of reduction could be reduced in the presence of inflammable air (hydrogen). Thus

(Rii) Metal calx + inflammable air → metal + water
                      heat

Finally, there was the classic method of reducing calxes, long known to smiths and other metallurgists

(Riii) Metal calx + charcoal → metal + fixed air
                    heat

How could one make sense of these reactions, either on Lavoisier's approach or on the basis of commitment to [6b]?

Notice first that, prior to the understanding of the composition of water,

74. On Kirwan's views, whether phlogiston combines with vital air to form water or fixed air depends on the temperature. At high heats, Kirwan agrees that water is always formed, but "...it cannot fairly be inferred that water results from their union in any lower heat; on the contrary, it appears that another compound of both, viz. fixed air, is then formed" (1782; 1789 43).

Lavoisier faced a serious difficulty with respect to (Rii). The initial description of reactions of this type was not, in fact, as I have given it. The "dew" that formed on the interior of the vessel was ignored, and chemists reported what they had found as

(Rii′)   Metal calx + inflammable air → metal
heat

But for the need to concede that something is absorbed in calcination, that is to accept [4], such experiments would have been perfect for the phlogistonians: if one takes a metal calx to consist of metal − phlogiston, then, by identifying phlogiston with inflammable air, experiments of type (Rii′) are immediately accounted for. Nevertheless, even after the phlogiston theory has been modified to accept [4], [6], and [6b], phlogistonians can account for these reactions by supposing that inflammable air is phlogiston; that the phlogiston unites to the calx, restoring the metal and releasing the compound Z. Even if this compound cannot be traced, it can be supposed that the phlogiston is enough to restore the metal, even if it does not emerge in its pure state but rather combined with Z (thus the resultant metals would be treated as having impurities). By contrast, Lavoisier is committed to supposing that the calx loses its vital air in the reduction and that both this and the inflammable air somehow disappear.

After a careful search for the products of the reaction reveals the presence of water, and after the acknowledgment of

[9]   Inflammable air + vital air → water
high heat

the epistemic situation changes. Lavoisier is now in a position to account for experiments of type (Rii) by deploying [9] to claim that the inflammable air combines with the vital air of the calx, leaving metal plus water. However, there is a phlogistonian explanation that will work equally well, namely that based on the following principles adopted by Cavendish:

[8a]   Metal calx = (metal − phlogiston) + water
Inflammable air + vital air = water
Inflammable air = phlogiston

The transition between Lavoisier's difficulty with respect to the reduction of calxes in inflammable air and his ability to fashion a response to the problem is apparent in Kirwan's reflections on this type of reaction:

> The second proof which I alleged in favour of the existence of phlogiston in metals, was deduced from the reduction of their calces to a metallic state, when heated in inflammable air, and the concomitant absorption of that air: to elude this proof, Mr. Lavoisier replied, that metallic calces, when heated, give out pure air, and that this air, meeting the inflammable air, formed

water. As most of these calces were heated to redness in Dr. Priestley's experiments, I allow water to have been formed by part of the inflammable air, while another part united to the calces, and therefore this experiment is not now as conclusive as it was when I alleged it, the composition of water being then unknown; . . . (1782; 1789 183)[75]

The difficulty posed by this experiment for the notion (which Kirwan favors) that calxes are formed when the exiting metal combines with incoming oxygen to form fixed air leads Kirwan to suppose that some calxes contain water, rather than fixed air. However, as experiments of type (Riii) reveal, the fixed air option seems initially to have some attractive features.

The first point to note is that, from Lavoisier's point of view, there are two types of experiment which are very different phenomenologically, both of which fall under (Riii). Everyone will agree that reactions in which discrete pieces of charcoal are introduced and placed beside or underneath the metal calx belong to this category. But there are also experiments which might look to the unwary like those belonging to (Ri) in which charcoal (or other forms of carbon) is nonetheless present. On Lavoisier's account, the preparation of metals and metallic calxes frequently leaves impurities behind. When calxes containing carbon impurities are heated, the reduction reaction should not be assigned to some new category

(Riv) Metal calx → metal + fixed air
heat

but recognized as an instance of (Riii).

The possibility of obtaining fixed air as one of the products when metallic calxes are *apparently* heated alone lends obvious support to a scheme of phlogistonian explanation based on another alternative:

[8b] Calx = (metal − phlogiston) + fixed air
Fixed air = inflammable air + vital air
Inflammable air = phlogiston

Kirwan's version of the phlogistonian approach tries to combine [8a] and [8b], supposing that inflammable air and vital air produce water at high heat and fixed air at lower temperatures. Consequently, he holds that those calxes formed at lower temperatures contain fixed air, while those that are only formed at higher temperatures contain water.

We can now recognize three different approaches to explaining reduction reactions. On Lavoisier's story, (Ri) reactions simply involve the decompo-

75. Kirwan goes on to refer to another experiment in which he believes that Lavoisier's explanation will not work. In this instance, the heat involved was lower, and Kirwan expresses skepticism about the possibility of forming water from inflammable and vital airs at these lower temperatures. Lavoisier counters Kirwan's claims about reactions of these gases at lower temperatures; see Lavoisier in Kirwan (1789 57).

sition of a calx into its constituents, the metal and vital air; (Rii) reactions involve the combination of inflammable air with the vital air of the calx to produce water and leave the metal. In all cases where carbon is present, either introduced as an observable reducing agent or contained as an impurity in the metal, the vital air released combines with the carbon to produce fixed air. Cavendish, advancing explanations based on [8a], must agree with Lavoisier's analysis of type (Riii), supposing that the production of fixed air results from decomposition of the water in the calx, with the carbon uniting with the vital air to produce fixed air and the phlogiston combining with and thus restoring the metal. In (Ri) reactions, he will propose that phlogiston from the water unites to the metal, releasing vital air, and, in (Rii) reactions, the calx absorbs the phlogiston (inflammable air) releasing its water and reviving the metal. Kirwan's account views some reactions as falling under (Riv) and revealing the presence of fixed air in some calxes. Other calxes contain water, and their reductions are treated as Cavendish would explain them.

It should now be evident what Lavoisier had to do to defend the superiority of his own scheme of interpretation. The suggestion that calxes contain fixed air (one which Lavoisier himself had endorsed at an early stage of his tortuous route to the approach of the 1780s) can only be turned back by providing clear experimental evidence of the presence of residual carbon in those metallic calxes that give rise, under reduction, to fixed air. Cavendish's preferred hypothesis, [8a], requires considerable ingenuity for Lavoisier to launch an experimental critique. Weighing the reactants and products in experiments of type (Ri) will be inconclusive. Lavoisier solves his problem by considering a different mode of calcination.

The famous "gun barrel" experiment is performed by placing iron filings in a gun barrel, heating to red heat, and passing water through the gun barrel. The products consist of the black calx of iron and inflammable air. All parties to the dispute can agree on the following description of the experiment:

[10]   Iron + water → black calx of iron + inflammable air
              red heat

Cavendish and other phlogistonians will suppose that what occurs in this experiment is that the inflammable air (phlogiston) is released from the iron to leave the "basis of the iron" (iron − phlogiston), and that this basis combines with water to yield the black calx. Lavoisier proposes that the water is decomposed to form vital air, which combines with the iron to form the calx, and inflammable air, which is released. At a purely qualitative level, nothing tells for or against either of these interpretations.

However, Lavoisier was able to show that an amount of oxygen whose weight equaled the gain in weight of the calx would combine with the inflammable air collected in the experiment to yield water, and that the weight of this water is the weight of the water lost in the experiment. He could thus provide an instantiation of CHEMICAL COMBINATION that would explain the *quantitative relations* among the products and reactants. On the phlogistonian scheme, however, there is no reason why *the precise amount of phlogiston* released by the metal should combine with vital air to yield *the amount of water lost.* Cavendish and other phlogistonians face the difficult open problem of accounting for a mysterious coincidence.

To appreciate the force of this problem, consider the transactions that are supposed to take place in the gun barrel experiment. Allegedly the calx is formed through the release of phlogiston and the absorption of water. These processes are supposed to be coupled in such a way that the amount of water absorbed in the formation of the calx contains just as much phlogiston (= inflammable air) as is released by the metal. *But the exiting phlogiston is not supposed to interact in any way with the incoming water.* However ideas about affinities are developed (in the writings of any of the participants to the controversy) nobody hypothesizes that X can displace X from a compound between X and Y. As a result, the option of suggesting that the exiting inflammable air combines with the vital air of the incoming water, releasing just the right amount of inflammable air, is simply unavailable. Lavoisier's experiment generates an open problem to which there seems to be no solution within the constraints adopted by his phlogistonian opponents.

The attack on Kirwan's more flexible version of the phlogiston theory proceeds on a number of fronts. As we have seen, a central assumption made by Kirwan is that it is possible to form fixed air by combining inflammable air and vital air at temperatures below those required for the formation of water. To this, Lavoisier and Fourcroy both respond by asking for an experimental demonstration that fixed air can indeed be formed in this way (see Fourcroy in Kirwan 1789, 219). They thus identify another open problem that phlogistonians fail to resolve. But, equally importantly, they undercut the motivation for holding that fixed air is a constituent of calxes by appealing to experiments that show how preparations of metals which take precautions to prevent contamination with carbon afford calxes that yield no fixed air when reduced. Fourcroy offers a diagnosis of the conditions under which we should expect to find carbon impurities and, consequently, the production of fixed air on the reduction of calxes:

> The carbonic acid which Mr. Hermstad obtained from the oxide of manganese distilled with iron, is produced only from that which the oxide absorbed from the atmosphere, or from the charcoal contained in the iron, which burns by means of the oxygen disengaged from the manganese. If red precipitate [the red calx of mercury: PK], thus treated with iron, sometimes affords various doses of this acid, and sometimes affords none at all, this depends on the state of the oxide of the mercury, which is more or less oxided, or has been either recently made or long exposed to the air, whether it contains nitrous acid, or is entirely deprived of it, and, lastly, upon the nature of the iron, which retains more or less charcoal. (In Kirwan 1789, 221–222)

By appealing to a variety of experiments involving different courses of preparation for the reactants, Fourcroy thus draws attention to the *variability* of the production of fixed air (or "carbonic acid") as an effect to be explained, one for which he and his colleagues can account. It is, of course, a consequence of the explanation offered that *pure* metallic calxes will yield no fixed air on reduction. In the concluding section of the French reply to Kirwan, Guyton de Morveau elaborates the point by drawing on an experiment of Priestley's.

> It is known that no carbonic acid is afforded in reductions, except when the acidifying vital air is resumed by charcoal, and that its quantity when produced, is in proportion to the small quantity of charcoal which the vital air accidentally meets with. The latter experiments of Dr. Priestley have put these truths out of all doubt.[76] He calcined iron in vital air by means of the burning glass; he reduced it by means of very dry hydrogenous gas upon very dry mercury; he found water in a quantity corresponding to the weight which the iron had lost, and to the weight of gas absorbed; the remainder was hydrogenous gas as before, and did not contain any fixed air at all. (In Kirwan 1789 293)

These responses do not simply clear up the problem posed for Lavoisier's chemistry by the existence of experiments which appear to be of type (Riv). They directly challenge Kirwan's own account. For recall, the composition of the calx is supposed to depend *not on the types of precautions in preparing reactants on which the French chemists insist but on the temperature at which calcination occurs.* If it can be shown (by experiments carried out by phlogistonians, no less!) that variations in the production of fixed air are not associated with the temperature of calcination, then Kirwan's attempt to combine [8a] and [8b] runs into inconsistency.

I have traced one line—albeit an important one—through the phlogistonian controversy of the 1780s. We have seen how the general approach to reasoning adopted in earlier sections—elimination of alternatives through the production of inconsistency, forced retreats that open problems, employment of background constraints—is concretely instantiated in the arguments of the participants. Yet the story I have told is, for all its greater complexity than Legend's tales of single crucial experiments, far too simple. The experiments on calcination and reduction that I have considered were only a part of the full body of experiments that both sides invoked. I claim that the rest of the story is more of the same. Lavoisier and his colleagues successfully turn back phlogistonian arguments, and they develop difficulties with phlogistonian accounts of other types of reactions. I shall close this section with far briefer discussions of these other aspects of the debate.

Acid-metal reactions took up just as much space in the chemical exchanges as did the results about calcination and reduction. Participants agreed that iron combines with nitrous acid (in water solution, to prevent a reaction that Lavoisier describes as "too tumultuous" (1782 516)) to yield the calx of the metal, together with nitrous air and water. On Lavoisier's interpretation, nitrous acid is composed of nitrous air, water, and oxygen (vital air); the reaction separates these constituents and joins the oxygen to the metal to form the calx. Kirwan, committed to the notion that the calx formed contains

---

76. At this point, Guyton refers to (Priestley 1775 III 82).

fixed air, is constrained by other experiments on the formation of nitrous air and nitrous acid, to offer the folowing scheme of interpretation:

[11]   Nitrous acid = water + fixed air + nitrous basis + phlogiston
       Nitrous air = nitrous basis + phlogiston
       Calx = basis of iron (iron − phlogiston) + fixed air

This scheme is not only more cumbersome than that proposed by Lavoisier and his colleagues. Once again, Kirwan makes himself vulnerable by taking on an unsolved problem. To make the accounting balance, he needs to identify fixed air as a constituent of nitrous acid, thereby inviting requests to show that nitrous acid contains fixed air (or carbonic acid). As in the case of reduction reactions that yielded fixed air, the French chemists insist on precautions in preparing reactants. Admitting that small amounts of carbonic acid can be discharged from nitrous acid at the beginning of reactions, Berthollet notes that the reaction can be stopped so that pure acid is obtained:

> All the nitre which is decomposed after this period, affords no carbonic acid; the presence of this acid is therefore nothing but an accident, and we shall hereafter explain why this accident takes place.
>
> By following the different decompositions, we find, in them all, the two principles we have acknowledged, and nothing else. (Berthollet in Kirwan 1789 118)

As with calcination and reduction reactions, Kirwan is squeezed. His approach to calcination, founded on [8b], commits him to finding fixed air in nitrous acid, and, when faced with Berthollet's results, he is again forced to inconsistency.

By the mid-1780s, champions of phlogiston were clearly on the defensive. Nevertheless, the controversy was not simply a matter of seeing whether their framework could respond to a mounting number of difficulties. Kirwan, Cavendish, and Priestley try to show that Lavoisier's doctrine is problematic, and part of the force of his case consists in the ability to turn back the challenges. One important line of objection concerns the table of affinities which Lavoisier had published in 1782. According to Kirwan, Lavoisier's account of the affinities of various substances for vital air is inconsistent with various experimental results. Allegedly, substances that have lesser affinity for oxygen (according to the table) are capable of combining with the oxygen attached to a substance with greater affinity for oxygen (1782; 1789 41). Lavoisier responds to this by acknowledging that his table of affinities is defective, insisting that any such table has to be understood as valid for only one temperature, and pointing out that the difficulties urged by Kirwan tell more seriously against the phlogistonian approach (Lavoisier in Kirwan 1789 45–55). As I mentioned earlier, the enunciation of principles of affinity that will cover all the instances of chemical combination is an unsolved problem for *all* chemists of the period, and Lavoisier contents himself with showing that he can do at least as well as his rivals.

Anyone who reads through the exchanges among Lavoisier, Priestley, Fourcroy, Guyton de Morveau, Kirwan, Cavendish, and Berthollet should be struck by the extraordinary difficulty of keeping in mind all the constraints assembled by experiment. Each interpretative move has consequences that ramify in ways that are very hard to foresee: introduce X as a constituent of Y and how will one deal with experiment 57? Lavoisier's own painful struggle to fashion a coherent interpretative scheme has been chronicled by Holmes, and it is worth closing by reflecting on the fact that a clear line of argumentation, emerging from the social process of debate and exchange, might, at an earlier stage, be cognitively impossible for the pioneering investigator who initiates that process. Holmes notes that, in 1776, Lavoisier regarded himself as discovering that respiration converts common air to fixed air—even though he had recorded this fact in 1774 and 1775. Holmes writes:

> How is it possible that he *discovered* this phenomenon only in the course of an experiment he performed between April and October 1776? I myself once regarded such a situation as so implausible that in an earlier draft of the present book I suggested that Lavoisier was imposing an ideal reconstruction, sacrificing historical veracity to the desire to develop a logical argument. Further examination, however, has led me to suppose that he was probably describing his mental development exactly as it appeared to him. His respiration experiment of April is consistent with the assumption that respiration only absorbs something from the air; for in order to reconstitute the original common air he had merely added to the residual air the air derived from mercury calx. His apparent neglect of his own prior knowledge that respiration also produces fixed air is another manifestation of a mental characteristic we have repeatedly observed. *When he concentrated his attention closely on a particular set of phenomena or relationships, he was apt to lose sight of other phenomena which, from a greater distance, appear obviously essential aspects of the problem he was attacking.* (Holmes 1985 65–66; final italics mine; see also 79)

If I have a quarrel with this exemplary analysis, it lies in the hint that the mental trait was an idiosyncrasy of Lavoisier's. Rather, I suggest, it was an endemic feature of the problem situation. Between 1772 and 1783, *nobody* could see how to fit together all the constraints and to manage a coherent interpretation. By chasing down numerous blind alleys (in ways that Holmes describes beautifully) Lavoisier was ultimately able to produce a wide-ranging argument, consisting of the logically simple steps characterized in my account of scientific reasoning, taken across a domain that was initially unsurveyable. That argument was appreciated by many of his contemporaries and (rightly!) led them to abandon the phlogiston theory in favor of the new chemistry. The entire episode serves historians and philosophers of science as a reminder that there is no need to invoke underdetermination, shifting standards, or conceptual incommensurability. To paraphrase one of Lavoisier's co-workers, *Nous n'avons pas besoin de ces hypotheses-là.* The actual arguments of scientists are much more interesting.

10. Worries

I shall conclude by looking, far too briefly, at some obvious concerns about the account of individual reasoning that I have offered. These concerns derive from three sources: first, the recognition that there are existing philosophical accounts of individual reasoning that treat scientific decision making far more precisely than I have been able to do; second, the possibility that my efforts at responding to underdetermination arguments only tackle relatively easy instances (albeit those that have figured in traditional thinking about scientific change) and that a more severe challenge can be generated by considering debates about the legitimacy of appealing to experimental evidence; third, and perhaps most dangerous, the intuitive sense that the processes for modifying practice that I have described would be unlikely to generate the kind of progress for which I have campaigned in Chapters 4 and 5.

10.1 The Charms of Bayesianism

The most prominent general account of scientific reasoning is that provided by contemporary Bayesians. Individual scientists should assign probabilities to hypotheses (probabilities which represent their degrees of belief in those hypotheses), and use Bayesian conditionalization to adjust those probabilities in the light of the evidence statements they come to accept in light of their encounters with nature.[77] It is both necessary and sufficient for rationality that degrees of credence be probabilities, subject to the proviso that no candidate hypothesis be assigned probability zero. Necessity is a consequence of the Dutch book theorem, and sufficiency stems from painful awareness that sources of formal constraints on prior probabilites are extremely hard to come by. Bayesians thus allow for the possibility that a rational agent might assign extremely low prior probability to an hypothesis that is, by our intuitive lights, the most plausible available. They take comfort in theorems showing the "washing out of the priors": given enough evidence, subjects who start from very different prior probabilities will have posterior probabilities that are arbitrarily close.

As several recent treatments have demonstrated (Rosenkrantz 1977, Horwich 1982, Howson and Urbach 1989, Earman 1992), Bayesianism has many virtues. It offers a unified account of the confirmation of hypotheses that can resolve some issues of underdetermination, can yield solutions to traditional logical puzzles about confirmation, and can explain the differential force of different types of evidence in *some* historical cases of scientific reasoning.[78]

77. This is the simplest version of Bayesianism, one which assumes that statements induced by perception receive probability one. For more complex accounts that allow "observation reports" to have probabilities less than one, see (Jeffrey 1965), and, for discussion, (Field 1978, Howson and Urbach 1989, Earman 1992).

78. The most impressive example here is Earman's account of evidence for general

Moreover, Bayesianism is clear, precise, and unified: there are results about proper reasoning and there is a single perspective from which these results flow. I anticipate the criticism that, with these virtues, Bayesianism is superior—perhaps as a descriptive theory, perhaps as a normative account—to the cumbersome approach of earlier sections.

However, the Bayesian perspective has important shortcomings. First, there is good psychological evidence that people do not naturally engage in Bayesian calculations (Tversky and Kahneman 1973, 1974, Nisbett and Ross 1980) and that explicit assignments of probability are sometimes quite different from relative intensities with which beliefs are held (see Goldman 1986 chapter 15, especially section 2).

However, this in no way undercuts the possibility that Bayesians might offer an excellent normative account. There is some residual appeal in the notion that the reasoning of scientists involves a fumbling grasp of cognitive strategies that the Bayesian methodologist brings to light, and that scientific reasoning would be improved if the heuristic procedures that approximate Bayesian patterns of reasoning were replaced by their precise counterparts.

There is an important similarity between my commendation of eliminative induction and the Bayesian suggestions.[79] I conceive of scientists using background constraints to devise a space of hypotheses and using statements generated from interactions with nature to eliminate all but one of the candidates. I allow that there may be residual doubt—possibly quite significant—about the constraints that are used in generating the space. The Bayesian will represent the same process by supposing that there is an initial assignment of nonzero probabilities to the hypotheses that comprises the space, together with the assignment of a probability to the "catchall" hypothesis that asserts the falsity of *all* hypotheses in the space. This adds to my representation both an assessment of the relative merits of the rivals and a quantitative evaluation of the possibility that the space has been wrongly constructed. As the work of elimination proceeds, the Bayesian apparatus can record just how the fortunes of the rival hypotheses are changing. So it would seem to refine the strategy that I view scientists as adopting, thus supporting the view that actual scientific reasoning is a fumbling approximation to something clearer and more sophisticated.

My claim, however, is that the extra detail is both unnecessary and arbitrary. Imagine two scientists reasoning about any of the questions that have occupied us in this chapter: one follows the eliminative strategy that I have recommended, the other pursues the Bayesian policy of assigning precise probabilities to candidate hypotheses. In what ways is the reasoning of the latter superior? Given the absence of constraints on prior probabilities, it is quite possible that the Bayesian arrives at some extremely unintuitive initial assignment, with the result that, when his colleague has come to accept the

---

relativity (see Earman 1992). Bayesians also struggle with the logical problem of old evidence (see Glymour 1980, Garber 1982, Howson and Urbach 1989, Earman 1992).

79. This is noted in (Earman 1992).

last remaining candidate, the Bayesian still assigns that hypothesis a very low probability. To be sure, there are convergence theorems about the long run—but, as writers from Keynes on have pointedly remarked, we want to achieve correct beliefs in the span of human lifetimes.

Critics of Bayesianism paint for us the picture of bizarre assignments of prior probabilities whose effects cannot be overcome sufficiently quickly, but it is important to recognize that the effects are *symptoms* of the problem. The root difficulty is that one *ought* not to partition the space of candidates in vast numbers of the ways for which the Bayesian allows. Furthermore, having recognized this, one can also see that *any* Bayesian partitioning of that space imposes an arbitrary and unmotivated structure. Only in special cases can responsible assignments of probabilities be made.[80]

Bayesianism also has trouble as an account of scientific reasoning because of its preoccupation with epistemically perfect situations, cases in which there is no inconsistency between statements that the scientist has reason to accept, no tensions that result from the demands that an open problem be brought within the scope of the accepted schemata. Just as Bayesians need some machinery for distinguishing reasonable from unreasonable prior probabilities, so too they need an apparatus to cope with those instances in which epistemic costs must be counted and escape trees explored. As we have seen in the reasoning of Darwin and Lavoisier, the hard work of reasoning does not consist in performing calculations of probabilities but in bringing into surveyable form the consequences of pursuing various responses to inconsistencies and open problems. As noted previously (footnote 48), a Bayesian might attempt to use the apparatus of decision theory to try to make more abstract and precise my account of epistemic costs and the ways in which such costs are generated. Yet, I suggest, any such account would be parasitic on the kinds of investigations I have outlined in general, and begun in particular instances. Without the notions I have introduced in discussing underdetermination problems or some surrogates for them, Bayesian accounts will always be remote from the intricate reasoning of scientific texts (let alone from the complexes of processes that underlie such texts). Moreover, the decision-theoretic apparatus would, I suspect, introduce a need for arbitrary precision, unmotivated specifications of probabilities.

Nevertheless, there are at least two ways in which Bayesian conceptions prove valuable. First, there are surely occasions on which scientists can use *non-Bayesian* forms of reasoning to assign responsible probabilities to candidate hypotheses and on which they can then proceed by using Bayesian

---

80. Wesley Salmon has articulated a position which combines the use of Bayesian conditionalization with the employment of plausibility arguments to set the prior probabilities (1967, 1982). He calls this "objective Bayesianism," and, in some scientific contexts, it may provide a useful refinement of the approach to reasoning that I endorse in the chapter. However it seems to me that such contexts are relatively rare. Philosophical purposes may sometimes be served by idealizing scientific contexts and using objective Bayesianism to sort out tricky methodological issues. My discussion of the role of optimality models in evolution and ecology (1988) is an effort in this direction.

conditionalization. Under these circumstances, the Bayesian apparatus can be grafted on to my preferred framework with improvement to both. Second, there are epistemological purposes for which idealization is an appropriate strategy. The precision of Bayesian accounts is welcome in posing and solving some methodological problems. One example of this is, I hope, the next chapter, where I shall suppose that scientists *can be idealized* as Bayesian decision makers. My excuse for this idealization is that, in attempting to study the results of interactions among cognitive agents, it is convenient to have a quantitative representation of each individual. The precision of Bayesianism may be artificial, but when we need precision we have nothing that is preferable to tolerating the artificiality.

10.2 The Birth Pangs of the New Science

In the previous section I claimed that the case for Kuhnian underdetermination could not be sustained by the example of the chemical revolution, even though this case has appeared to many to be a principal inspiration for Kuhnian epistemological relativism. Those mindful of recent work in the history and sociology of science will recognize that the impossibility of closing scientific debate by the employment of good reasoning has become almost a commonplace (see, for example, Shapin 1982). They will naturally wonder whether my reactionary claims are based on a careful choice of examples. So Kuhn was unlucky about the case of Lavoisier—never mind! there are still numerous other instances to show the shifts of standards, the loss of data, the possibilities of alternative responses to the same situation, on which Legend's detractors love to insist.

I explain the confidence of relativists rather differently. Without an account of individual reasoning that made contact with the actual arguments of actual scientists, sociological relativism could readily be fueled by citing the *general* possibility of underdetermination and gesturing at the difficulties with which scientific debates are often resolved.[81] Those who win by default are vulnerable when real opponents show up. So, I suggest, applying a realistic analysis of individual scientific reasoning to the favored examples will reveal that the claims of relativists have been premature. The examples of Lavoisier and Darwin are typical.

Nevertheless, one example is more threatening than the others. In their comprehensive study of the Boyle-Hobbes debate, Shapin and Schaffer have demonstrated that the traditional dismissal of Hobbes as a tedious crank whose opposition to the new program of experimental science could reasonably be

---

81. Of course, there have been philosophical attempts, most notably those of Laudan and Shapere, to provide views of scientific reasoning that would subvert relativist claims. Although I believe that these attempts have yielded important insights, I think that they have been unconvincing because of failure to reconstruct historical examples in some detail. The general account of the earlier sections of this chapter is an effort to improve on Laudan and Shapere; the discussions of Sections 8 and 9 try to do the necessary historical work.

ignored cannot be sustained (Shapin and Schaffer 1985).[82] Shapin and Schaffer reveal Hobbes as a subtle thinker who simultaneously opposed the conception of knowledge articulated by Boyle, the social arrangements for the acquisition of knowledge adopted by the early Royal Society, the use of the air pump as a tool for promoting knowledge, and the specific claims made by Boyle about the "spring of the air." Hobbes succeeds in advancing serious criticisms, especially of Boyle's apparently hyperpositivistic methodological views (Shapin and Schaffer 1985 81ff.). Thus, in this instance, it is initially plausible to claim that there is (Kuhnian) underdetermination, based on commitment to different views of knowledge, and, ultimately, to different "forms of life," so that there was no possibility of resolving the dispute *and thus endorsing the practices of the new experimental science* on the basis of superior reasoning.[83]

Nevertheless, I think that Shapin and Schaffer only succeed in exposing how complex and difficult it was to defend the central claims of the new experimental science. Hobbes's arguments about the proper form of social arrangement for obtaining knowledge raise questions about authority, testimony, and trust which will occupy us in the next chapter.[84] What I hope to do here is to show how the approach to the reasoning of individual scientists that I have offered in earlier sections enables us to see how to criticize the claim that differences in conceptions of knowledge and method made the debate between Boyle and Hobbes a standoff.

Shapin and Schaffer take for granted the general doctrine of underdetermination, supposing that there is no way in which opponents could be forced to concede Boyle's experimental results and that there are always ways of avoiding Boyle's explanations of those results. The former point is made forcefully by invoking Harry Collins's concept of the "experimenter's regress" (Collins 1985) and connecting it to the historical material. Shapin and Schaffer write:

> The claims Boyle made about his phenomena could be turned into matters of fact [i.e., accepted within the community: PK] by replication of the pump. But then other experimenters had to be able to judge when such replication

82. For a characteristically forthright presentation of the traditional view about Hobbes, see (Conant 1957 57).

83. The Wittgensteinian phrase "forms of life" is used by Shapin and Schaffer, whose central thesis is that the debate involved *not only* alternative conceptions of knowledge and of knowledge-gaining practice *but also* commitments to different ideas about the proper form of society. It is interesting to note that *Hobbes* saw the issue in a similar fashion: his own explanation for the dismissal of his ideas identifies "the hatred of Hobbes [*odium Hobbii*]" (Shapin and Schaffer 1985 379), which we might suppose to descend from the widespread opposition to his political theory. However, whether there is a tight connection or merely guilt by association is a question worth exploring.

84. I shall not attempt to reconstruct the arguments between Hobbes and Boyle about the appropriate way in which to organize cognitive labor. However, I believe that the conceptual apparatus that I assemble in Chapter 8 can be used to resolve such issues, both for historical debates and for present disputes.

had been accomplished. The only way to do this was to use Boyle's phenomena as *calibrations* of their own machines. (226)

They also note that, even if Hobbes were to accept the Boylean findings as genuine, he could—and did—reject the hypotheses that were supposed to explain those findings.

> He contended that, whatever hypothetical cause or state of nature Boyle adduced to explain his experimentally produced phenomena, an alternative and superior explanation could be proffered and was, in fact, already available. In particular, Hobbes stipulated that Boyle's explanations invoked vacuism. Hobbes's alternatives proceeded from plenism. (111)

This second type of challenge to Boyle leads very directly to the methodological differences between the two protagonists. For Hobbes could argue that any explanatory endeavor must deploy hypotheses that are "conceivable, that is, not absurd" (in Shapin and Schaffer 1985 362), and that plenist explanations stand out as superior on this score.

Shapin and Schaffer's conclusions here rest on the tacit supposition that any consistent position is defensible: they do not explore the kinds of difficulties that Hobbes might encounter in pursuing the lines of resistance here attributed to him, because they recognize that there is a consistent "way of going on" and they assume that there are no epistemic criteria that can be invoked to make more fine-grained evaluations.[85] The task of this chapter and its predecessor has been to show first that there is a basis for such criteria (the versions of the external standard) and second that some further standards (which are, we hope, in accordance with this basis[86]) are actually deployed within scientific debates. Before we concede that the Boyle-Hobbes debate is a standoff, we ought to see whether a more developed account of individual scientific reasoning will break the stalemate.

The issues involved are precisely those considered in earlier sections. Just as Galileo's telescope and Galileo's phenomena, Lavoisier's techniques for purifying reactants and Lavoisier's interpretations of experiments are bound up together, so too the idea of a properly working air pump is intertwined with Boyle's results. Strictly speaking, the epistemic threat here is not that of *regress* but of *circularity:* can we find a way to legitimize the employment of an instrument or a technique without presupposing the correctness of the controversial results to which it gives rise? Our previous discussion of examples suggests a way of breaking in to the circle. One establishes the reliability of an instrument (or technique) by connecting its performance to procedures that can be carried out *independently;* one shows the dependence of particular variations in the performance of the instrument on changes in the design or manufacture which can be understood by deploying independently accepted schemata. Lavoisier and his colleagues could show how dif-

85. More exactly, they believe that the only such criteria are "local," so that while discriminations can be made by those who are prepared to commit themselves in particular ways, clashes among rival forms of life cannot be settled by the invocation of such criteria.

86. The third and final subsection will take up the worry that this is *mere* hope.

ferent amounts of *independently recognized* carbon contamination yielded proportionate amounts of fixed air. Boyle could show how the deviations from his expected findings varied with the stopping power of the substances which he used to try to block leaks—and the variation in quality of plugs could be assessed and explained in uncontroversial terms. It is entirely characteristic of Boyle's argumentation to point out the ways in which various technological improvements—which can be assessed independently—yield a more complete collapse of the column of a Torricelli barometer placed within the globe above the pump.

Consider now Hobbes's attempts to respond to Boyle by offering alternative hypotheses about his experimental findings. Hobbes maintains that the space above the pump is not a vacuum, but a plenum, and that the various Boylean effects—the falling of mercury columns, the ability of the stopper to suspend weights, the death of animals placed within the globe—can be explained in terms of the violent circulation of the air. Shapin and Schaffer do not inquire whether there is a coherent explanatory practice—a set of explanatory schemata, in my terms—that we could ascribe to Hobbes and that would articulate these alleged explanations. Nor do they note Hobbes's difficulty in reconciling his basic explanatory idea (the rapid circulatory motion of the air particles) with Boyle's findings about the behavior of feathers in the globe above the pump. (Boyle discovered no signs of any "wind," even when he introduced vanes to agitate the residual air within the globe.)

Finally, Hobbes's standards for the legitimacy of explanation and his application of those standards in the case at hand were subject to important critiques. Seventeenth century natural philosophy brimmed over with controversies about the force of metaphysical arguments about the conceivability or absurdity of various entities (including vacua). Here we *may* be able to endorse the idea that the questions whether explanation requires conceivability and whether vacua are absurd could not be resolved by reasoning available to the participants. However, admitting a standoff on such questions does nothing to help Hobbes for his only means of offsetting the apparent epistemic gains of the Boylean enterprise (the explanations of the barometers, the suspended weights, the mice, the feathers, in terms of the evacuation of the globe) is to insist on the illegitimacy of explanations that appeal to vacua. If that issue is rightly seen as undecided, and if there is no promising line of escape from the problems that plenists face in coping with the variety of Boylean phenomena, then there is no underdetermination. Hobbes was insightful. He raised criticisms that could lead to refinements of the experimentalist project. Nonetheless, Hobbes was wrong.

Much more could, and should, be said. I have simply tried to indicate a line of response to the intricate and challenging claims advanced by Shapin and Schaffer. Working out the details must be a task for another occasion.

10.3 The Threat of Skepticism

The deepest worry about the account I have offered centers on the relationship between my account of individual reasoning and my account of scientific progress. In earlier chapters I have defended the old-fashioned idea that scientists aim at (and sometimes attain) significant truth about nature. If individuals modify their practices according to the procedures I have described in this chapter, what reasons do we have for thinking that they will be able to achieve this aim? Is there any guarantee that the procedures actually followed will deliver what my account of progress promises?

Some preliminary issues need to be disentangled here. First, my story about scientific progress supposes that the sequence of *consensus* practices is progressive. What I have focused on here is the reasoning used in modifying *individual* practices. It is in principle possible that while such reasoning is not well adapted for modifying individual practices in a progressive fashion, *communities* in which individuals modified their practices in these ways would nonetheless make progress. Although I believe that there is an important point here, I do not think that skeptical questions can be completely dodged by recognizing that the individual is only one part of a community, since, as I have suggested discussing large scientific changes, the result of interaction among scientists is frequently the construction of a line of reasoning that becomes available to all and that generates communitywide acceptance of some modification of practice. The account I have offered in this chapter (especially in the discussions of Darwin and Lavoisier) attempts to analyze the character of this kind of reasoning, and it would thus be a dangerous concession to allow that reasoning of this type was ill adapted for the promotion of progress.

A second issue concerns the exact nature of the question that is being posed. Do we seek "reasons for thinking" that our reasoning is likely to advance our epistemic ends? Do we want "guarantees"? Wants are one thing, reasonable expectations quite another. The fate of classic Cartesian epistemology should make clear that we cannot hope for a demonstration that a favored set of methods for generating beliefs about nature is bound to succeed. So, if we frame our question as, Can we show, without assuming anything about nature, that these procedures for modifying practice are bound to yield epistemically valuable practices (practices that contain correct schemata, pose genuinely significant questions, and offer true answers to them, and so forth)?, then the response has to be negative. There are three loci at which the question can be weakened: (a) we can allow that some of our current beliefs about nature may be used in providing the demonstration, (b) we can suppose that the goal is only to show that the procedures make it *likely* (rather than inevitable) that good changes occur, (c) we can demand only that practices *improve* rather than that they *attain* our epistemic goals.

The problem which we are confronting is a generic version of Hume's problem. Hume took induction as an exemplar of scientific reasoning and asked whether we could demonstrate (without making substantive assumptions about nature) that induction inevitably yields true conclusions. That question cannot be answered for the simple reason that the forms of induction we use sometimes yield false conclusions, and we might have inhabited a world in which these forms often yielded false conclusions. Many philosophical responses to Hume choose to weaken the question at one or both the loci (b) and (c). They retain the notion that we cannot presuppose anything about the character of nature, but suggest that we only have to show that induction is *likely* to succeed, or that induction will be as likely to succeed as other methods, or that induction will work *in the long run*.[87] None of the resultant problems is any more tractable than that posed by Hume, and for a systematic reason. Classical enterprises of justification establish conclusions by reasoning according to approved canons from accepted premises. Skeptics can win easy victories by severely limiting the set of accepted premises and the approved canons of reasoning. In particular, if the skeptic challenges us to justify a form of reasoning that is involved in producing all our beliefs about nature and imposes a condition of noncircularity to the effect that we must not make use of any information grounded in the form of reasoning in question, then we find ourselves with resources that are too weak to meet the challenge.

The motivations for *not* weakening the skeptical question at locus (a) are relatively clear. Were we to rely on our current beliefs about nature in justifying the procedures of reasoning through which we arrived at those beliefs, there is a serious danger that the entire enterprise would be infected with error. Perhaps we are merely engaged in self-congratulation, when, all the while, faulty methods are being validated by the flawed conclusions to which they give rise. Naturalists have to insist that this is a genuine possibility, which cannot be excluded by invoking some set of a priori principles and rules of inference that are beyond criticism. We should know, in advance of skeptical embarrassments, that some forms of the problem of classical justification are solvable and others are not. Unanswerable challenges can be generated by calling into question single claims, bodies of doctrine, or methods of modifying practice that are involved in all empirical investigations. Even though weakening the skeptical question at locus (a) recognizes that there are unanswerable forms of skepticism, there is no alternative to the project of using what we think we know to appraise the methods which we take to be reliable.

So I urge a naturalistic response to skeptical questions which commends projects of *local* self-criticism. The task is to scrutinize and to improve our favored methods of individual reasoning by drawing on current beliefs about nature and our relationship to it. Waiving concerns about the relationship between individual reasoning and consensus practice, we can conceive of the situation as follows: Contemporary science develops out of far more primitive practices by employing the forms of reasoning I have discussed in this chapter. Ultimately, the lineage may stem from some ur-practice, adopted by our hominid ancestors, who classified nature in a particular fashion and were disposed to project from their past experiences in a particular way. Using the

87. See, for example, the efforts of Feigl (1950), Reichenbach (1959), and Salmon (1963) to vindicate induction. There are well-known difficulties with these endeavors, stemming from the possibility that many alternative methods can differ radically from induction in the conclusions that they recommend in the arbitrarily large short run, and yet converge to the inductive expectations. As Keynes pointed out, there is also little comfort in the notion that induction will eventually succeed.

eliminative strategy discussed previously, subject to the principle of unification, the classificatory scheme of the initial language has been modified,
explanatory schemata introduced, and ever more elaborate sets of questions,
statements, techniques, and instruments generated. Given what we believe
about the starting point, the processes of modification, and the world that is
the object of investigation, can we show that the history is an optimistic story
of self-correction?

According to the hopeful picture, we began with rudimentary representations of nature and primitive notions of how to modify those representations
and gradually replaced them with cognitively superior representations and
strategies. But perhaps we began in so unfortunate a state that we are in-
capable of working ourselves into any accurate representations of nature. Or
maybe there are constraints on the processes of modification that prevent us
from making any significant improvements. Live versions of skepticism proceed by taking these possibilities seriously.

In some quarters it is fashionable to think that there is an easy response
to the first type of skeptical worry, that there is "encouragement in Darwin"
(Quine 1970a; see also Peirce 1958 VII 29–30, Ruse 1986, Rescher 1989a).
If our initial cognitive equipment were as poor as the skeptic portrays it as
being, then, the suggestion runs, our ancestors would have been eliminated
by natural selection. They weren't, so it wasn't. Darwinian evolutionary theory seems to support the notion that our initial ways of classifying stimuli
must correspond to objective regularities in nature, and our modes of reasoning must work reliably in producing accurate representations.

Sadly, invoking natural selection will not do the intended job. Human
brains have been assembled, over evolutionary time, out of structures originally selected for properties far removed from the capacity for pursuing
scientific investigations. To the counter that our brains must be minimally
competent at representing nature and reasoning about it there are three
obvious replies. One, in the spirit of the idea that the brain is the product of
evolutionary tinkering, simply denies that this minimal competence will take
us very far in establishing the reliability of the historical process out of which
contemporary scientific beliefs have emerged. The second, scrutinizing the
argument from selection in the style of contemporary neo-Darwinism, notes
that the selection pressures felt by organisms are dependent on the costs and
benefits of various consequences. We think of hominids on the savannah
requiring an accurate way to discriminate leopards and conclude that parts
of the ancestral schemes of representation, having evolved under strong selection, must accurately depict the environment. Yet, where selection is in-
tense in the way it is here, the penalties are only severe for failures to recognize
present predators. The hominid representation can be quite at odds with
natural regularities, lumping all kinds of harmless things with potential dangers, provided that false positives are evolutionarily inconsequential and provided that the representation always cues the subject to danger (Stich 1984,
1990). The third, reflecting on the most detailed available suggestion for
understanding the evolution of human reasoning, notes that selection seems

to have favored an ability to think about *social* situations and relations. Apparently we are far more successful at solving problems in reasoning when their topics are socially significant than when they deal with features of the asocial world (Cosmides and Tooby 1988).

It would be wrong to conclude that the appeal to Darwin actually reinforces skepticism. Our current knowledge of human cognitive evolution is too rudimentary to allow for much more than speculation. Instead of arriving at conclusions on the basis of analyses that have more kinship with Kipling than with Darwin, the appropriate response is neither optimism nor pessimism but agnosticism. Our ignorance is, however, remediable, and we can hope that more detailed accounts of human cognitive abilities and comparative studies of related organisms might adjudicate this form of skeptical question.

Yet whether matters turn out well or badly on the question of our initial state, the crucial issue surely concerns the possibility of successful correction of practice. How much improvement might we expect from continually deploying the strategies discussed in this chapter? Consider, for example, the use of eliminative induction. Here the relevant skeptical worry focuses on our aptitude for framing inductive problems. We might succeed in reviewing what we take to be the relevant variety underlying an inductive generalization, eliminating all the hypotheses we represent as rivals, and still be wrong because of the inadequacies of our ideas about possible hypotheses and dependencies in nature. The right response to this kind of skeptical concern seems to me to be prefigured in my discussion of the "pessimistic induction from the history of science." First, there is the clear possibility that we are relatively good at framing inductive problems with respect to some types of phenomena and rather bad at coping with others. Through psychological investigations and scrutiny of the historical record of the sciences, we can even hope to draw useful distinctions. Second, even before engaging in detailed studies of particular areas and inductive practices, we have ordinary empirical grounds for thinking that the varieties within our ordinary classificatory groups are well understood: our success in comprehending the kinds of factors that affect the behavior of metals, the colors of birds, the activities of members of our own species is reflected in our own ability to produce variation and in nature's apparent inability to surprise us with novel variants.

However, there is a far more serious skeptical worry generated by the thought that the history of science is filled with debates that could not have been settled by deploying the types of reasoning that I have canvassed. Skeptics contend that, at a number of times in the historical process that has led to current science, there have been alternative possibilities for modifying practice, each of which might with equal justice have been adopted. We can imagine several possible histories of science, yielding divergent conceptions of nature and rival sets of canons of reasoning, in each of which the protagonists retrospectively praise past decisions as exercises in self-correction. Because nothing distinguishes the actual course of events from these potential histories, there is no basis for concluding that the actual evolution of science is self-correcting while the others are not.

This species of skepticism is founded on the idea that underdetermination is omnipresent. It can only be resolved through examination of individual instances in which alternative modifications of practice are alleged to be equally defensible. Unlike more general forms of skepticism, this line of attack recognizes the importance of empirical information in assessing the status of our methods of reasoning and offers the deadly suggestion that *when judged from the scientific perspective to which they have given rise, the methods actually employed by scientists are too weak to show the correctness of that perspective.*

Precisely because I view this type of skepticism as potentially damaging, I have tried, in this chapter, to show that we can give a realistic account of scientific reasoning that resolves those situations in which underdetermination is supposed to be present. Although it may initially seem that studies of the overthrow of the phlogiston theory or the triumph of Darwinism are remote from fundamental epistemological problems, I believe that misjudgments about these cases foster the most biting form of skepticism. Our failure to provide transcendental guarantees of the effectiveness of scientific reasoning should not worry us. The thought that reflections on the growth of science undermine the credentials of the enterprise should.

If my attempts to classify skeptical questions are successful then the appropriate strategy is "divide and conquer." Certain global worries should be dismissed as on a par with requests to trisect angles with ruler and compass or to show the consistency of first-order arithmetic within itself. Other concerns must be addressed by patiently studying the available historical, psychological, and sociological evidence to see whether, and to what extent, the kinds of reasoning I have discussed are likely to yield epistemically valuable practices, given our best information about ourselves and the world with which we interact. Quite evidently, I have only scratched the surface of this general problem, but I hope that one particular aspect of it has been dealt with more thoroughly. Despite the currency of the idea that historical choices have been underdetermined by reason and evidence, a closer look tells us otherwise.

8

The Organization of Cognitive Labor

1. Introduction

The general problem of social epistemology, as I conceive it, is to identify the properties of epistemically well-designed social systems, that is, to specify the conditions under which a group of individuals, operating according to various rules for modifying their individual practices, succeed, through their interactions, in generating a progressive sequence of consensus practices. According to this conception, social structures are viewed as relations among individuals: thus my departure from the tradition of epistemological theorizing remains relatively conservative.[1] If we remind ourselves of the framework introduced in Chapter 3, then the general problem can be resolved into more specific instances. If people join the scientific community through the socialization process I have sketched; if they form their divergent individual practices partly through interacting with nature and thinking by themselves, partly by borrowing from and lending to others; if what is transmitted to the next generation is shaped by their individual decisions and by their interactions, how will the whole system best work to promote a progressive sequence of consensus practices? Is it possible that the individual ways of responding to nature matter far less than the coordination, cooperation, and competition that go on among the individuals?

In what follows, I shall focus on two clusters of problems. The first set is concerned with scientists' responses to others. The second considers the effects of individual efforts on communitywide belief. So, in the earlier sections of this chapter, I shall look at action *on* the individual. Later sections will study the effects that stem *from* the doings of the individuals who make up a scientific community.

---

1. More radical suggestions for socializing epistemology can be found in (Fuller 1988), (Rouse 1987), (Longino 1990), and the writings of Bruno Latour. Many of the previous chapters tackle the ideas and arguments that lead these writers to break with traditional thinking about human knowledge. A different line of objection, posed forcefully to me by Steven Shapin, is that the species of methodological individualism that I deploy in articulating my version of social epistemology cannot be sustained. For the moment, I shall rest content with challenging those who believe that my individualistic framework is too narrow to offer examples of social aspects of knowledge that cannot be accommodated within it.

In considering the first group of problems I conceive of trust in others as essential to scientific activity.[2] For an active researcher within a scientific community, there are important questions about the assignment of trust: Whom should one trust? When should one trust others more than oneself? When is it worth risking the errors that others might make? I shall explore some ways in which researchers might address these questions and consider the impact of their decisions on community-wide research.

The primary topic of the second group of questions is the distribution of effort within scientific communities. The framework of Chapter 3 depicts individuals with different individual practices pursuing diverse projects. To inquire about the structure of a well-ordered scientific community is to ask how these projects should be pursued so as best to promote the community project. How much division of effort is desirable? How can diversity be maintained in a scientific community? How should consensus be formed?

My general approach to both sets of questions is to draw on what we believe about the available possibilities for individual reasoning and the co-ordination of individual effort. Reflecting on the history of science and on current science, we can seek to identify recurrent problem situations that scientists face. At this point, there are two types of inquiry that are worth pursuing: first, we want to know what, given the range of possibilities, is the best approach to the problem situation in which we are interested; second, we should scrutinize which of the available combinations of individual decision procedures and sets of social relations would move the community closer to or further away from the optimal approach. More pedantically, given a problem situation, we seek

1. The optimum community response, A_{opt}, conceived as a distribution of the efforts of the individual members.
2. For each combination of individual decision rules and social relations, $\langle \{D_i\}, \{R^s_{ji}\} \rangle$, a representation of the success of that combination in terms of its distance from the expected effect of A_{opt}.

Through the examination of a range of instances of recurrent problem situations, we can hope to understand the impacts that various systems of social and individual decisions have on the growth of science.

The inquiries just envisaged are often too ambitious. Optimal solutions to decision problems are usually only available in highly simplified situations. Much of this chapter is devoted to examining the kinds of equilibrium distributions of epistemic effort that would emerge under various types of social

2. This point is elaborated extensively by Shapin in forthcoming writings. He is largely concerned with broader issues of trust than those that occupy me here, particularly the honesty of potential informants. As will become apparent, I shall be concerned more with assessing competence than evaluating honesty, but this should not be taken to imply that issues about sincerity are unimportant, either in Shapin's prime historical context (seventeenth-century Europe) or in the context of contemporary science. Ironically, many of the recent instances of scientific fraud reveal the extent to which scientists operate on trust.

conditions and to exploring whether these distributions enable us to avoid certain types of cognitive disasters. Instead of thinking about how best to achieve a cognitive goal, we can consider whether one type of social arrangement avoids a pit into which another falls.

Although my analyses are very abstract, they ultimately connect with concrete issues about science policy. I do not find it troubling that questions of social epistemology might have practical consequences—enabling us, for example, to consider the merits of rival systems for awarding grants or refereeing scientific contributions. Philosophy of science should earn its way by trying to draw specific morals for the organization of scientific research. But I would caution against overinterpreting my results: although my analyses reveal neglected epistemic possibilities, they are far too idealized to enable us to be confident in reaching conclusions about practical strategies for (say) funding research. *Perhaps* that can come later.

In pursuing these problems, I shall employ an analytic idiom inspired by Bayesian decision theory, microeconomics, and population biology. The advantage of this idiom is that it enables me to formulate my problems with some precision, and that precision is important for both identifying consequences and disclosing previously hidden assumptions. Precision is bought at the cost of realism.[3] My toy scientists do not behave like real scientists, and my toy communities are not real communities.

We can think of the problems that concern me as including those that would face a philosopher-monarch, interested in organizing the scientific work force so as to promote the collective achievement of significant truth. Science, of course, has no such benevolent dictator. In consequence, individual scientists face coordination problems. If we suppose that they internalize the (fictitious) monarch's values, how will they fare? If we assume instead that they are motivated in baser ways or that they are locked into systems of authority and deference, will they necessarily do worse than a society of unrelated individuals, each of whom is pure of heart?

The principal moral of my discussions is a cautionary reminder: do not think that you can identify very general features of scientific life—reliance on authority, competition, desire for credit—as epistemically good or bad. Much thinking about the growth of science is permeated by the thought that once scientists are shown to be motivated by various types of social concerns, something epistemically dreadful has been established. On the contrary, as I shall repeatedly emphasize, particular kinds of social arrangements make good epistemic use of the grubbiest motives.

Beyond this, I want to note that the details often matter. Although we shall see that competition is frequently helpful in enabling a community to

3. *Some* degree of formalism is needed in the text to enable readers to understand the kinds of conclusions I draw about idealized situations. But I have tried to banish the most technical points into separate "Technical Discussions." The main conclusions can be understood by omitting these, and this might prove most satisfactory on a first reading. Those interested in seeing how the conclusions are derived, and under what kinds of circumstances they obtain, can then return to the Technical Discussions.

achieve valuable cognitive diversity, there are also cases in which competition is impotent or in which it generates unwelcome homogeneity. Only close attention to the pressures actually present in scientific situations will teach us whether the social systems we employ do a good job of coordinating the efforts of individuals.

My investigations should be viewed as mapping a space within which identifications of the epistemically important characteristics of scientific communities can be made. They suggest obvious empirical questions: Can real scientists be construed as hybrid agents, torn between their epistemic and nonepistemic goals, and, if so, how much weight do they give to defending truth as opposed to receiving credit? How do scientists actually evaluate the chances that a proposal for modifying practice will succeed? I do not suppose that answering these types of questions will be easy. However, if my approach is right, then they are the kinds of questions that should be asked in developing a more substantive account of the growth of scientific knowledge in communities of scientists.

2. Authority

Reliance on authority affects all our cognitive lives, and, for present purposes, we can distinguish three ways in which it permeates the cognitive lives of scientists. First, there is the general epistemic dependence on the past that figures in everyone's early intellectual ontogeny. We absorb the lore of our predecessors through the teaching of parents and other authorities. Second, at the time of entry into the scientific community, novices endorse a communitywide conception of legitimate epistemic authority. Certain people are to be trusted to decide on certain issues, and the novice must accept whatever agreements they reach on those issues. Third, during the course of individual research, scientists interact with one another, adopting the claims made by *some* of their colleagues, investigating the proposals of others, ignoring the suggestions of yet others, when the claims, proposals, and suggestions in question go beyond what is agreed upon by the pertinent community.

A major theme of the foregoing chapters is that the overarching kinds of authority that fall under the first two types do not trap us in inevitable error. My accounts of progress and of individual reasoning attempt to develop the idea that we work our way free of the mistakes of earlier generations through further encounters with nature. But if this optimistic picture is to be sustained, it must be the case that the *third* type of attribution of authority does not interfere with the process of self-correction, but works constructively to further the community project. I shall be concerned with this type of attribution of authority, the differential assessment of peers.

To bring the problem into focus it is worth recalling some episodes from recent science. In the spring of 1989, two electrochemists, Stanley Pons and Martin Fleischmann, held a celebrated press conference, at which they announced the possibility of obtaining cold fusion on a tabletop. Prior to the announcement that claim was incredible: according to the consensus practice of nuclear physics, fusion cannot be achieved at room temperatures in the fashion that Pons and Fleischmann described. However, the payoff if Pons and Fleischmann were right would be enormous, and immediately after the press release laboratory telephones began to ring. Electrochemists knew and respected Pons and Fleischmann and, accordingly, took their apparently incredible claim seriously. Outsiders from physics had typically never heard of either Pons or Fleischmann. Many of the telephone calls they placed posed the same questions: Who are Pons and Fleischmann? Can they be trusted? When told of the high standing that Pons and Fleischmann had within the electrochemical community, the interested physicists began to consider the possibility that the outlandish finding might be right. So a significant expenditure of scientific effort was begun, as numerous phsyicists and chemists tried to replicate the Pons-Fleischmann experiment.

Contrast this situation with another. During the past twenty years, self-styled creation scientists have made periodic announcements about the co-presence of dinosaur and human tracks in the same strata and about the existence of human artefacts which, when dated by standard techniques, yield ages comparable to those of supposedly ancient rocks. Their pronouncements challenge paleontology just as Pons and Fleischmann questioned our understanding of nuclear fusion. But Duane Gish, Henry Morris, and their colleagues at the Institute for Creation Research do not inspire the same dedicated investigations. They catch the ear of the scientific community only when they are able to threaten, only when they have demonstrated an ability to influence legislators and publishers, making it necessary for scientists to divert time from profitable research to the enterprise of rebutting creationism.

The differences between the two cases are readily traced to differences in authority. Pons and Fleischmann had considerable authority among electrochemists, and they obtained authority among physicists because they had authority for some people who had authority for some physicists. Gish and Morris have no authority among paleontologists, nor do they have authority with anyone who has authority for paleontologists. However, though this contrast may throw into relief some basic features of the attribution and withholding of authority in science, more mundane illustrations are useful for bringing other points to our attention.

Consider anyscientist working in anylab. During the course of a day's work there will be numerous opportunities for relying on others. Some will be taken; others will not. Particular parts of the day's project will be assigned to technicians, graduate students, support staff. The scientist will typically perform other tasks herself. In addition, there may well be special opportunities to redesign some aspect of the course of research. Perhaps a journal arrives with an article which, if sound, would enable a time-consuming procedure to be abbreviated. Or a grant proposal, sent for review, may suggest an alternative sequence of experiments. Or a new catalogue may offer a novel version of a relevant piece of apparatus. All these opportunities, both those pursued or dismissed in the quotidian assignment of jobs and the more special chances, call for a decision by anyscientist: Should I do this myself or rely on someone else? Can I trust X to do A? Is this procedure/instrument/technique reliable?

Understanding the role of authority in science requires us to probe these decisions. How should they be made? What kinds of individual decision rules and social relations facilitate the making of such decisions in ways that will profit the community to which anyscientist belongs? I shall start at the most general level, investigating the conditions under which cooperation among scientists, the relying on others or the borrowing from others, is worthwhile.

3. Cooperation

The most obvious advantages of deference to authority are that it enables individual scientists to pursue their epistemic projects more rapidly and makes feasible investigations that would be impossible for a single individual. Suppose that a scientist is dedicated to a particular inquiry: the scientist's overriding concern is to bring this inquiry to a conclusion by discovering the true answer to a particular question. In terms that I shall employ more systematically in later sections, this imagined scientist is a *pure epistemic agent,* one for whom the primary goal is to reach an epistemically valuable state. I further suppose that the scientist has total resources (time, energy, money) E and needs k items of information. Let the cost of acquiring each directly be C, the cost of acquiring each from an authority be c. The scientist's project is individually impossible but cooperatively feasible just in case:

$$kc < E < kC. \tag{3.1}$$

We can assume that the investigation consists of a period in which the needed information is acquired followed by a period in which the scientist attempts to put the information to use. The chances of success in the project can be written as the product of two probabilities: the probability that the information acquired is correct and the probability of deployment of correct information at the second stage. The latter can be written as $F(E -$ acquisition costs), where F is nondecreasing and $F(x) = 0$ if $x < 0$.[4]

Consider the simplest case in which $k = 1$. You have two choices: you can do the work yourself or you can rely on authority. Which decision is preferable? It depends on two error rates, yours and that of your potential authority. Assume that borrowing is cheaper than doing the work yourself, $c < C$. Let your error rate be p, the potential authority's error rate q.[5] As

---

4. Of course, the function F cannot be entirely arbitrary, but must meet the conditions on probability (thus, for example, taking only values in $[0, 1]$).

5. A cautionary note on notation. In the sections that follow, numerous probabilities and other parameters will be introduced. Given a limited lexicon, the same symbols have sometimes been assigned different interpretations in different sections. Although I have tried to maintain uniformity in notation for related problems, the reader should not assume that symbols employed in sections dealing with very different topics have the same meanings.

noted, I suppose that your decision is motivated by the desire to bring this project to a correct conclusion. Then you should rely on authority if

$$(1 - p) \cdot F(E - C) < (1 - q) \cdot F(E - c). \tag{3.2}$$

This inequality is automatically satisfied if the potential authority is more expert at the relevant task than you are—$p > q$—for recall that $c < C$ and F is nondecreasing. But even if you are more reliable than the potential authority it may still be worth your while to take the risk of borrowing.

To see this, imagine that $F(x)$ is (a) 0 when $x < 0$, (b) mx when $0 < mx < 1$, (c) 1 when $mx > 1$ (where m is a constant representing the rate at which resources increase the probability of finding a solution). Assume also that $E - C > 0$ (it would be possible for you to do the work yourself and still have a chance of succeeding in the project), and that $m(E - c) < 1$ (the project is sufficiently complex that you cannot have a probability > 1 of success, even if you borrow). Borrowing is epistemically preferable if

$$q < [(C - c) + p(E - c)]/(E - c). \tag{3.3}$$

Even when you are perfect ($p = 0$), if the costs of borrowing are negligible ($c = 0$), you can tolerate a maximum error rate of C/E in your potential authority, and C/E may be sizable if you would have to expend a lot of your resources on acquiring the information directly.[6]

The case in which the probability function F is linear is intermediate between two others. If a new piece of information would yield rapid initial returns, then, while it is still sometimes worth borrowing from others, the maximum tolerable error rate is decreased. By contrast, if you would have to expend considerable efforts, once the information is acquired, in learning how to use it most effectively in your project, then it will sometimes be worth borrowing even from extremely unreliable sources.

Up to this point, I have been assuming that the scientist's decision is dominated by a particular type of epistemic intention: the goal is that the individual scientist will solve the problem at hand. It is possible to imagine an even more epistemically devoted scientist, one who cares only that the problem be solved and who is prepared to pool resources with others in a richer cooperative effort. A scientist of this altruistic bent would engage in a different type of decision making, surveying the scientific community to ensure that her own efforts worked with those of others in advancing the community's understanding.

6. As noted in the text, the discussion proceeds on the assumption that the project is relatively complex. If we relax this assumption, there are two subcases:

(a) The project is so straightforward that you can attain probability 1 of getting a solution whether you borrow or not; $m(E - C) > 1$ (a fortiori, $m(E - c) > 1$); unsurprisingly, the maximum tolerable error rate is now p; i.e., it is only worth borrowing from those who are more reliable than you.

(b) You can achieve probability 1 of succeeding if you borrow, but not if you do not; $m(E - c) > 1$, $m(E - C) < 1$; now the maximum tolerable error rate is $(1 - m(E - C)) + m(E - C)p$; it is easy to see that this value is greater than p, so that, here, it is worth borrowing from people who are less reliable than you are.

When I consider cognitive diversity in later sections, I shall, in effect, be exploring the kinds of ideal distributions of effort that altruistic epistemically pure agents would aim to achieve. For the moment, however, I am interested in looking at the cooperative inclinations of a different type of agent, an *epistemically sullied* agent, one who is driven not only by a desire to solve the problem, but also by the quest for priority which Merton (1973) emphasizes.

Consider the predicament of an epistemically sullied scientist, X, engaged in a research project, when some peer, Y, announces a result that would, if correct, provide a way to simplify the investigation. X reasons as follows: "Suppose I make use of Y's result; there are two possibilities, Y is wrong or Y is right; if Y is wrong, then there is no chance of my solving the problem; if Y is right then I will have a probability $F(E - c)$ of arriving at a solution; my rivals, whose resources are equal to mine, who also borrow from Y will have an equal chance of solving the problem; if n of us borrow from Y and $N - n$ do not, then the expected number of problem solvers is $nF(E - c) + (N - n)(1 - p)F(E - C)$, so that, given that I produce a solution, my chances of being the first are $1/[nF(E - c) + (N - n)(1 - p)F(E - C)]$. Conversely, suppose that I do not borrow from Y; again, there are two possibilities, Y is wrong or Y is right; either way, I have a chance of getting the correct result if I set out to get the needed piece of information myself, and, given that I get the right result, then there's a chance of $F(E - C)$ that I will produce a correct solution to the whole problem; if Y is right, and n members of the community borrow from Y then the expected number of those who arrive at a correct solution is $nF(E - c) + (N - n)(1 - p)F(E - C)$ as before; if Y is wrong, then the expected number of competitors is $(N - n)(1 - p) F(E - C)$; either way, my chance of being the first solver is inversely as the expected number of solvers."

This, I suggest, is eminently sensible reasoning for someone whose goal is to be the first solver of a scientific problem, a *scientific entrepreneur*. It involves a number of important assumptions. First, my imagined X makes no distinction of talent: all the N scientists who are engaged in trying to solve the problem are envisaged as equally likely to succeed. *One* way of relaxing this assumption would be to suppose that X reasons not about the actual community, but about a "virtual community" in which talented scientists are allowed to count more than once. (Imagine that a community of five sound but undistinguished scientists and one superstar is treated as of size ten, with the superstar being regarded as equivalent to five "regular" scientists.) Second, X takes the probability that he, *or any of the others*, will achieve the correct version of Y's result, if they do not borrow, to be the same value $1 - p$. In fact, X need merely assume that his own error rate is the average of those whom he takes to be in competition with him. Finally, X does not allow for any kind of partial credit: there is no payoff for being second or for showing that Y's result is indeed right.

At this stage I want to point to a moral that will become familiar later. From the community perspective, it is likely that sullied scientists will do better than the epistemically pure. This is because a pure community heads toward cognitive uniformity: either all the members find it worth borrowing from Y or they do not. By contrast, as Technical Discussion 1 shows, in the sullied community, there are ample opportunities for division of cognitive labor. Some follow the strategy of aiming for a quick victory by borrowing from $Y;$ others work independently. In this way, the sullied community hedges its bets.[7] That is, intuitively, a good thing. Later, I shall try to supply some arguments that would underwrite intuitions of this general type.

So far, I have assumed an unrealistic condition of symmetry among my imaginary scientists. All are supposed to have the same resources at their disposal, to assess the error rates of themselves and others in the same ways. By introducing asymmetries into the treatment of the entrepreneur's predicament, we make it easier for different members of the community to pursue different strategies. This is readily evident in instances in which, for one member of the community p is zero, while, for another, $p > q$. (Intuitively, one is well qualified to pursue the type of work that would lead to acquisition of the information Y promises, while the other is not expert in such matters.) The latter will find it profitable to borrow from Y, while the former will only borrow if the problem is sufficiently complex ($F(E - C)$ is sufficiently low).

There is a different type of asymmetry, one stemming from differences in resources. As Technical Discussion 1 shows, we can expect those who have fewer resources sometimes to take far greater risks in borrowing information than would be justified by a sober epistemic assessment. If the scientific community is divided into those who have relatively large amounts of resources and those who have less, then we may expect that the latter group will include some members who pursue what seem to be rather unpromising ways of solving problems, while the resource-rich adopt more conservative strategies. As I have already intimated, it is likely that this type of cognitive diversity is no bad thing from the community perspective.

Technical Discussion 1

Borrowing for Sullied Agents

The decision tree in Figure 8.1 captures (the first version of) THE SCIENTIFIC ENTREPRENEUR'S PREDICAMENT.

The expected utility of borrowing from Y is

$$U \cdot (1 - q)(F(E - c))/(nF(E - c) + (N - n)(1 - p)F(E - C))$$

where U is the payoff for being the first solver.

The expected utility of not borrowing and obtaining the information by oneself is

---

7. So presumably would a community of epistemic altruists, for they would adjust their behavior to what others were doing and attempt to maximize the community's epistemic utility.

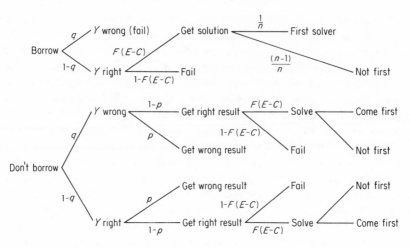

Figure 8.1. The Scientific Entrepreneur's Predicament.

$$U\{q/(N - n) + (1 - q)(1 - p)F(E - C)/(nF(E - c) + (N - n)$$
$$(1 - p)F(E - C))\}.$$

The community of scientists would reach a stable state—involving a division of the population into n who borrow and $N - n$ who do the work themselves—if these expected utilities were equal. This would occur if

$$n = ((1-q)F(E-c) - (1-p)F(E-C))/(F(E-c) - (1-p) \qquad (3.4)$$
$$F(E - C)).$$

If q becomes too large (the potential source is too unreliable), then the numerator will be negative; in this case there will be no borrowing, and everyone in the community will do the work of obtaining the information. Unless q is zero, the numerator is always strictly less than the denominator; hence, provided that the community is not too small, there will always be some people who do not borrow. The condition for there to be some borrowers is that the numerator should be positive, that is,

$$q < (F(E - c) - (1 - p)F(E - C))/F(E - c). \qquad (3.5)$$

If doing the work oneself is very expensive in terms of resources, and if the problem of applying the information is sufficiently complex, then $F(E - C)$ will be small in comparison with $F(E - c)$, and, in these circumstances, very high error rates in the potential source can be tolerated.

Suppose, for simplicity's sake, that F is linear provided $0 < x < K$, and that $E < K$ (where K is a constant). Assume $p = 0$ (the potential users of Y's information are perfectly reliable at obtaining such information by themselves), that $c = 0$ (the costs of borrowing are negligible), and that $C = rE$ where $0 < r < 1$. Then, if $r > q$, some members of the community will borrow from Y, and there will be an internal equilibrium state at which the number of borrowers is given by

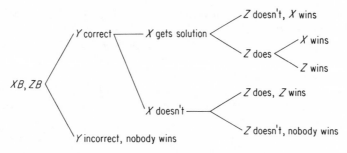

Figure 8.2

$$n = 1 - \frac{q}{r}. \tag{3.6}$$

Provided that $q > 0$ (Y is not perfectly reliable) and the community is sufficiently large, there will be some members of the community who do the work themselves.

Asymmetry in Resources

Imagine that there are two competing scientists X and Z, and that a third scientist, Y, announces a result that, if correct, would simplify the project on which both X and Z are engaged. Each of the competitors has two strategies: borrow (B) and do the work independently (I). I shall suppose that both assess their own error rates and each other's error rate at the same value p, that they both evaluate Y's error rate as q, that the costs, C, c, are the same for both, but they have different resources, E_1 for X and E_2 for Z. The payoffs to each are dependent on what the other does, and X and Z are effectively playing a two-person game. For each pairwise combination of strategies, the payoffs are determined by considering decision trees that trace the possible outcomes. Figure 8.2 is the tree for the combination (B, B).

Following this approach, we achieve the following payoff matrix:

		Z	
		B	I
			$q(1-p)F(E_2-C)^8$
X	B	$(1-q)F(E_1-c)\{1-F(E_2-c)/2\}$	$(1-q)F(E_1 - c)$
		$(1-q)F(E_2-c)$	$q(1-p)F(E_2- C)\{1-F(E_1-C)/2\}$
	I	$q(1-p)F(E_1-C)$	$q(1-p)F(E_1-C)\{1 - F(E_2-C)/2\}$

8. Here I assume that Z has no chance of achieving the first solution if Y is correct. Intuitively, X's head start from borrowing will make it impossible for Z to catch up. This contrasts with the analysis offered earlier, in which I did allow for the possibility that a nonborrower might be the first solver, even in instances where the borrowed result was correct.

Understanding the outcomes of the competitive interaction between X and Z involves identifying the equilibria for this game. The major features of the situation can be discerned from considering the special case in which F is linear, with $F(x) = x/E_1$ for $0 < x < E$, $p = 0$, and $c = 0$. In this case, the payoff matrix becomes:

$$
\begin{array}{cc}
(1-q)E_2/2E_1 & q(E_2-C)/E_1 \\
(1-q)(1-E_2/2E_1) & (1-q) \\
(1-q)E_2/E_1 & q(E_2-C)(E_1+C)/2E_1^2 \\
q(E_1-C)/E_1 & q(E_1-C)(2E_1-E_2+C)/2E_1^2
\end{array}
$$

It is not hard to show that there are instances in which both scientists prefer to borrow (BB is an equilibrium), and instances in which both prefer to work independently of Y (II is an equilibrium). The only instances in which the combination (BI) (X borrows, Z works independently) is an equilibrium are the trivial cases in which $C = 0$. However, it is possible for the scientists to achieve the combination (IB) (X works independently, Z borrows). This occurs if Y's error rate is substantial, but not too high; if Z's resources are fairly small in comparison to those of X; and if the work required to obtain the information independently would consume most of Z's resources.

---

4. Attributing Authority

The discussion of the last section rested on the idea that scientists are able to assess the error rates of others, or, to put it more positively, that they are able to judge how reliable other members of the community are. How are such estimates made? How are authorities evaluated?

Sometimes, like instruments, potential authorities can be calibrated directly. We compare the output of an authority with our own opinions on topics where there is overlap. This is a temptingly simple suggestion, but I propose that, from the beginning, we think about authority in a broader way.

We are interested in a function $a(X, Y)$ that measures Y's authority for X, or, more exactly, X's assessment of the probability that what Y says will be true.[9] Here X is an individual scientist and Y may be another scientist, a research team, a journal, a series of scientific monographs, or some other composite entity on which scientists may potentially depend. (There are often

---

9. More exactly, we are concerned with X's assessments of the probability that an arbitrary statement (belonging to a particular class—e.g., reports about a particular kind of topic) will be true, *given simply that Y produced or endorsed it.* As we shall see in Section 11, the probabilities you assign to some statements are computed partly on the basis of the content of those statements and partly by consideration of the authority of those who produce them. This is most evident in cases where erstwhile trustworthy people announce findings that, given the prevailing body of beliefs, seem highly implausible.

serious questions about the entity whose authority is to be assessed: when an unreliable scientist publishes in a highly prestigious journal, known for its strict refereeing, the authority attributed to the article may be an amalgam of the authority of scientist and journal.) Attributions of authority are rarely uniform across topic. I shall assume, in what follows, that the assignments of authority are relativized to a range of issues with respect to which the potential authority's deliverances can be assigned the same reliability.

An idealized treatment of authority that includes obvious social factors can proceed by breaking a scientist's credibility into two parts: there is *unearned* authority that stems from the scientist's social position (either within the community of scientists or in the wider society), the type of authority that arises from being associated with a major institution or from having been trained by a prominent figure; this contrasts with *earned* authority, that credibility assigned by reflection on the scientist's performances or through consideration of others' opinions of those performances. So I propose that we consider communities of scientists in which individuals evaluate one another in accordance with the equation

$$a(X, Y) = w(X, Y)a_u(X, Y) + (1 - w(X, Y))a_e(X, Y) \qquad (4.1)$$

(where $w(X, Y)$ is a weight function, whose value indicates the relative importance X takes unearned authority to have in evaluating Y). I shall consider different possibilities for weighing unearned authority (different values of w) and different methods of computing earned authority (different measures of a_e).

Is this approach to authority complete? Perhaps the assessment of the reliability of others depends not just on such large social factors as prestige within the community but on the personal relations between evaluator and potential authority as well.[10] This could easily be incorporated by amending the basic equation to

$$a(X, Y) = w_1(X, Y)a_u(X, Y) + w_2(X, Y)a_p(X, Y) \qquad (4.2)$$
$$+ w_3(X, Y)a_e(X, Y)$$

where $w_1 + w_2 + w_3 = 1$. This amendment would be useful in studying the ways in which differences in personal connections might be reflected in the pursuit of different lines of research in different subgroups. Initially, however, my principal concern will be to examine the balancing of considerations of track record (earned authority) against other forms of attribution, and, for these purposes, there is no loss in generality in collapsing the a_u and a_p components. So I shall use the simpler version throughout the following sections. However, in Section 8, I shall look at a special type of personal

---

10. This point is clear in Shapin's studies of the importance of personal testimony about the probity of informants and would surely have to be treated if we were to respond to concerns about the honesty of others. Annedore Schulze has also emphasized to me the role of personal assessments of the character of others in the decision making of scientists working in small groups. I am indebted to both Shapin and Schulze for helpful discussion of these matters.

factor in the attribution of authority, considering the ways in which our perceptions of others might partially depend on their view of us.

Before leaving the topic of authority, it is important to distinguish between authority and credit. A number of writers have emphasized the role that the search for credit plays in the conduct of scientific research (Latour and Woolgar 1979, chapter 5, Hull 1988). As I have already emphasized, authority is topic-relative. A scientist's credit, on the other hand, seems to be based on the *overall* assessment of that scientist's contributions. Thus it is very easy for a scientist to have high authority within a particular area, his area of expertise, but to have little credit. Scientific communities are full of respectable members whose deliverances about their assigned projects are dependable but who are viewed by their colleagues as pedestrian. In many instances, the authority structure of a community is that of an inverted pyramid: almost all those who have been trained, and who have survived their novitiate, have fairly high authority (with respect to the topics about which they make pronouncements). Nevertheless, the same communities can be sharply pyramidal in terms of credit, with a tiny fraction of the members aspiring to the highest levels of reputation (and concomitant resources). As I shall suggest in the next section, there is an intimate relation between credit and a particular type of authority, but the notions are not generally interchangeable.

5. Direct Calibration

Let us now explore the consequences of various ways of attributing authority to others. The best hope for minimizing the effects of the social structure is to suppose that members of the community always set w to be 0 and calculate earned authority through direct calibration. Under these circumstances, $a(X, Y)$ is simply Y's truth ratio with respect to the sample of statements about which X has an independent opinion: in effect, X uses the straight rule to project the probability that Y's claims will be correct from the frequency with which Y asserts the truth, by X's lights, within the class of Y's statements available for appraisal by X.[11] So, for example, if X believes $p_1, p_2, - p_3,$

---

11. This is gross oversimplification of a complex practice. Notice first that if the entire corpus of Y's public pronouncements is considered it is likely to contain so many banalities that the truth ratio will automatically be very high. By relativizing to a range of issues on which Y's pronouncements are considered, I hope to prevent this difficulty. Intuitively, X looks to Y's original contributions on a particular topic: so, for example, in assessing Pons and Fleischmann, chemists recalled their particular, original claims within electrochemistry and asked how many of these had been subsequently validated. In practice, the complex assertions that scientists make are often used by their peers in evaluations of authority by effectively *atomizing* their published and circulated work. Whole papers, or even series of papers, are identified with a single, central claim, and track records are judged by considering how many of these claims are right (from the perspective of the assessor).

However, other considerations enter in. Trustworthiness is often appraised by considering the experimental and technical skills of the person in question and, in some sciences, inspecting the visual representations that are produced. Hence, even given the relativization

p_4, and $- p_5$, and Y is recognized by X to assert $-p_3$, $- p_4$, $- p_5$, p_6, and $- p_7$, then $a(X, Y) = 2/3$.

Is this at all realistic? Do scientists ever engage in this type of computation? I believe that they do. The physicists who telephoned their electrochemist friends to ask about Pons and Fleischmann typically wanted to know how frequently the defenders of cold fusion had reported correct results about similar matters. The electrochemists who responded were often in a position to provide a minireview of the records of both Pons and Fleischmann.[12]

The simplest type of case in which direct calibration functions involves the reporting of experimental results. Here the difficulties with individuating and counting the deliverances of the target (mentioned in note 11) can be addressed relatively straightforwardly because we have a preferred way of classifying the reports that are of interest: we want to know how frequently, in giving experimental reports of such-and-such a kind, our potential informant gives correct reports. However, there are other contexts in which direct calibration can also be used. So we might ask, of a prominent scientist who has a history of making large claims about promising lines of research in a field, how often her suggestions about which directions to follow have proved fruitful. Here we discover that connection between authority and credit at which I hinted in the last section. One way to obtain substantial credit within a scientific community is to acquire the reputation for knowing "where the field should be going," and this is sometimes done through making claims about the important lines of research. As one's peers compare these claims with their own judgments about what has proved valuable, they engage in direct calibration that confers authority with respect to the development of the field, and this type of authority is an important determinant of credit.[13]

Deference to authorities might affect a scientific field by blocking the spread of new ideas within the community, so that proponents of heterodoxy would ipso facto lose credibility. A simple, somewhat artificial, way to focus this worry is to consider what I shall call the *alliance-splitting problem*.

Imagine that we have three scientists *A, B,* and *C. C* advocates a finding that would challenge the accepted ideas of the community to which all three belong. In terms of direct calibration, *B* and *C* are natural allies, in the sense

---

to topic that I introduce in my analysis, the idea of computing truth ratios oversimplifies the practice of direct assessment. I use it here as a way of contrasting the evaluation of track records with the attribution of authority on the basis of social position (itself an eqully complicated business), and, for present purposes, oversimplification enables sharp presentation of the issues arising from the contrast. (I am grateful to Michael DePaul for correspondence in which he raised interesting questions about the actual appeal to track records.)

12. Here I am indebted to an insightful presentation by Jan Talbot in a symposium on cold fusion held at UCSD in the spring of 1989. Electrochemists, but not physicists, could (and did) calibrate Pons and Fleischmann directly.

13. I am grateful to Ernan McMullin for bringing home to me the differences between attributions of authority about experimental findings and attributions of authority about more "theoretical" matters. The discussion of the text is intended to note the differences rather than to engage in the important work of analyzing them.

that B ascribes to C a higher truth ratio than B ascribes to A. A is a potential authority who claims that C's alleged finding is bogus. Under what conditions can the presence of A split the alliance between B and C?

If B assigns no weight to unearned authority ($w(B, A) = w(B, C) = 0$), and if B computes earned authority through direct calibration, then, by our assumption about truth ratios, $a(B, C) > a(B, A)$. Suppose that B is in no position to investigate C's finding directly, that B must arrive at a decision for or against C's finding, that A and C are the only members of the community who express opinions about C's finding (the only potential authorities on whom B can rely). Imagine further that B follows the decision rule

(R)   If you have to make a decision with respect to p, you are unable to make that decision through independent inquiry, and there are exactly two potential authorities whose opinions about p conflict, then you should follow the judgment of the person to whom you assign higher authority.

Then, plainly, B will endorse C's finding. Under these (admittedly artificial) conditions, we achieve the traditional epistemologist's utopia: social factors play no important role and, in effect, B's individual reasoning underlies her judgments.

Despite its obvious unreality, this is a useful point from which to begin, for it provides a baseline against which the disturbing effects of appeals to authority may be measured. When authority is measured by direct calibration, other informants are essentially extensions of the individual who appraises them. Let us now look at more interesting ways of assessing the authority of others, using the alliance-splitting test as an assay for identifying the epistemological roles of various social relations.

6. Prestige Effects

The most obvious way in which the presence of authorities can break up natural alliances is through prestige effects. Suppose that our community is one in which everyone who has undergone a reputable training program is always assigned a value of unearned authority above some threshold. People who have been associated with privileged institutions are attributed greater unearned authority. Under these conditions, strong natural alliances can be broken.

Consider, once again, our three scientists A, B, and C. As before, C announces a new, controversial finding and A dismisses it as flawed. The public assertions on questions that concern the three are as follows:

A	$-p_1$	p_2	p_3		
B			$-p_3$	p_4	p_5
C	p_1	$-p_2$	$-p_3$	p_4	p_5

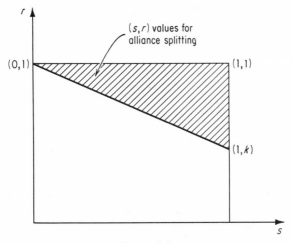

Figure 8.3

This is an extreme situation in which B and C are perfect allies, agreeing on all statements about which each has an independent opinion. B and A, on the other hand, *disagree* with respect to the only statement on which each has an independent opinion. However, because of A's prestigious position, $a_u(B, A) = 1$, and, perhaps because of the limited basis on which B can directly calibrate A, perhaps because of A's prominence, $w(B, A)$, the weight B assigns to A's unearned authority in assessing A, is some relatively high number r. C is a perfectly respectable member of the community who receives the threshold value for unearned authority k. Let $w(B, C) = s$. Earned authority is calculated through direct calibration.

B's assignments of authority are easily seen to be

$$a(B, A) = r \qquad (6.1)$$
$$a(B, C) = sk + (1 - s).$$

So the natural alliance between B and C is split if

$$r > sk + (1 - s) \quad \text{or} \quad r + s(1 - k) > 1. \qquad (6.2)$$

As Figure 8.3 reveals, (s, r) has to lie in the triangle whose vertices are the points $(0, 1)$, $(1, 1)$, $(1, k)$. Intuitively, there are many ways in which A's prestige can split even a perfect natural alliance: all that is required is that the weights B assigns to unearned authority be sufficiently high and the threshold value of unearned authority for a respectable member of the community sufficiently low. Under these conditions, if B is in the predicament of deciding the merits of C's finding subject to the constraints on the artificial decision situation I have described—in particular using rule (R)—then A's prestige will block the acceptance of C's challenge to orthodoxy. As we might have expected, and as Bacon and Descartes feared, appeals to authority can help preserve the status quo.[14]

So far there are few surprises. Matters become more intriguing as we add detail.

7. Indirect Calibration

There are numerous people whose advice might be valuable for a scientist but whom the scientist is unable to rate directly. When you need conclusions outside your speciality, direct calibration of potential informants is unreliable, perhaps even impossible, since the sample of statements available for computing a truth ratio is small, possibly empty. Under such circumstances, attributions of authority must proceed either by use of unearned authority or by what I shall call *indirect calibration*—using the judgments of those who have already been assessed in evaluating others.[15]

I shall simplify the general situation by focusing on cases in which the only type of indirect calibration uses scientists who are directly calibrated to evaluate those who are indirectly calibrated (the maximum length of paths in the directed graphs is 2) and by ignoring possibilities of mutual assessment and iterated adjustments. So we can write the earned authority of Y for X as

$$a_e(X, Y) = \sum_i a(X, Z_i) \cdot t(Z_i, Y)/\sum a(X, Z_i) \qquad (7.1)$$

where $a(U, V)$ is the authority *already* assigned to V by U, $t(U, V)$ is the truth ratio of V as measured by U, and the summation is over all the paths that X uses in evaluating Y. This equation can be extended to include direct calibration as a possible component of total earned authority, if we treat X–X–Y as a degenerate two-step path leading from X to Y. In elaborating this idea I shall *not* assume that $a(U, U)$ is always 1. Indeed, people can sometimes

14. Careful qualification is needed here. As Richard Foley pointed out to me, the algebraic approach that I adopt does not mandate conclusions that authority will inevitably have a cramping, conservative effect. It is possible to apply my basic algebraic approach to *different* problem situations, within which appeals to authority can *aid* the introduction of new ideas into the community: simply suppose that those most likely to make innovations are those with highest authority. I have chosen to apply the algebra within a particular problem situation, that in which innovations are likely to be produced by mavericks with low authority, because this seems to be a common scenario in the development of at least some sciences and because it highlights the apparent problems with deference to authority. As we saw at the end of Section 3, and as we shall see again later, the pressures to take risks are strongest on those who are at the fringes of scientific communities. Hence it seems to me that the alliance-splitting scenario serves as a useful assay for examining the effects of authority.

15. I owe to Steven Shapin the observation that Michael Polanyi had already recognized the importance of indirect assessment of other scientists. See (Polanyi 1958 217).

be very unconfident about their direct assessments of their fellows, which is, of course, why indirect calibration is used in the first place.[16]

Let us now consider our earlier example. We have three scientists whose publicly expressed beliefs are as follows:

$$
\begin{array}{lccccc}
A & -p_1 & p_2 & p_3 & & \\
B & & & -p_3 & p_4 & p_5 \\
C & p_1 & -p_2 & -p_3 & p_4 & p_5
\end{array}
$$

Let us suppose that A is a prestigious scientist whom B uses in assessing C. As before, let $w(B, A) = r$, $a_u(B, A) = 1$, $w(B, C) = s$, $a_u(B, C) = k$. If C's authority is evaluated through indirect calibration, we shall also need to consider B's confidence in her own reliability as a direct assessor of C. Suppose, then, that $a(B, B) = h$. Under these conditions, we have:

$$a(B, A) = r \tag{7.2}$$

$$a_e(B, C) = (r \cdot 0 + h \cdot 1)/(r + h) = h/(r + h) \tag{7.3}$$

$$a(B, C) = sk + (1 - s)h/(r + h). \tag{7.4}$$

Plainly, the value of $a(B, C)$ is reduced from that obtained through direct calibration $(sk + (1 - s))$, unless B gives no weight to A's unearned authority $(r = 0)$.

Under what conditions will the alliance between B and C be split? Suppose first that $h = 1$, the case in which B is confident in her own power to assess C and, in consequence, the situation most favorable for resisting alliance splitting. The appliance is split if

$$a(B, A) > a(B, C) \tag{7.5}$$

that is

$$r^2 + r(1 - sk) > (1 - s) + sk. \tag{7.6}$$

As Figure 8.4 makes clear, the region of (s, r) points that correspond to alliance splitting is now increased: in particular, even when $s = 0$ (unearned authority is discounted in the evaluation of C), the alliance is split if $r > 0.62$. At the opposite extreme, if B is extremely unconfident about her ability to evaluate C directly $(h = 0)$ then the alliance is split if $r > sk$. It follows that if B assigns equal weight to unearned authority in evaluating A and C, and C is a reputable, but not high-ranking, member of the community, then the B–C alliance will be split.

The example I have been discussing reveals complete disagreement between the evaluator B and the putative authority A and perfect harmony

16. Here again, issues about the *kinds* of statements that are being assessed arise. In using Y to rate Z, $a(X, Y)$ must be computed by considering Y's reliability in judgments about the reliability of others with respect to some particular range of scientific topics. $a(X, X)$ is X's measure of X's ability to rate others on that range of topics. If X believes that her beliefs about this topic are likely to be wrong, then she may assign herself quite a low value for $a(X, X)$.

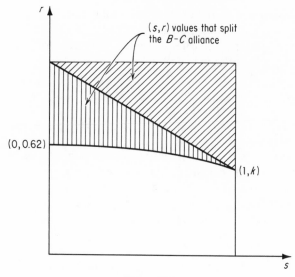

Figure 8.4

between B and the subve sive C. Nonetheless, if B is led to give some weight to A's prestigious position the B-C alliance can be split, and it can be split more readily if B calibrates C indirectly. Because of the perfect disagreement between A and B the alliance cannot be split unless *some* weight is given to A's unearned authority. Let us now consider a more realistic situation, one in which B, who diverges from A in some but not all ways, evaluates a relative newcomer C with whose views B is in sympathy.

Suppose that the public assertions of A, B, and C are as follows:

$$
\begin{array}{lllllllll}
A & p_1 & -p_2 & p_3 & -p_4 & -p_5 & p_6 & p_7 & p_8 \\
B & p_1 & p_2 & p_3 & & p_5 & p_6 & & \\
C & & p_2 & & p_4 & & & &
\end{array}
$$

Were B to calibrate C directly, ignoring unearned authority, then $a(B, C) = 1$, and B and C would form an unsplittable alliance. But, recognizing that C is relatively unknown and that the truth ratio is computed on a slender basis, B might choose to calibrate C indirectly by referring to the judgment of the authority A of whom C is severely critical. Imagine that unearned authority is neglected in assessing both A and C ($w(B, C) = w(B, A) = 0$). Let $a(B, B)$, B's confidence in her power to evaluate C directly, be h. Then the attributions of authority are

$$a(B, A) = 0.6$$

$$a(B, C) = ((0.6 \times 0) + (h \times 1))/(0.6 + h) = h/(0.6 + h).$$

The alliance will thus be split if $h < 0.9$. Unless B is extraordinarily confident of her direct assessments, indirect calibration alone, without invocation of unearned authority, suffices to resist the challenges of the maverick C.

8. Backscratching

In Section 4 I considered the possibility of introducing a term into the authority equation to represent the personal interactions between X and Y. Although we can often lump "personal" authority with unearned authority, there is one interesting case that deserves special attention. Any account of the attribution of authority in science should recognize the possibility of backscratching, a process in which your positive opinion of me raises my respect for you, which, in turn, may further increase your attribution of authority to me. (Conversely, your negative view of me would decrease my opinion of you, leading to a further lowering of your regard for me.) Is it possible that this type of mutual adjustment of attribution of authority should lead to a runaway process in which we ultimately view one another either as totally authoritative (so that the authority function takes value one) or as utterly unreliable (the authority function takes value zero)? Perhaps the personal aspects of attributions of authority could swamp the effects of social standing and of agreement in belief.

Amending the basic equation for attributing authority, we can write:

$$a(X,\ Y) = w_1(X,\ Y)a_u(X,\ Y) + w_2(X,\ Y)a(Y,\ X) + \qquad (8.1)$$
$$w_3(X,\ Y)\, a_e(X,\ Y)$$

$$a(Y,\ X) = w_1(Y,\ X)a_u(Y,\ X) + w_2(Y,\ X)a(X,\ Y) + \qquad (8.2)$$
$$w_3(Y,\ X)\, a_e(Y,\ X)$$

Let us assume that $w_i(X,\ Y) = w_i(Y,\ X) = w_i$.[17] Let us also suppose that degrees of authority are adjusted through a number of "rounds," and that the values of a_u and a_e remain constant through the process. Using indices to mark the rounds, we get:

$$a_n(X,\ Y) = w_1\, a_u(X,\ Y) + w_2\, a_{n-1}(Y,\ X) + w_3\, a_e(X,\ Y) \qquad (8.3)$$

$$a_{n-1}(Y,\ X) = w_1\, a_u(Y,\ X) + w_2\, a_{n-2}(X,\ Y) + w_3\, a_e(Y,\ X). \qquad (8.4)$$

An internal equilibrium would require $a_n = a_{n-2}$. So $a_n(X,\ Y)$ would satisfy:

$$a_n(X,\ Y) = w_1\, a_u(X,\ Y) + w_2(w_1\, a_u(Y,\ X) + w_2\, a_n(X,\ Y) + \qquad (8.5)$$
$$w_3\, a_e(Y,\ X)) + w_3\, a_e(X,\ Y).$$

That is:

$$a_n(X,\ Y) = \{w_1(a_u(X,\ Y) + w_2\, a_u(Y,\ X)) + \qquad (8.6)$$
$$w_3(a_e(X,\ Y) + w_2\, a_e(Y,\ X))\}/(1 - w_2^2).$$

The process could only be runaway if the right-hand side of this equation were less than zero or greater than one. Assuming that $w_2 < 1$, it is plain

---

17. This is a nontrivial assumption, for there is no reason to think that people involved in mutual assessment would give the same weight to all three factors. However, it helps to simplify the algebraic treatment of the problem, and I shall forego discussion of the more general case here.

that the right-hand side cannot be negative, and not hard to see that it cannot exceed one. For we maximize the numerator by setting $a_u(X, Y)$, $a_u(Y, X)$, $a_e(X, Y)$, and $a_e(Y, X)$ all equal to one. Thus the value of the numerator is less than or equal to:

$$w_1(1 + w_2) + w_3(1 + w_2) = w_1 + w_3 + w_2(w_1 + w_3) =$$
$$w_1 + w_3 + w_2(1 - w_2^2) = w_1 + w_2 + w_3 - w_2^2 = 1 - w_2^2.$$

Hence the right-hand side cannot be larger than one. In consequence, there will not be a runaway process, provided that $w_2 < 1$.

I do not wish to claim that the approach I have chosen (which embodies a highly artificial symmetry between X and Y) is the only possible way to model the effects of backscratching. However, I think it is noteworthy that some rather straightforward assumptions yield scenarios in which backscratching doesn't have the consequence that mutual admiration societies will protect members' opinions against criticism.

Technical Discussion 2

Three Interesting Illustrative Scenarios

(1) Imagine that calibration is direct. Since X and Y evaluate each other by focusing on the same class of propositions and looking for the subclass on which they agree, $a_e(X, Y) = a_e(Y, X)$. If $w_1 = 0$, then it is easy to see that, whatever the value of w_2, $a(X, Y) = a_e(X, Y)$.

(2) Suppose that Y is a recently trained newcomer ($a_u(X, Y) = 1/2$), and that X is established ($a_u(Y, X) = 1$). As before, earned authority is computed by direct calibration. Let $a_e(X, Y) = a_e(Y, X) = d$, and suppose that $w_1 = w_3 = (1 - w_2)/2$. Under these assumptions, the equilibrium value of $a(X, Y)$ is $(1 + d)/2 - 1/4(1 + w_2)$. It is interesting to compare this value with that which would have been assigned in the absence of backscratching. To make the comparison, let us keep the same values for earned and unearned authority, and suppose that $w_2 = 0$, while w_1 and w_3 retain their previous proportions (that is, in the present instance, $w_1 = w_3 = 1/2$). Then the value without backscratching is $a_u(X, Y)/2 + a_e(X, Y)/2 = (1 + 2d)/4$. Thus the change in attributed authority due to backscratching is $1/4 - 1/4(1 + w_2) = w_2/4(1 + w_2)$. This has a maximum value of $1/8$.

(3) Imagine that $w_3 = 0$, and, as in the second scenario, that $a_u(X, Y) = 1/2$ and $a_u(Y, X) = 1$. Now the value of $a(X, Y)$ is $1 - (1 + w_2)/2$. Again, we can compare this with the value that would have obtained in the absence of backscratching, that is, $1/2$. As with the second scenario, the authority that X attributes to Y is boosted by backscratching, and the maximum effect of backscratching is to raise it from $1/2$ to $3/4$.

We can represent the effects of backscratching for the case in which X is an established figure, Y a beginner, by summing up these findings in the following table:

	No Backscratching	Backscratching
Direct calibration	d	d
Earned/unearned mix	$(1 + 2d)/4$	$(1 + d)/2 - 1/4(1 + w_2)$
Unearned authority	$1/2$	$1 - 1/2(1 + w_2)$

In general, the change in attributed authority effected by backscratching can be represented as follows: Suppose that, if the weights are w_1, w_2, and w_3, then, in the absence of backscratching, the weights would be $w_1/(1 - w_2)$, $w_3/(1 - w_2)$. The change in the authority X attributes to Y due to the influence of backscratching is:

$$\Delta a(X, Y) = \{w_1(a_u(X, Y) + w_2 a_u(Y, X)) + w_3(a_e(X, Y) + w_2 a_e(Y, X))\}/$$
$$(1 - w_2^2) - w_1 a_u(X, Y)/(1 - w_2) - w_3 a_e(X, Y)/(1 - w_2)$$
$$= w_1 w_2(a_u(X, Y) - a_u(Y, X))/(1 - w_2^2).$$

This reveals clearly that the effects of backscratching depend on the assignment of weight to unearned authority.

9. The Entrepreneurial Predicament Revisited

After looking at some of the ways in which scientists might attribute authority to others, I now want to return to the approach to cooperation and the borrowing of others' ideas begun in Section 3. My first goal will be to understand how scientists would be expected to behave in circumstances in which the *perceptions* of others were given overriding importance. In Section 3, the focus was initially on epistemic values and pure agents who honored these values: scientists were supposed to plan their courses of action so as to maximize their chances of being right. Later, in setting up the entrepreneur's predicament, I supposed that the utilities were fixed by whether one is the *first*,to be right. In integrating the behavior of scientists in cooperative-competitive situations with the treatment of authority and credit, the first step will be to move further away from the purely epistemic. Let us now suppose that the utility of a course of action is determined by the *credit* it will bring, where credit is a function of the perceptions of others. What matters, then, is being *perceived* as the first to be right. The twin dangers are no longer being wrong and being pipped to the post, but being perceived to be wrong and being perceived to have been beaten.

It will be simplest to begin with the situation in which a scientist is working on a project without competition from others and to reconstruct the entrepreneur's predicament. The type of situation I shall consider is one which David Hull regards as central to scientific activity (Hull 1988). As he reminds us, there is too much published literature for scientists to engage in thorough search, and too little time for checking if scientists hope to make contributions of their own: "Scientists incorporate into their own work those findings that support it, usually without checking" (348). To obtain the credit that Hull

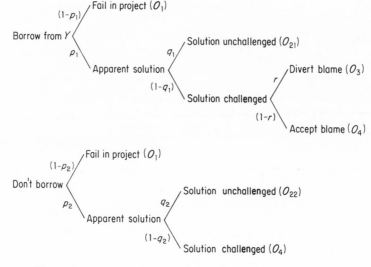

Figure 8.5. Decision Situation for Scientific Entrepreneurs.

(like other authors as diverse as Hagstrom and Latour) sees as one of the main motivations for devoting a large proportion of a human life to difficult, technical problems, scientists have to publish and market their ideas: "With the possibility of credit comes the possibility of blame. Scientists cannot spend very much time checking the work of others if they are to make contributions. They reserve checking for those findings that bear most closely on their own research, chiefly those that threaten it" (394). Hull thus offers us the picture of scientific entrepreneurs, operating within a credit economy.

As in Section 3, we suppose that X is engaged in some research project and that Y introduces something that could be used to reduce X's question to some simpler issue, if it were accurate. Should X adopt Y's innovation and pursue the apparently simpler project? Assuming that X has no competition with respect to this project and that the task is simply to receive credit by discovering a solution that is accepted by others, we can represent X's situation as in Figure 8.5.

This tree represents a simplified conception of the possible futures for a scientific entrepreneur whose possible strategies are to take over Y's potentially helpful result or to ignore it and continue with the present course of research. The new version of the predicament differs from that of Section 3 in two respects: I have left out of consideration the pressures that may be set up by the presence of competitors who are working on the same project, and I have explicitly couched the outcomes in terms of *others' reactions* to what X does rather than in terms of X's attaining a correct solution.

Writing the utility of O_i as u_i, I shall make the following assumptions:

(a) $u_1 = 0$ Failure in the project has utility 0.
(b) $u_{21} > 0$ Unchallenged apparent solutions have positive utility.

(c) $u_{22} > u_{21}$ There is greater utility in achieving an apparent solution without Y's help than with Y's help.

(d) $u_3 < u_{21}$ The utility of achieving an unchallenged solution is greater than that of being challenged and successfully diverting blame.

(e) $u_3 > u_4$ The utility of diverting blame is greater than that of receiving blame.

(f) $u_4 < 0$ Being blamed has negative utility.

(g) $p_1 > p_2$ Use of Y's idea would increase the chance of achieving an apparent solution.

As Technical Discussion 3 reveals, many interesting scenarios satisfy these plausible conditions. Sometimes there is just the kind of pressure on scientists to borrow that Hull emphasizes. On other occasions, scientists engage in the type of careful scrutiny that Legend hypothesized. As we shall discover later (§12), if competition from rivals is not too severe, if Y's innovation promises increased probability of success, and if there is no benefit in building on a flawed result, the strategies of borrowing from Y and ignoring Y will be inferior to the strategy of checking Y's proposal. Yet even when Legend delivers the right conclusions, they do not emerge for the traditionally accepted reasons.

Technical Discussion 3

Deviations from the Null Situation

Define the *null situation* by the following conditions:

$$q_1 = q_2 = q \tag{9.1}$$

The probability of challenges is independent of whether one borrows from Y.

$$r = [u_{22} - u_{21}]/u_{22} \tag{9.2}$$

The probability of diverting blame under challenge is the proportion of one's contribution to the apparent solution, measured by the marginal extra credit one would receive for an original solution.

$$p_1/p_2 = u_{22}/u_{21} \tag{9.3}$$

The credit assigned for original or partially borrowed success is directly proportional to the probabilities of success given borrowing or independent work.

$$u_3 = 0 \tag{9.4}$$

Dodging blame if the solution is challenged is a null outcome.

The expected utilities of the two strategies are:

$$p_1qu_{21} + p_1(1 - q)(1 - r) u_4 \tag{9.5}$$

$$p_2qu_{22} + p_2(1 - q) u_4. \tag{9.6}$$

By (9.3), the first terms are equal and the ratio of the second terms is:

$$u_{22}(1 - r)/u_{21}. \tag{9.7}$$

By (9.2), this is 1. Hence, given our additional assumptions, (9.5) and (9.6) represent the expected utilities of each strategy and take on the same value. In the null situation, borrowing from Y is as good as but no better than continuing with independent research.

Thinking about the null situation is valuable because it enables us to recognize various scenarios which would realize Hull's expectations about the behavior of scientists and various scenarios that would defeat those expectations. Consider, first, deviations from (9.2)–(9.4) that would *favor* borrowing from Y. Borrowing is preferable if we hold the constraints on the null situation constant except for replacing one or more of (9.2)–(9.4) with its counterpart:

$$r > (u_{22} - u_{21})/u_{21} \tag{9.2'}$$

$$p_1/p_2 > u_{22}/u_{21} \tag{9.3'}$$

$$u_3 > 0. \tag{9.4'}$$

(9.2') might obtain if Y is unpopular in the community or is regarded as unreliable, so that an entrepreneur might expect to be able to divert blame if matters go awry. (9.3') could hold because the scientist is engaged in a currently stymied research project of great significance, to which Y's suggestion promises new life—p_2 is effectively 0, p_1 is nonnegligible, and, while u_{21} $< u_{22}$, these values are of the same order of magnitude. Finally, (9.4') might be true because the scientist can expect to receive partial credit for a contribution to a challenged solution, even if what was borrowed from Y has to be discarded.

However, it should also be clear that there are situations in which simply taking over Y's results *without checking* is inferior to ignoring them. All that is required is for one or more of (9.2)–(9.4) to be replaced by their counterparts from:

$$r < (u_{22} - u_{21})/u_{22} \tag{9.2''}$$

$$p_1/p_2 < u_{22}/u_{21} \tag{9.3''}$$

$$u_3 < 0. \tag{9.4''}$$

The situations in which these hold are the inverses of those that support Hull's conclusions.

---

10. Entrepreneurs, Authority, and Credit

I now want to consider the ways in which differences in authority affect the decision making of scientific entrepreneurs. Technical Discussion 4 reveals how ease of borrowing can be profoundly affected by considerations of relative rank. Both the traditional conception (according to which scientists only "borrow" results that they have independently checked) and Hull's emphasis on the acquisition of credit offer too uniform a picture of scientific behavior. Both capture something important that can be understood more exactly by probing the conditions of scientific decision making, seeing how all kinds of social and epistemic factors interact with one another.

Technical Discussion 4 extends the analysis by allowing for effects of competition. Competition situations will now be marked by interesting asymmetries: the chances of receiving credit as the first discoverer of a solution to a problem will depend on one's standing in the community. Scientific entrepreneurs thus have to evaluate not only their own authority and that of their potential sources, but also that of their competitors. In order to keep the analysis relatively simple, I deal only with the case in which Y announces a result that could be used by either X or Z, who are the only competitors on a project. Suppose that X has higher authority than Y and that Y has higher authority than Z. As at the end of Section 3, Technical Discussion 4 examines the two-person game that X and Z play against one another. Social competition has appeared to promote welcome diversity within a scientific community. The present analysis shows that authority effects can weaken the competitive pressure. Where they do favor diversity, the asymmetries produced are different from those generated by differences in resources.

Technical Discussion 4

Authority and the Entrepreneur's Decision

Start with the predicament described in the last section and with the uncontroversial constraints (a)–(g). Suppose that earned authority is measured through direct calibration, and that there is no backscratching effect. For the moment, I shall retain (9.3) and (9.4), but replace (9.1) and (9.2), so that we have:

$$q_1/q_2 = a(X, Y)/a(X, X) \tag{10.1}$$

the probabilities of successful challenges to apparent solutions are in the ratio of the degrees of authority of X and Y (as measured by X, the entrepreneur whose problem this is)

$$r = a^*(X)/(a^*(X) + a^*(Y)) \tag{10.2}$$

{where $a^*(V)$ is a measure of V's average authority within the community}

the chances of diverting blame are proportional to X's relative authority with respect to Y within the community

$$p_1/p_2 = u_{22}/u_{21} \tag{10.3}$$

$$u_3 = 0. \tag{10.4}$$

Subject to these conditions, the general inequality for the preferability of borrowing from Y is

$$p_1 u_{21}(q_1 - q_2) + \{p_1(1 - q_1)(1 - r) - p_2(1 - q_2)\}u_4 > 0. \quad (10.5)$$

It should not be too hard to see that there are *many* ways of satisfying or of violating this inequality. I shall consider a few special scenarios that are particularly interesting.

Suppose first that the measure of authority within the community, a^*, used in calculating the probability of diverting blame, r, is simply the *unearned* authority of the target individual, and that the probabilities of resisting challenge are the minimum degrees of authority of the participants in the attempted solution, where these are computed by giving zero weight to unearned authority and using direct calibration. Then, if $t(X,Y)$ is the truth ratio that X ascribes to Y, we have:

$$r = a_u(X, X)/(a_u(X, X) + a_u(X, Y)) \quad (10.6)$$

$$q_1 = t(X, Y) \qquad q_2 = t(X, X) = 1. \quad (10.7)$$

This represents a situation in which X, the scientist making the decision, is extremely self-confident, attributing to himself an authority of one. But suppose that X has relatively low status in the community, while Y is highly prestigious, so that:

$$a_u(X, Y) = 1. \quad (10.8)$$

(10.5) now becomes:

$$p_1 u_{21}(q_1 - 1) + p_1(1 - q_1)(1 - r)u_4 > 0. \quad (10.9)$$

But unless X would rate Y as perfectly reliable, $q_1 < 1$; hence, the first term is negative, and, because $u_4 < 0$ (see (f)), so is the second. Hence the inequality is unsatisfiable, and it cannot pay X to borrow from Y. This is the *case of the self-confident prole and the unreliable aristo.*

Conversely, if X is a low-ranking individual within the community, then she may have little to lose through blame if a solution to the problem in hand is successfully challenged, so that u_4 is small. Suppose that she is also unconfident, taking the community's appraisal of her into account in assigning authority to herself. Under these conditions it is possible that $a(X, Y) > a(X, X)$, so that (neglecting the term in the small u_4) (10.5) reduces to:

$$p_1 u_{21}\{a(X, Y) - a(X, X)\} > 0, \quad (10.10)$$

which is satisfied. Call this the *case of the intellectually downtrodden.*

So far I have supposed that the credit that scientists receive is proportional to their contributions, an idea reflected in:

$$p_1/p_2 = u_{22}/u_{21}, \quad (10.3)$$

which supposes that benefits received with or without help from another are inversely proportional to the probabilities of success. It is natural to consider replacing (10.3) with an expression of Merton's "Matthew effect" (1973 443–447):

$$u_{21} = a_u(X, X) \cdot v/(a_u(X, X) + a_u(X, Y)) \quad u_{22} = v. \quad (10.11)$$

According to (10.11), when two scientists combine to produce an accepted solution, credit is distributed according to their prior prestige. "To him who hath, it shall be given."

Under these conditions, the inequality that must hold for borrowing to be preferable:

$$p_1 q_1 u_{21} + p_1(1 - q_1)(1 - r) u_4 > p_2 q_2 u_{22} + p_2(1 - q_2)u_4. \quad (10.12)$$

becomes

$$p_1\{q_1 v \cdot a_u(X, X)/(a_u(X, X) + a_u(X, Y)) + \quad (10.13)$$
$$(1 - q_1)u_4 \cdot a_u(X, Y)/[a_u(X, X) + a_u(X, Y)] > p_2\{q_2 v + (1 - q_2)u_4\}.$$

Suppose further that the costs of being blamed if the apparent solution fails are proportional to one's prestige:

$$u_4 = -kv \cdot a_u(X, X) \quad \text{where} \quad k > 0. \quad (10.14)$$

Consider three subcases: (i) X is high-ranking, Y is low-ranking ($a_u(X, X) = 1$, $a_u(X, Y) = 0$); (ii) X is low-ranking, Y is high-ranking ($a_u(X, X) = 0$, $a_u(X, Y) = 1$); (iii) X and Y are of equal rank ($a_u(X, X) = a_u(X, Y)$).

In the first case, (10.14) yields:

$$p_1/p_2 > (q_2(1 + k) - k)/q_1. \quad (10.15)$$

When k is small, the q_i are computed according to (10.2), and overall authority is measured by direct calibration, this can be further reduced to:

$$p_1 \cdot t(X, Y) > p_2, \quad (10.16)$$

which is easily satisfiable if Y's innovation is potentially helpful and Y does not appear to be too unreliable.

When the prestige of X and Y is reversed (as in (ii)), (10.14) becomes

$$0 > p_2 q_2 v \quad (10.17)$$

which is unsatisfiable. Assigning a low non-zero value to $a(X, X)$ generates a condition for borrowing that can be satisfied, but only if the p_1/p_2 ratio is extremely high.

Finally, if X and Y share the same rank with $a_u(X, X) = a_u(X, Y) = h$, then (10.14) yields

$$p_1/p_2 > 2(q_2(1 + kh) - kh)/(q_1(1 + kh) - kh). \quad (10.18)$$

When kh is small (either because the costs of blame are not too large or because the individuals are sufficiently low-ranking), and the q_i are measured through direct calibration (as with (i)), borrowing will be preferable if

$$p_1 \cdot t(X, Y) > 2p_2. \quad (10.19)$$

Comparing this with (10.16), we see that the effect of the leveling of rank has been to double the value that the quantity $p_1 \cdot t(X, Y)/p_2$ must exceed:

the high-ranking may borrow much more easily from the low-ranking, and, when equals borrow from one another, twice as much reliability is demanded from the source or the probability of success through borrowing must be twice as high (or there must be some combination of circumstances that produces the same effect).

The Game for Competing Entrepreneurs

Suppose X and Z compete, and that both have the opportunity to borrow from Y. We can represent the agents' assessments of payoffs from each combination of strategies by decision trees. Consider the tree for the case in which both X and Z borrow from Y (BB) in Figure 8.6. Here, p_1 is the chance (for each) of producing an apparent solution based on Y's work, and p_2 is the chance (for each) of producing an apparent solution by working independently of Y. Reasons analogous to those canvassed in Section 3 might lead us to assume that $p_1 > p_2$, and I shall start by making this supposition (although I shall scrutinize it later). If we simplify the situation by supposing that, if both succeed in producing a solution, then both will produce the same solution ($k = 0$), then the payoffs are:

For X: $u_{21}\{p_1(1 - p_1)a(X, Y)a(X, X) + p_1^2 a^*(X)/(a^*(X) + a^*(Z))\}$
For Z: $u_{21}\{p_1(1 - p_1)a(Z, Y)a(Z, Z) + p_1^2 a^*(Z)/(a^*(X) + a^*(Z))\}$.

Using similar reasoning with respect to the other combinations, we get the following assignments of payoffs:
For (BI) (X borrows, Z works independently):

X: $u_{21}\{p_1(1 - p_2)a(X, Y)a(X, X) + p_1 p_2 a^*(X)/(a^*(X) + a^*(Z))\}$
Z: $u_{22}\{p_2(1 - p_1)a(Z, Z)^2 + p_1 p_2 a^*(Z)/(a^*(X) + a^*(Z))\}$

For (IB):

X: $u_{22}\{p_2(1 - p_1)a(X, X)^2 + p_1 p_2 a^*(X)/(a^*(X) + a^*(Z))\}$
Z: $u_{21}\{p_1(1 - p_2)a(Z, Y)a(Z, Z) + p_1 p_2 a^*(Z)/(a^*(X) + a^*(Z))\}$

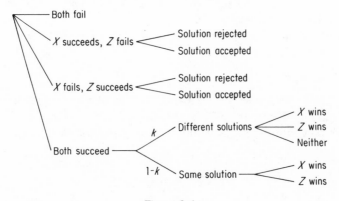

Figure 8.6

For (II):

$$X: u_{22}\{p_2(1 - p_2)a(X, X)^2 + p_2^2 a^*(X)/(a^*(X) + a^*(Z))\}$$
$$Z: u_{22}\{p_2(1 - p_2)a(Z, Z)^2 + p_2^2 a^*(Z)/(a^*(X) + a^*(Z))\}$$

It is not hard to show that there are choices of the parameters u_{21}, u_{22}, $a(X, X)$, and so forth, that will yield equilibria at (BB) and (II). The interesting question is whether the asymmetry in status can produce a situation in which there is an asymmetric equilibrium.

To answer this question it will help to make some special assumptions, consonant with but not determined by my qualitative description of the situation. Let us assume self-confidence for both scientists: $a(X, X) = a(Z, Z) = 1$. Relative status in the community will be appraised by supposing that $a^*(X) = 1$, $a^*(Z) = \frac{1}{2}$. I shall assume that $a(X, Y) = a(Z, Y) = r$, where $\frac{1}{2} < r < 1$, and that $u_{22} = ku_{21}$ where $k > 1$. Normalizing the utilities so that $u_{21} = 1$, the payoff matrix becomes

	B	I
	$rp_1(1 - p_1) + p_1^2/3$	$k(p_2(1 - p_1) + p_1p_2/3)$
B		
	$rp_1(1 - p_1) + 2p_1^2/3$	$rp_1(1 - p_2) + 2p_1p_2$
	$rp_1(1 - p_2) + p_1p_2/3$	$k(p_2(1 - p_2) + p_2^2/3)$
I		
	$k(p_2(1 - p_1) + 2p_1p_2/3)$	$k(p_2(1 - p_2) + 2p_2^2/3)$

Recall from the end of Section 3 that differences in resources could yield *IB* as an equilibrium, given appropriate relations among error rates and resources. Here, however, given the assumptions made so far, *IB cannot* be an equilibrium. For *IB* to be an equilibrium would require

$$kp_2 - rp_1 > 2p_1(p_1 - p_2)/3(1 - p_1) \quad \text{and} \tag{10.20}$$

$$kp_2 - rp_1 < p_2(p_1 - p_2)/3(1 - p_2) \tag{10.21}$$

If $p_1 > p_2$ these are not jointly satisfiable.

Plainly, the absence of the asymmetric equilibrium *IB* results from the supposition that borrowing from *Y* increases the chances of producing an apparent solution to the problem that confronts both *X* and *Z*. Given our supposition about the attributions of authority to *X*, *Y*, and *Z*, the supposition is quite reasonable, but it is important to recognize that it is very different from the assumptions that guided our earlier discussion. For, at the end of Section 3, it was assumed that *Y* was relatively unreliable. However, supposing that $p_2 > p_1$ is of no avail, since in that case the left-hand side of (10.21) is positive while the right-hand side is negative. In consequence, even if we waive the assumption that $p_1 < p_2$, it is impossible to have an equilibrium at *IB*.

By contrast, it is possible for *BI* to be an equilibrium.[18] The pertinent conditions are:

18. Again this contrasts with the example studied in Section 3, where *BI* could only be an equilibrium under trivial conditions.

$$kp_2 - rp_1 < 2p_2(p_1 - p_2)/3(1 - p_2) \qquad (10.22)$$

$$kp_2 - rp_1 > p_1(p_1 - p_2)/3(1 - p_1) \qquad (10.23)$$

These are jointly satisfiable provided that $p_2 > p_1/(2 - p_1)$. However, the constraints on k require that $k - 1$ be relatively small. Hence, we can obtain an asymmetric equilibrium only when the relative chances of success through working independently are substantial (so that there is a boost, but not an enormous boost, from borrowing) and if the differences in payoff from working independently and using Y's result are relatively slight.

---

11. The Community Response to Innovation

I have been exploring ways in which individual scientists might attribute authority to others, and the effects of such attributions in a range of rather artificial examples. It is now time to return to the type of case that figured prominently in my initial discussion of reliance on authority. We imagine that someone in the scientific community announces a new finding that is at odds with current thinking. How should a well-designed community respond? What kinds of decision rules and social arrangements will lead scientists to distribute themselves so as to approximate the community optimum?

Imagine a community of N scientists, one of whose members announces a challenging new finding. From the viewpoint of orthodoxy it is improbable that that finding is correct. Think of yourself as a philosopher-monarch with power to adjust the labor of the scientific work force as you see fit. We now envisage three possible strategies for each worker: a scientist can adopt the new finding, ignore it, or attempt to replicate it. A community strategy consists of an assignment of each worker to one of these options. As a philosopher-monarch you are interested in which community strategy gives you the best chance of attaining the community ends of science.

Much depends at this point on how we envisage consensus practice as being determined, an important issue that we shall explore briefly in Section 23. To begin with, let us suppose that having everyone adopt the new finding would automatically add it to consensus practice, that having everyone ignore it would automatically exclude it from consensus practice, and that a single successful replication would suffice to introduce it into consensus practice.[19] Initially, I examine a simplified version of the community optimization problem, in which we reduce the possible community strategies to three:

(*I*) Everyone ignores the challenging finding.
(*M*) Everyone modifies his or her views to embrace the new finding.
(*R*) All but one person ignore the finding; the single exception attempts to replicate it.

---

19. I shall relax this highly unrealistic assumption later in this section.

The more limited question I shall consider is, Under what conditions is R a better strategy than both I and M?

In advance of the finding, you would have assigned a probability of 0 to the statement asserted by the alleged finder C. Now you use your assessment of C's authority to grant a probability p that the statement is correct.[20] You also assign a probability q that the potential replicator will work reliably, endorsing C's result if it is correct and rejecting it if it is false. The expected utilities of the strategies can be written as

$$U(I) = -pv \tag{11.1}$$

$$U(M) = pu - (1 - p)w \tag{11.2}$$

$$U(R) = pqu - p(1 - q)v - (1 - p)(1 - q)w - e \tag{11.3}$$

where $-v$ is the cost of ignoring the challenge (if true), $-w$ is the cost of building on it (if false), u is the benefit of building on it (if true), and $-e$ is the cost resulting from the loss of work that the replicator would otherwise have done. I shall assume that u, v, and w are all appreciably larger than e, and that $w > u > v$.

I is preferable to M if $p < w/(u + v + w)$. Typically I will be preferable to M because p will be small, $p < \frac{1}{3}$, and, by the assumption that $w > u > v$, $w/(u + v + w) > \frac{1}{3}$. Moreover, if the replicator is highly reliable (q close to 1), then R will be preferable to M even when p is relatively high. For R is preferable to M if

$$p < (qw - e)/(qw + (1 - q)(u + v)) \tag{11.4}$$

and when q is close to 1 and e is small in comparison with w, the right-hand side is close to 1.

The interesting comparison is between R and I. R is preferable to I just in case

$$q > (e + (1 - p)w)/(p(u + v) + (1 - p)w). \tag{11.5}$$

This sets quite a strong condition on the reliability of the potential replicator. Intuitively, if p is very small and if the costs of building on error are very severe (w is large), the potential replicator had better be extremely reliable. Furthermore, if the right-hand side is greater than 1, the inequality is unsatisfiable however reliable the potential replicator may be. This occurs if

$$p < e/(u + v). \tag{11.6}$$

---

20. As remarked in note 9, the probability is computed by considering two factors: the content of the statement and the trustworthiness of the source. The weighting of the prior probability (based on content, and, in the case at hand, 0) and the probability based on utterance by the source (the reliability of the source on this range of topics) depend on your assessment of the reliability of the two channels. How likely is it that the traditional wisdom is mistaken? What is the chance that the assessment of the authority of the source is suspect? The ratios of the probabilities assigned in answering these questions give the relative weights assigned to the content-based probability and the authority-based probability. See p. 336.

How should you, the philosopher-monarch, compute values of p and q? Before C announced his finding you would have attributed to C a certain authority $a(C)$, to be bestowed on any result he asserted, and, on the basis of your prior commitments, you would have assigned this particular result the probability 0. The probability you *now* assign to C's finding should be a weighted average of these, with the weights expressing your relative confidence in your prior judgment about C and in the body of scientific lore that underwrote your dismissal of any such result as that he has now announced. So it is reasonable to write

$$p = z \cdot a(C) \tag{11.7}$$

where z is between 0 and 1, and is ever closer to 0 the more firmly entrenched the parts of scientific practice that C's claim calls into question. I shall suppose that q is simply your assessment of the authority of the potential replicator.

When well-established parts of science are called into question, it seems reasonable to think that the past track record of those parts of science should be weighted at least twice as heavily as judgments about the authority of a single individual. Thus z should be less than ⅓, and, even given a perfect potential replicator, I is preferable to M.

Consider now a community that meets the *Pons conditions:* (a) $a(C)$ is close to 1 (the person who announces the finding has high authority), (b) there are potential replicators in the community for whom q is close to 1 and e is negligible (people of high authority who could turn to replication without serious cost to the community), (c) u and v are of the order of $w/2$ (there are serious epistemic gains from incorporating the finding if it is correct). R will be preferable to I if

$$z > e/w \tag{11.8}$$

(for we require that q, which is approximately 1, be greater than $e + (1 - z)w/zw + (1 - z)w$). Hence, since e is negligible, even a very small weight given to prior judgments about the high authority of the finder will make it profitable for the community to expend some effort in replicating rather than simply ignoring the finding.

By contrast, if the community meets the *Gish conditions,* in which $a(C) = 0$ (the authority of the finder, *at least with respect to the kind of statement in question,* is zero), then ignoring the alleged finding is preferable no matter how small the costs of replication or how high the reliability of a potential replicator. For recall that the condition for R to be preferable to I is unsatisfiable if

$$p < e/(u + v). \tag{11.6}$$

If $a(C) = 0$ then p $(= z \cdot a(C))$ is automatically zero for any choice of z. Since e, u, v are all positive, the inequality is satisfied.

Intuitively, our examination of the simplified problem tells us that attempts at replication are frequently (although not always) a good thing for the community. But how much effort should be spent on replication efforts? Let us turn now to the community optimization problem where we explicitly consider different assignments of scientists to ignore, modify, or try to replicate. In light of the preceding discussion, we can ignore the issue of how to distribute the work force between modifying and ignoring. Instead, we can represent each of the strategies with which we are concerned by a function f that assigns each member of the domain 1 if that scientist works on replication and 0 otherwise. In general, of course, the reliability of our potential replicators will vary, but the discussion is greatly simplified by supposing that each member of the community has probability q of delivering the correct result in an effort at replication. Viewing the community as homogeneous in this way, each strategy can be represented as R_n (n attempt to replicate; the rest do not).

Technical Discussion 5 examines some special cases, which show that the optimum number of workers assigned to replication efforts decreases as the probability of the finding decreases and the costs of building on it (if it is wrong) increase. This is readily comprehensible. Under such circumstances, it is reasonable to impose a more stringent condition, namely to require that r out of a *smaller pool of potential replicators* should manage to replicate the finding. Decreasing the number of those assigned to replication efforts does just that.

Technical Discussion 5

Optimal Number of Replicators

The optimal community strategy depends upon the procedure for forming consensus. Assume that the finding will be added to consensus practice just in case r of those attempting to replicate it succeed in doing so (r need not be independent of n: so, for example, community policy might be to accept the finding if it is replicated by half those who attempt to do so, provided that this number is at least two). The possible outcomes and their associated probabilities can be represented as in Figure 8.7. Given that the finding will

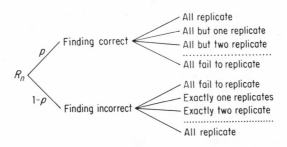

Figure 8.7

be accepted if r of the potential replicators succeed in replicating it (and will be decisively rejected if r of the potential replicators refute it), we can compute the following probabilities:

P_1 = probability of acceptance if finding correct
$= p\{q^n + nq^{n-1}(1-q) + \ldots\ldots + n!q^r(1-q)^{n-r}/(n-r)!r!\}$

P_2 = probability of acceptance if finding incorrect
$= (1-p)\{(1-q)^n + n(1-q)^{n-1}q + \ldots\ldots + n!(1-q)^r q^{n-r}/(n-r)!r!\}$

P_3 = probability of rejection if finding correct
$= p\{n!q^{r-1}(1-q)^{n-r+1}/(n-r+1)!(r-1)! + \ldots\ldots + (1-q)^n\}.$

The expected utility of the strategy R_n can then be written as

$$U(R_n) = P_1 u - P_2 w - P_3 v - ne.$$

An extremely well-designed community would maximize $U(R_n)$ with respect to both n and r. However, I shall suppose that the consensus formation system applies globally, covering a whole range of situations, and that it is not tuned to this particular problem. Let us therefore take the value of r as antecedently fixed, and consider how n should be adjusted so as to maximize $U(R_n)$.

To calculate the pertinent value of n, we consider the consequences of changing n to $n + 1$. When n changes to $n + 1$, the changes in P_1, P_2, and P_3 are given by:

$$\Delta P_1 = pn!q^r(1-q)^{n+1-r}/(r-1)!(n+1-r)!$$
$$\Delta P_2 = (1-p)n!(1-q)^r q^{n+1-r}/(r-1)!(n+1-r)!$$
$$\Delta P_3 = -pn!q^r(1-q)^{n+1-r}/(r-1)!(n+1-r)!$$

The change in the expected utility of R_n is

$$\Delta U(R_n) = n!\{pq^r(1-q)^{n+1-r}(u+v) - (1-p)(1-q)^r q^{n+1-r}w\}/(r-1)!(n+1-r)! - e.$$

Assuming that the costs of losing the work of a single scientist in the quest for replication are negligible, $e = 0$, then n should be chosen to satisfy

$$(1-q)^{n+1-r}/q^{n+1-r} = (1-q)^r(1-p)w/q^r p(u+v)$$

I shall assume, in what follows, that $q = \frac{1}{2}$. Let $1 - q/q = x$, $1 - p/p = y$, $w/u + v = z$. Then, taking logarithms (to whatever base is most convenient), the optimal value of n is

$$n_{opt} = [(2r-1)\log x + \log y + \log z]/\log x. \qquad (11.9)$$

Consider some special cases. (a) When $w = u + v$ ($z = 1$), and $p = \frac{1}{2}$ ($y = 1$), then, whatever the value of q, $n_{opt} = 2r - 1$. (b) When $w = u + v$, $p = \frac{1}{5}$ ($y = 4$), and $q = \frac{4}{5}$ ($x = \frac{1}{4}$), then $n_{opt} = 2r - 2$. (c) When $w = 4(u + v)$, $p = \frac{1}{5}$, $q = \frac{4}{5}$, then $n_{opt} = 2r - 3$.

Because we suppose that the finding is at odds with orthodoxy, p will be small, so that y will be greater than 1 ($\log y > 0$). Since the potential replicators in the community will typically be reliable, $q > \frac{1}{2}$, and hence $x < 1$ ($\log x < 0$). We have already assumed that $w > u > v$; w need not be greater than $u + v$, and z will be greater or less than 1 (z greater than or less than 0) according as w is greater or less than $u + v$. Putting these facts together, we see that n_{opt} decreases as y and w increase.

---

12. Individual Responses to Innovation

Having explored what the community would prefer to see happen in response to a challenging finding, let us now consider whether scientists, attributing authority in various ways and employing various decision rules, could work their way to the type of strategy that the community favors. Imagine, then, that you belong to a community of scientists and that some scientist, X, has announced a surprisingly heterodox finding. I shall suppose that you have two possible responses: you can ignore X and go on with your research project, or you can attempt to replicate X's result, deferring your current project. It is possible that there are conditions under which a third strategy, that of incorporating X's finding into your own research work without any antecedent check, might also become tempting—for reasons that we have explored in earlier sections—but I shall suppose that these do not obtain in the present case.

The strategy of ignoring X is basically a null option. By pursuing it you do not change your future well-being, so that the extra utility that accrues to you is 0. If you try to replicate, on the other hand, there are potential gains and losses. If you are the first to replicate then there will be some benefit u^*—possibly epistemic, possibly social, possibly mixed (the attainment of truth, the acquisition of status, the recognition of having discovered something important). If you are the first to show that X is wrong, then you will gain v^*, the utility accruing to the first refuter (again, this may be epistemic, social, or mixed). But, by diverting your effort from the ongoing research project, there will be costs of delay, d. On the basis of prior assessment of the authority of X and prior reasons for assigning probability 0 to X's challenging finding, you now suppose that there is a probability p that X is right. You are confident that, in any attempt at replication, you will judge correctly. However, to receive the benefits you have to be the *first* replicator or refuter. Your chances depend on the extent of the competition. Suppose that you conceive of all your colleagues as equally likely to win the race to replicate or refute. Then, if $n - 1$ join you in trying to replicate, the probability of your being the first is $1/n$. The expected extra utility of trying to replicate is therefore

$$pu^*/n + (1 - p)v^*/n - d.$$

Trying to replicate X is preferable just in case this quantity is positive; that is

$$p > (nd - v^*)/(u^* - v^*) \quad \text{if} \quad u^* > v^* \quad \text{or} \qquad (12.1)$$

$$p < (v^* - nd)/(v^* - u^*) \quad \text{if} \quad v^* > u^*. \tag{12.2}$$

Assume that the new finding is sufficiently important that it would gain high credit for the first replicator and epistemic dividends for all researchers in the field, so that $u^* > v^*$. If you and all your colleagues decide in the same way, then at least one person will try to replicate X's result if

$$p > ((d - v^*)/(u^* - v^*)). \tag{12.3}$$

(This is the condition under which, in a community all of whose members ignore the finding, it becomes preferable for a single person to switch to trying to replicate.) The community can reach this desirable outcome, even when the probability assigned to the finding after the announcement by X is low, provided that $d - v^*$ is small in comparison with $u - v^*$. This occurs trivially when the delay costs are smaller than the gains of first refutation: $d < v^*$. Surely many of the scientists who heard Pons and Fleischmann's claims about cold fusion were engaged in projects that could be delayed without serious cost. For them there were two significant questions: (i) Should a value greater than zero be given to the probability that the claim about cold fusion is correct? (ii) Would there be significant benefits from refuting Pons and Fleischmann if they were incorrect?

Both questions involve issues of authority. Assuming that the probability assigned to the experimental claim weighs the prior authority of the experimenters against the traditional considerations that underlie the antecedent view that there is zero probability of obtaining fusion on a table-top at room temperature, then, as before, we may write

$$p = z \cdot a(Y, PF) \tag{12.4}$$

where Y is the scientist making the probability judgment and PF is the composite entity Pons-Fleischmann. To obtain a nonzero probability, all that is required is that nonzero weight be assigned to the prior authority of the experimenters and that they have nonzero prior authority. All that was needed from the telephone conversations *on this score* was the assurance that Pons and Fleischmann were reputable, not outsiders like Gish and Morris.

The second issue is more subtle. Each scientist could assess for herself the costs of delay in current research. The benefits of refuting a challenge to orthodoxy depend, however, on the authority of the challengers within the community. If nobody is prepared to defer to these challengers on this issue, then, supposing that their challenge is wrong, little or no epistemic damage will be done. Similarly, if the challengers are already seen as having low authority, exposing the flaws in the present challenge will not redound to the credit of the refuter. So v^* will be a nondecreasing function of the perceived authority of the challengers within the community.[21] Suppose, for the sake of simplicity, that

---

21. Michael DePaul pointed out to me the possibility that the utility of a successful refutation might depend not only on the perceived authority of the challengers but also on the number of people engaged in attempts to replicate. Thus, once the challenging finding

$$v^* = V \cdot a^*(X) \tag{12.5}$$

where V is a constant, X is the challenger, and a^* is some averaging function over $a(Y, X)$ for the scientists Y who belong to the community. V will surely be large, for the benefits of exposing errors made by those to whom everybody assigns an authority of 1 are substantial. The crucial inequality for deciding to replicate is thus

$$d < a^*(PF) \cdot V. \tag{12.6}$$

When telephone calls reveal that other scientists Y give a sizable value to $a(Y, PF)$, or when it is found that Pons and Fleischmann have high unearned authority, scientists whose delay costs are small will make the decision to investigate their finding.

Throughout earlier sections, we have faced the worry that appealing to authority could retard desirable changes in science, making it difficult for correct challenges to orthodoxy to be appreciated. Indeed, as we saw earlier, reliance on authority in assessing the credibility of third parties can lead scientists to take a jaundiced view of novel claims. I have explored a highly idealized model in which the community preference is for some efforts at replication, and I have suggested that a variety of epistemic or social forces could lead scientists to modify their research in ways that satisfy this *generic* preference. Moral: the impact of appeals to authority has to be evaluated by recognizing the other forces that come into play in scientific decision making— even when a high weight is given to unearned authority, even when calibration is indirect, a community is not doomed to stagnation if there are sufficient rewards (social or epistemic) for entrepreneurs and critics.

But this moral only considers the problem in a relatively gross way. If we have a community in which the new finding will be adopted into consensus practice if just one person succeeds in replicating it, then, as we have seen, it is quite likely that the pressures on individual scientists will lead someone to attempt replication. As yet, however, we have no assurance that, in communities with more complex consensus-forming mechanisms, *enough* member scientists will attempt replication, or, in general, that the distribution of potential replicators will bear any relationship to the community optimal distribution. Technical Discussion 6 explores the much harder problem of how multiple replication might be achieved. It brings to light a serious difficulty.

---

is taken up, there is a possibility of bandwagon effects, even runaway bandwagons. This can easily be modeled by supposing that

$$v^* = V \cdot a^*(X)(1 + kn)$$

where v^*, V, a^* are as in the text, n is the number of scientists attempting to replicate/ refute the challenging finding, and k is a constant. Since my first concern is with the conditions under which *someone* replicates, I am effectively looking at the decision problem for someone when $n = 0$, so that there is no difference between the approach in the text and that which would see utility as an increasing function of the number of other replicator/ refuters. The problem of adjusting the number of potential replicators will be considered later.

In a community in which findings are adopted into consensus practice if only one member succeeds in replicating, the first scientist knows that there is no danger that her labor will be wasted through the failure of anyone else to pursue the possibility of replication. However, if a single successful replication would not be sufficiently impressive to secure community decision, the potential replicators have to rely on the formation of a critical mass of size r. Thus part of their decision making must involve making judgments of the likely behavior of others, so that they can estimate the expected size of the group of replicators. We can imagine that this estimation is done through informal polling, and that the telephone calls that followed the Pons-Fleischmann announcement were also intended to discover whether a sufficiently large number of people took the alleged result seriously enough to devote their time to investigating it.

The picture that emerges from the preceding discussions of competition, cooperation, and authority is a complicated one. Despite the apparently harmful effects of deference to authority in inhibiting the reception of maverick ideas, the more extensive discussions of the past two sections indicate ways in which authority structures within a scientific community might foster attention to novel, antecedently implausible, results. The exact relation between the actual distribution of effort and the community optimum will depend on the values of various parameters, but, even in communities in which intuitively worrying forms of authority play a major role in the assessments scientists make of their fellows, it is quite possible for those parameters to be set in ways that make for good epistemic design. To understand how well a particular community is likely to do, one cannot rest with simple descriptions of it as "authoritarian" or "individualistic." There is no substitute for looking at the types of models I have surveyed.

Technical Discussion 6

Let us now consider a community in which the finding will be adopted in consensus practice if and only if at least r members succeed in replicating it. I imagine that each member of the community calculates the expected utility of trying to replicate by making the following assumptions: *self-confidence,* If the finding is correct then I [the subject in question] will succeed in replicating it; *homogeneity,* All other potential replicators have probability q of getting the correct outcome from an attempt to replicate; *purity,* Credit will only accrue to me if I get the right answer, I am the first to do so, and enough others do so to make the finding part of consensus practice.

The agent reasons about the case in which $n - 1$ others attempt to replicate, (where $n > r$), by computing the expected utility of trying to replicate according to the tree in Figure 8.8.

The expected utility of trying to replicate $U(R)$ is given by

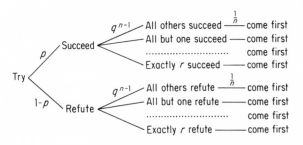

Figure 8.8

$$U(R) = \{pu^* + (1 - p)v^*\}Q - d \qquad (12.7)$$

where

$$Q = \sum_{j=r}^{n-1} (n - 1)!q^{j-1}(1 - q)^{n-j}/(n - j)!j!.$$

Suppose now that the community manages to reach the critical number of r replicators.[22] The number of those members of the community who attempt to replicate will be given by the value of n such that

$$\sum_{j=r}^{n} n!q^{j-1}(1 - q)^{n+1-j}/(n + 1 - j)!j! < d/\{pu^* + (1 - p)v^*\} \qquad (12.8)$$

$$d/\{pu^* + (1 - p)v^*\} < \sum_{j=r}^{n-1} (n - 1)!q^{j-1}(1 - q)^{n-j}/(n - j)!j!$$

The community is well designed only if this value of n is close to n_{opt}, computed in the last section.[23] More pointedly, we can say that a system of rewards for member scientists—ways of assigning values to u^* and v^*—is epistemically well designed if it produces coincidence (approximate coincidence) between the value of n calculated here and n_{opt}. It should now be clear that it is in principle possible to assess the design of actual scientific communities with respect to the replication problem. For we could investigate their method of consensus formation; use it to compute the value of n_{opt} for variable p, u, v, and w; identify the ideal relationship between these parameters and u^*, v^*, and d, and then examine whether or not this relationship is borne out on actual occasions in which the replication problem arises. This would require the distribution of n_{opt} replicators to be *stable* when attained, and to be *attainable*. Satisfaction of (12.8) only entails that the community will remain at the optimum if it reaches it. It does not guarantee that the community can reach the optimum.

---

22. I shall consider how this might occur shortly.

23. This claim rests on the assumption that the values of p and q used in giving the community-level analysis are just those that the members of the community employ in making their decisions. If that is not so, then the notation needs to be changed so as to make the necessary disambiguation between the agent's perspective on these probabilities and the correct values.

13. Division of Cognitive Labor

At various points in the previous sections I have suggested that there are advantages for a scientific community in cognitive diversity. Intuitively, a community that is prepared to hedge its bets when the situation is unclear is likely to do better than a community that moves quickly to a state of uniform opinion. Much of the rest of this chapter will be devoted to exploring this intuitive idea and trying to understand the kinds of social arrangements that might foster welcome diversity.

The problem is easily illustrated by many of the examples that have been considered in earlier chapters. The *compromise model* of Chapter 6 made much of the idea that scientific debates are resolved through the public articulation and acceptance of a line of reasoning that takes considerable time to emerge, and that the working out of this line of reasoning depends crucially on the presence in the community of people who are prepared to work on and defend rival positions. So, for example, in the resolution of the Darwinian debate, and, even more, in the triumph of Lavoisier's new chemistry, it was important that opposing points of view were kept alive and that the objections they generated were used to refine the ultimately successful positions.[24] The example of "the great Devonian controversy," discussed in Chapter 6, makes even more obvious the value of preserving rival approaches in situations of unclarity, for here the ultimate solution emerged from interaction between the previously dominant rivals. Even the much-decried hyperspecialization of science may play a valuable role.[25]

How, then, is cognitive diversity maintained? Given epistemic purity—a community of agents dedicated to modifying their practices so as to achieve purely epistemic ends—it is possible, even likely, that the outcome of separate decisions will be an epistemically homogeneous community. Even if differences between the alternatives are small, if one is slightly more developed, or more successful in overcoming difficulties, then those who value truth (and the kindred epistemic virtues discussed in Chapter 4) will favor that. Consequently, in a community of clear-headed scientists, devoted to espousing epistemically virtuous individual practices, we may expect cognitive uniformity.[26]

24. Kuhn (1962/70 158–159) points out the importance for the scientific community of persistent opposition, and his (1977) advances the general idea of the value of cognitive diversity. Much of Feyerabend's earlier work is also devoted to celebrating the desirability of keeping minority approaches alive, using them to sharpen both the dominant views and one another (see Feyerabend 1963b, 1965, 1970). The root idea goes back at least to Mill (1859).

25. I am grateful to Kim Sterelny for this observation.

26. This will not occur if the scientists are altruistic. If they are prepared—and able— to adjust their cognitive commitments so as best to serve the community, then, of course, they will replicate the decision making that identifies the community optimum. Whether there are, in fact, any scientists so devoted to the community cognitive project as to waive any concerns for their own epistemic advancement in favor of adopting a minority approach ("because someone ought to do that") is a question I leave for the reader.

The following sections consider the desirability and possibility of cognitive diversity in a number of idealized situations. The principal theme that I shall pursue is that epistemically *sullied* communities often seem to fare relatively well. While communities of isolated knowers who proceeded in the ways commended by individualistic epistemologies would often be doomed to cognitive uniformity, social pressures and types of motivations often regarded as antipathetic to the progress of science will turn out to be helpful. However, as with the discussions of authority, competition, and cooperation, the picture is complex, and there is no substitute for examining the details.

Before I offer my analyses it is worth noting an alternative approach to the maintenance of cognitive diversity, one whose principal emphasis is on the psychology of individuals rather than on the competitive and cooperative relations among individuals. Perhaps we can appeal to the idea of cognitive variation among individuals to recognize the possibility that, faced with a murky situation, some scientists will prefer one approach, others another.[27] My own discussions of cognitive variation in earlier chapters should make it apparent that I am not unsympathetic to this idea, but I do not think that it can be the entire story about the maintenance of cognitive diversity. In the first place, there are numerous instances in which a scientist's persistence in the face of severe difficulties seems best explained by the characteristics of the social system—its framework of rewards and sanctions, its competitive pressures—rather than by cognitive differences. Owen's opposition to Darwin and the persistence of both Murchison and De La Beche in defending their rival interpretations of the Devon strata are caused, in significant part, by the social positions and competitive relationships among the actors. Other instances can readily be garnered from the historical examples I have discussed earlier (see, in particular, Chapters 2, 6, and 7). Second, although cognitive variation among individuals may play an important role once rival approaches have established themselves as genuine contenders, the route to establishment often passes through extremely unpromising terrain. Those who articulate initially implausible alternatives are kept going, at least in part, by various kinds of nonepistemic pressures: fascination with a particular idea, devotion to a national tradition, desire to make a name for oneself. These kinds of motivations can, I suggest, be clearly discerned in the long intellectual voyages that led to Copernicus's *De Revolutionibus*, Lavoisier's *Traité*, and Darwin's *Origin*.

14. Alternative Methods: Community Decision

As will become clear in later sections, the problem of understanding how communities ought to address intertheoretic debate—or, in my terms, situ-

---

27. In an extremely interesting paper (Solomon 1992), Miriam Solomon pursues an approach of this type, arguing that scientists who use the Tversky-Kahneman "salience heuristic" and who find different things salient will differ in their cognitive commitments. For reasons offered in the text, I see Solomon's analysis as complementary to my own.

ations in which there are rival proposals for modifying the language, schemata, questions, and statements of consensus practice—is rather complex. I shall therefore begin with a question about division of labor which is relatively more tractable. This question focuses on choice among rival methods for solving an experimental problem. It can be introduced with a fanciful version of a scenario, familiar from the history of recent science.[28]

Once there was a very important molecule (VIM). Many people in the chemical community wanted to know the structure of VIM. Two methods for fathoming the structure were available. Method I involved using X-ray crystallography, inspecting the resultant photographs, and using them to eliminate possibilities about bonding patterns. Method II involved guesswork and the building of tinker-toy models. Everybody agreed that the chances that an individual would discover the structure of VIM by using method I were greater than the chances that that individual would discover the structure by using method II. Since all members of the community were epistemically pure each chemist used method I. They are still working on the problem.

The community goal is to fathom the structure of VIM as quickly as possible. Suppose that each method is associated with a probability function $p(n)$ representing the chance that the method will deliver an answer if n workers are assigned to it. Assume further that any answer delivered is recognizably either correct or incorrect. Imagine also that the relations between the probability functions represent their behavior over any time intervals we might consider—so, for any time interval t, the probabilities that method I delivers an answer within t and that method II delivers an answer within t are in the same ratios as the functions p.[29] N workers are available for distribution between the two methods. The community optimum distribution is given by having n workers use method I and $N - n$ use method II so as to maximize the probability that the structure of VIM will be discovered, that is, to maximize

$$p_1(n) + p_2(N - n) - \text{Prob(both methods deliver)}.$$

I'll assume, for simplicity's sake, that the probability that both methods will deliver the correct answer is zero. In general, this is a nontrivial assumption, but, for present purposes, let us assume that X-ray crystallography is very bad at finding the structure of molecules that can be figured out by tinker-

28. The illustration is obviously prompted by James Watson's account of the race for the structure of DNA (Watson 1967). Of course, many aspects of Watson's personal history have been disputed (see Sayre 1975, Olby 1974), but there is little doubt that it captures some important features of competitive life in science (see the reviews in Stent's critical edition of (Watson 1967)). Hence, Watson's account provides a useful way of making vivid a recurrent type of scientific situation.

29. These functions measure the chance that a method will deliver a *correct* answer, for an assignment of n workers, given that the world is as represented by the community's current knowledge about the molecule. Thus, for example, if little is known about VIM and if both methods have been pursued by similar numbers of workers for fathoming a large class of molecules, with method I proving successful much more frequently than method II, whatever the number of workers assigned, $p_1(n) > p_2(n)$.

toy building and vice versa. (The effect of this assumption is solely to simplify the algebra. Qualitatively similar conclusions can be obtained, at far greater length, if it is not made.)

Evidently, the solution to the problem depends on the form of the functions $p(n)$. I shall call these *return functions*, since they measure the return in probability of reaching the goal for an investment of n workers. I shall take these functions to be subject to the following constraints: they should increase monotonically with n, they should be zero when n is zero, and they should tend asymptotically to some value p when n goes to infinity (p represents the intrinsic prospects of the method, the probability of its success when we abstract from limitations of human effort).[30] Given the simplifying assumption that both methods cannot work, we know that the values of the asymptotes, p_1 and p_2, must sum to less than 1. These constraints leave a lot of room for choice of functions. To make one point explicit, it's quite possible that the forms of the functions should be different for the two methods. (Imagine that one method responds much more quickly than the other to the efforts of workers.)

I shall consider two possibilities for the functions. Suppose first that $p_i(n) = p_i(1 - e^{-kn})$. Then $p_1(n) + p_2(N - n)$ is maximized when

$$n = (kN + \ln p_1 - \ln p_2)/2 \, k \qquad (14.1)$$

Thus, when $\ln p_1 - \ln p_2 > kN$, the optimal distribution is to assign all the available workers to method I. Notice, however, that even when method I has more intrinsic promise than method II ($p_1 > p_2$), there is a range of conditions—when $\ln p_1 - \ln p_2 < kN$—under which the optimal distribution is to divide the community. Intuitively, a genuine division of cognitive labor would be best for the community if there is a large available work force (N is large), or if the methods respond quickly to the injection of effort (k is not too small), or if the difference in intrinsic promise between the methods is not too great (p_1 and p_2 are fairly close). The inequality given previously represents the ways in which tradeoffs are made.

The functions considered in the last paragraph have the property that the rate of increase in $p(n)$ is maximal when n is small. This idea may not be at all realistic. Perhaps the chances of achieving an answer by following a given method increase quite slowly at first, then go up rapidly once a critical mass of workers has accumulated, and eventually increase very slowly as saturation is approached. We can describe this behavior by mimicking the logistic growth equation of population biology, supposing that the p_i are given by

$$p_i(n) = p_i(3 \, n^2 - 2 \, n^3/kN)/k^2 \, N^2 \qquad (n < kN) \qquad (14.2)$$
$$p_i(n) = p_i \qquad\qquad\qquad (n > kN).$$

---

30. I originally thought that these constraints would apply in all cases. However, as Stephen Stich pointed out to me, too many cooks may spoil the broth. Imagine, for example, that the method involves observing some sensitive organisms and that crowding in the field would disturb the organisms' normal behavior. A live example of this phenomenon is hinted at by Shirley Strum (1987 197–198). However, in many instances, I suspect that the probabilities do indeed increase monotonically with the number of workers.

If the probabilities are given by these functions, then there are various cases of interest, depending on the value of k. Provided that $k < 1/2$, it is possible to realize the intrinsic prospects of both methods, so the optimal distribution divides the work force. If $1/2 < k < 1$, it is not hard to show that the optimal value of n is less than kN. More precisely, n should be chosen so that

$$(p_1 - p_2) \, n^2 + (2 \, p_2 - kp_1 - kp_2) \, Nn + p_2 \, N^2(k - 1) = 0. \quad (14.3)$$

When k is greater than or equal to 1, n should be N (recall that method I is superior; that is, $p_1 > p_2$).

Let me give a qualitative interpretation of these findings. As in the previous case, k is a critical parameter, representing the responsiveness of the methods. If the methods are so responsive that the intrinsic prospects of both can be realized with the available work force, then it is easy to appreciate that the community epistemic interests are best served by dividing the labor. Even when k is between 1/2 and 1—so that it is possible to realize the intrinsic prospects of one method but not those of both—it may be better to divide the work force so that the prospects of neither method are realized. Provided that the difference between p_1 and p_2 is not too great, it will be better to assign a new worker to method II if method II already has sufficient devotees to offer a large return from a new investment, rather than to method I, if method I is nearly saturated. However, once k reaches 1, it is always better to assign all resources to the method whose intrinsic prospects are higher. For $k = 1$, there are barely enough workers to saturate one method, and for greater values it is not possible to achieve the intrinsic prospects of either. Under these conditions, a new worker is always more profitably assigned to the superior method (method I). The optimal distribution involves no genuine division.

Are there situations in which the return functions take the logistic form, in which the work force is relatively small, and in which the optimal distribution divides the community? Yes. Suppose that the $p_i(n)$ have different intrinsic maxima p_i *and* that they have different k_i, so that the method with smaller intrinsic prospects yields returns at a faster rate. Specifically, imagine that $p_1 = 2 \, p_2$, $k_1 = 4 \, k_2$; even more specifically, let $p_1 = 1/2$, $p_2 = 1/4$, $k_1 = 2$, $k_2 = 1/2$, $N = 6$. Under these circumstances, it is not hard to show that the optimal distribution is to assign three workers to method I and three to method II (a distribution I shall represent as $\langle 3, 3 \rangle$). I introduce this example to forestall the worry that genuine division of labor is only optimal when the work force contains superfluous people.

15. Alternative Methods: Individual Decision

Let us now consider how various types of communities would distribute themselves. We can begin with a community of the epistemically pure. On one simple understanding of epistemic purity, epistemically pure agents judge methods according to the intrinsic qualities of those methods, not according to what their fellows are doing. If we understand epistemic purity in this simple way, then it is easy to see that there can be discrepancies between the resultant distribution and the optimal distribution. We have examined a number of cases in which the optimal distribution takes the form $\langle n, m \rangle$ with both n, m nonzero. In all these cases, method I was taken to be intrinsically more likely to succeed—that is, $p_1 > p_2$—so that the simple notion of epistemic purity would generate the distribution $\langle N, 0 \rangle$.

But perhaps we should think of individual epistemic rationality a bit differently. Suppose it is a requirement of epistemic purity that a pure Bayesian agent maximize her chances of following a method that yields the answer. We can interpret the requirement in two ways:. (A) We imagine the agent making a decision in complete ignorance of what other members of the community are doing, so that the task is to choose i so that $p_i(1)$ is as large as possible. (B) We imagine that the agent knows the current distribution $\langle r, s \rangle$, so that method I is to be chosen just in case $p_1(r + 1) > p_2(s + 1)$. On either interpretation it is easy for there to be discrepancies between the resultant distribution and the community optimum.

Consider (A). Suppose that the return functions take either of the forms considered in the last section, with k the same for both methods and $p_1 > p_2$ Then, in both cases, $p_1(1) > p_2(1)$. Hence, each agent will choose method I, and the distribution will be; $\langle N, 0 \rangle$. But we have seen that there are choices of the parameters for which the optimal distribution is $\langle n, m \rangle$ with n, m nonzero—imagine, for example, that the community has enough workers to realize the intrinsic prospects of both methods. But the discrepancy can even arise without superfluous workers. In the numerical example of the end of the last section, $p_2(1) > p_1(1)$, so that our envisaged requirement would yield the distribution $\langle 0, 6 \rangle$, a distribution that not only is different from the optimal distribution but that actually *minimizes* the probability that the community will discover the answer!

Turn now to (B). Here, it is not hard to see that there will be a runaway process. If the current distribution gives a higher value to p_i then (according to (B)) a pure Bayesian agent should choose method i, so that the discrepancy is increased. Now imagine that decisions are made sequentially. If $p_1(1) > p_2(1)$ then the first agent chooses method I. But now, by the monotonicity of the p_i, $p_1(2) > p_1(1) > p_2(1)$, so the second agent chooses method I, and it is easy to see that the distribution is $\langle N, 0 \rangle$. (Again, the numerical example shows not only that we don't need superfluous workers, but that the actual distribution might even be $\langle 0, N \rangle$.) These results follow without making any assumptions that contravene the preconditions of the cases in which the optimal distribution gives a genuine division of cognitive labor. Hence, (B) also yields discrepancies between the distribution achieved and the optimum.

We have been imagining an epistemically pure community of chemists striving and failing to fathom VIM. I shall imagine that the optimal distribution for them involves a genuine division of labor (corresponding to one of those cases considered in the last section in which $p_1 > p_2$, $p_1(1) > p_2(1)$). They fail to achieve this, since all of them follow one of the dictates of epistemic purity that lead to the distribution $\langle N, 0 \rangle$. Moreover, because the structure of VIM can only be fathomed by method II, their inability, as a community, to hedge their bets is costly.

By contrast, in a neighboring nation, the chemical community is composed of ruthless egoists. Each of the members of this community makes decisions rationally, in the sense that actions are chosen to maximize the chances of achieving goals, but the goals are personal rather than epistemic. Those who elect to work on VIM do so because they believe that whoever discovers the structure of VIM will win a prize coveted by all. They make the simplifying (but not altogether implausible) assumption that, if a method succeeds, then each person pursuing that method has an equal chance of winning the prize. How should we expect the sullied community to distribute its effort?

Imagine that the community has reached a distribution $\langle n, N - n \rangle$. You are a scientist currently working on method I, and you ponder the possibility of switching to method II. The change would be good for you—given my assumption about your interests and aspirations—if it would increase the probability that you win the prize. Now the probability of your winning is the probability that someone in your group wins, divided by the number of group members. (Intuitively, by choosing a method you buy into a lottery that has a probability of paying up, a probability dependent on the number of ticketholders; your chance of collecting anything is the probability that the lottery pays up divided by the number of tickets.) Thus at $\langle n, N - n \rangle$, it will behoove a scientist working on method I to switch to method II if

$$p_2(N - n + 1)/(N - n + 1) > p_1(n)/n. \qquad (15.1)$$

To understand how our imaginary sullied agents might distribute themselves, we need to discover equilibria, points at which nobody is better off switching to the alternative method. Let us say that the distribution $\langle n, N - n \rangle$ is *stable downward* if $p_1(n)/n$ is greater than or equal to $p_2(N - n + 1)/(N - n + 1)$ and *unstable downward* otherwise. Similarly, $\langle n, N - n \rangle$ is *stable upward* if $p_1(n + 1)/(n + 1)$ is less than or equal to $p_2(N - n)/(N - n)$ and *unstable upward* otherwise. $\langle n, N - n \rangle$ is *bilaterally stable* just in case it is stable both upward and downward.

If a community of sullied scientists reaches a bilaterally stable distribution, then we can expect it to stay there. However, *stability* is one thing, *attainability* another. Even though a particular distribution might be maintained, once it had been achieved, it might prove impossible for a group of self-interested scientists to reach it. For any distribution $\langle n, N - n \rangle$ that is bilaterally stable, we can define its *zone of attraction* to be the set of distributions that collapse to $\langle n, N - n \rangle$. More precisely, say that $\langle m, N - m \rangle$ collapses up to $\langle n, N - n \rangle$ just in case $m < n$, and for each x, $m < x < n$, $\langle x, N - x \rangle$ is unstable upward, and analogously for collapsing downward. $\langle n, N - n \rangle$ is *attainable* if its zone of attraction contains all distributions.

The sullied community might work much better than the purists who failed to divide the labor and who are still struggling to fathom VIM. More exactly, there may be a distribution that is both stable and attainable and that offers a higher probability of community success than the distributions we considered for the epistemically pure. The very factors that are frequently thought of as interfering with the (epistemically well-designed) pursuit of science—the thirst for fame and fortune, for example—might actually play a constructive role in our community epistemic projects, enabling us, as a group, to do far better than we would have done had we behaved as independent epistemically pure individuals. Or, to draw the moral a bit differently, social institutions within science might take advantage of our personal foibles to channel our efforts toward community goals rather than the epistemic ends that we might set for ourselves as individuals.

But is the possibility genuine? Consider cases. The simplest is that in which the return functions are given by

$$p_i(n) = p_i(1 - e^{-kn}) \qquad \text{with } k \text{ large and } p_1 > p_2.[31]$$

There is a bilaterally stable distribution for the sullied community in the neighborhood of $\langle n^*, N - n^* \rangle$, with $n^* = p_1 N/(p_1 + p_2)$. The distribution is attainable. Moreover, if p_1 is only slightly larger than p_2, the distribution yields a probability of community success that is close to that given by the optimal distribution. Even if p_1 is substantially larger than p_2, provided only that $p_1 < p_2 e^{kN}$ [$(\ln p_1 - \ln p_2) < kN$], the distribution $\langle n^*, N - n^* \rangle$ will generate a probability of community success greater than that obtained from $\langle N, 0 \rangle$. *Moral:* there are conditions under which the sullied do better than their epistemically pure cousins, even conditions under which they come as close as you please to the ideal.

Life is more complicated if the return functions take the forms

$$p_i(n) = p_i(3n^2 - 2n^3/kN)/k^2 N^2 \quad \text{for} \quad n < kN \qquad (15.3)$$

$$p_i(n) = p_i \quad \text{for} \quad n > kN$$

$$\text{where} \quad p_2 < p_1 \quad \text{and} \quad k < p_2/(p_1 + p_2).$$

Under these conditions, it is possible to achieve the intrinsic prospects of both methods and *any* distribution $\langle n, N - n \rangle$ with $kN < n < N - kN$ is an optimal distribution. There is a bilaterally stable distribution $\langle n, N - n^* \rangle$, given by $n^* = p_1 N/(p_1 + p_2)$. Provided that $p_2/(p_1 + p_2) > k$, this bilaterally stable distribution will be optimal. So far, so good. So long as the intrinsic prospects of the inferior method are not too low, and the methods respond quickly to the assignment of workers, there will be an optimal division of cognitive labor which the community can *maintain*—if it can but reach it.

But there's the rub. The zone of attraction of the stable distribution

---

31. It's also necessary for the claims that follow to be true that p_2 not be too small. These vague conditions can be formulated more precisely by requiring that

$$\exp\{-kp_2 N/(p_1 + p_2)\} \ll 1.$$

includes all the optimal distributions, but it is quite possible that the community should get stuck at a suboptimal distribution, particularly at the extreme $\langle N, 0 \rangle$. Intuitively, if p_2 is too small or N too large, there may be no benefit in a maverick's abandoning method I.[32] The good news is that there are some instances of this general type in which the sullied community not only does better than its high-minded cousins but actually achieves a stable optimal division of cognitive labor. The bad news is that when the community is too big, self-interest leads it to the same suboptimal state as purity.

However, there is a remedy. The trouble with large communities (more exactly communities for which kN is too big) is that a single deserter from method I cannot contribute enough effort to method II to make that method profitable. Several people need to jump ship together. Imagine, then, that the community is divided into fiefdoms (laboratories) and that when the local chief (the lab director) decides to switch, the local peasantry (the graduate students) move too. Suppose that each lab contains q members and that the director can thus bring it about that x members of the community switch where x is less than or equal to q. Of course, if $q > kN$, then a single laboratory can realize the intrinsic prospects of method II, and it is easy to see that there are conditions under which a stable optimal distribution is attainable. If $q < kN$, there are interesting special cases. Flexibility in the community is promoted if q is greater than or equal to $3kN/4$. Moreover, if q is at least $3kN/4$, the stable optimal distribution is attainable if $p_2/p_1 > 8k/9$. Since we are assuming that k is small, this result shows that the sullied community can manage an optimal division of cognitive labor, even when the prospects of method II are markedly less than those of method I. *Moral:* A certain amount of local autocracy—lab directors who can control the allegiances of a number of workers—can enable the community to be more flexible than it would be otherwise.

My imaginary sullied community is, I think, a recognizable idealization of a subcommunity depicted in James Watson's famous book *The Double Helix*. If Watson's characters, *Watson* and *Crick,* are at one remove from actual scientists, then the members of my imagined community are at a further remove. However, the kinship is, I hope obvious: sullied communities may come close to an optimal distribution because some enterprising members see that it is to their advantage to try a method whose intrinsic prospects are relatively low. In Watson's account, *Watson* and *Crick* take that kind of gamble. The models I've constructed here might, whimsically, be called "Lucky Jim models."

---

32. The condition for its being profitable to switch from method I to method II, if everybody is using method I, is $p_2(3 - 2/kN) > p_1 k^2 N$. Since, $p_2 < p_1$ (by our assumption that method II is the inferior method), there's no chance of satisfying this if N is too large—specifically if $Nk^2 > 3$. On the other hand, the stable distribution is attainable under some conditions: (a) if $Nk < 4$ and $p_2(kN + 1)(kN - 2) > p_1 k^3 N^2$; (b) if $Nk > 4$, $Nk^2 < 3$, and $p_2(3kN - 2) > p_1 k^3 N^2$.

16. Approaches to the Problem of Theory Choice

The most forceful illustration of the value of division of cognitive labor occurs in the context of (what has traditionally been called) "theory choice." There are numerous occasions in the history of science and in contemporary science on which a community is faced with alternative ways of modifying its consensus practice, and on which the rivals are sufficiently close to make it inadvisable to take an immediate decision. Under these conditions, the community would be better served if efforts were expended in pursuing both (each) of the alternatives, and, if pursuit is impossible without cognitive commitment to the epistemic virtue of the option pursued, then it would be best for the community if some amended their individual practices in one of the proposed ways, others by adopting the alternative course(s).[33]

For the purposes of the following sections I shall retain the traditional idiom of theory choice, rather than attempting to survey the full range of instances for which my framework allows. In consequence, the paradigmatic situation will be one in which there are two rival "theories"—each conceived as a statement that might be incorporated within consensus practice—and that the community goal is to incorporate that theory (if either) which is true.

Precise description of cases of theory choice turns out to be tricky—which is why the last sections developed the more tractable "Lucky Jim" models. Crudely, the troubles stem from the existence of two sources of uncertainty: we need to take into account the probability that a theory will improve its apparent epistemic status and also the probability that, if it does so, it will be closer to the epistemic goal (truth). Nonetheless, if we are prepared to make some large idealizations, we can find analogues of the results of earlier sections.

Imagine that, at some moment in the history of some science, we have a pair of rival, incompatible theories, T_1 and T_2. Given the available evidence, the probability that T_i is true is q_i, and all members of the scientific community concerned with the theories recognize this. Suppose further that $q_1 + q_2 = 1$, that $q_1 > q_2$, but that q_1 is approximately equal to q_2.[34]

The community goal is to arrive at universal acceptance of the true theory, to eliminate the problems that currently beset this theory, and to develop the theory in its applications to both theoretical problems and practical matters.

33. Although it is in principle possible for a community to reach a good distribution by having its members pursue rival approaches in a detached fashion, I suspect that such alternatives are usually most successfully developed by their partisans. Those who want to draw a distinction between strategies of pursuit and strategies of belief seem to me to adopt a highly idealized picture of the cognitive capacities of scientists.

34. There are many different idealizations here: I assume that theories can be associated with definite probabilities on the basis of the available evidence, that there is universal recognition of the right probabilities, and that one of the theories is correct. (Usually, of course, the negation of a theory is not anything we would recognize as a sensible theory.) One first move toward greater realism would be to relax this last supposition, allowing that $q_1 + q_2 = r < 1$. See Section 22.

In pursuing this goal, the community can follow one of two generic strategies: (A) assign all the scientists to T_i, (B) assign n scientists to T_1, $N - n$ to T_2 (where $0 < n < N$). I shall consider the merits of these strategies from the perspective of a later stage in the history of the community—"the time of reckoning"—at which we assign epistemic utilities to various consequences.[35]

The possible outcomes at the time of reckoning are as follows: If everyone has been pursuing the true theory, then, I assume, that is the best of all possibilities, and, through its resolution of problems, and so forth, the community has amassed epistemic utility u_1. On the other hand, if everyone has been working on the false theory, then, I shall suppose, nothing has been accomplished and, indeed, epistemic mischief has been perpetrated by introducing false assumptions into problem formulations, for epistemic utility $-u_1$. (It might be more plausible to think of these utilities as added on to a basic utility that will be obtained by committing the scientific work force to *any* course of action. Work on any theory will yield some epistemic good, and the utilities I discuss can be seen as measuring the differential gains of accepting true or false theories, resolving disputes, and so forth). To understand the consequences of dividing the labor, we need to introduce the concept of a *conclusive state*, a situation in which the present standoff between T_1 and T_2 is resolved. I shall imagine that both the available theories are currently beset by anomalies, problems that it is necessary for them to overcome if they are to win unqualified acceptance (honoring the traditional idea that theories are born refuted). We would reach a conclusive state in favor of one of the theories, say T_1, just in case, at the time of reckoning, T_1 had managed to overcome its problems and T_2 had not, despite being given an opportunity to do so. The controversy between T_1 and T_2 is resolved if *both* theories are

35. Here it would be more realistic to allow for the possibility that the community aims to use the presently available theories in achieving a more adequate descendant theory that would be closer to the truth than either of those now available. (See, for example, Martin Rudwick's discussion of the great Devonian controversy in (Rudwick 1985), examined all too briefly in Chapter 6, and in particular, the diagram at 412–413, which reveals the ways in which later positions about the strata in question combined features of earlier theories.) Obviously, allowing for the possibility of attaining closer approximations to epistemic goals through synthesis of parts of current competitors would underscore the value of division of cognitive labor.

The idea of a time of reckoning is a useful fiction, comparable to that employed by population biologists in hypothesizing that the relative reproductive successes of different types can be measured by looking at the value of a particular quantity (e.g., the number of copulations achieved or the number of eggs fertilized) in a particular generation. In principle, one should consider the total histories of consequences resulting from the choice of particular strategies and the epistemic utilities associated with each. It is possible that there are some instances in which it would be better to pursue a strategy of kind A, investing all available resources in a currently more promising theory, and returning to its discarded rival later, if the promise is not realized, and that the epistemic benefits of this strategy do not show up on the simplified analysis I offer here. However, it is clear that there are other conditions in which this is not so, and the simplification involved in positing a time of reckoning does no harm.

pursued and if *one* overcomes its current anomalies, while the other does not.

Now let us make the very optimistic assumption that nature, while not forthcoming, is also not hostile: while correct theories may encounter anomalies, theories that successfully overcome *all* their anomalies are correct. Thus, if we reach a conclusive state, then there is no need to worry about false positive results—in a conclusive state we resolve the issue and we resolve it correctly. In light of this assumption, I shall assign epistemic utilities to the outcomes as follows: if division of labor by following one of the B strategies leads to a conclusive state, then we attain epistemic utility u_2 ($0 < u_2 < u_1$); if it does not, then the epistemic utility is 0 (we're still in the same predicament, although our labor may have given us a clearer view of the problems that each of the rivals faces).[36]

The expected utility of the more promising A strategy (assign all scientists to T_1) is easily computed. It is

$$q_1u_1 - q_2u_1. \tag{16.1}$$

To work out the expected utility of the strategy B_n (choose the distribution $\langle n, N - n \rangle$), we need to recognize that a conclusive state in favor of T_1 will be attained just in case (a) T_1 overcomes its current anomalies, (b) enough workers are assigned to T_2 to give T_2 a chance. I shall assume (somewhat arbitrarily) that the probability that (b) is the case is 1 if $N - n$ is larger than some value m and 0 otherwise. The probability that (a) obtains is the probability that T_1 is true multiplied by the probability that T_1 responds to the efforts of n scientists. Letting $p_i^*(n)$ be the probability that a *true* theory T_i responds to the assignment of n workers by overcoming its problems, we can write the expected utility of B_n, where $m < n < N - m$ (so that both theories are given a chance) as

$$q_1p_1^*(n)u_2 + q_2p_2^*(N - n)u_2.$$

Division of labor is thus preferable if there's an n, meeting the "give both a chance" constraint, $m < n < N - m$, such that

$$q_1p_1^*(n)u_2 + q_2p_2^*(N - n)u_2 > q_1u_1 - q_2u_1 \tag{16.2}$$

36. My claims about the u_i can easily be adjusted to reflect differences in views about the values of particular outcomes, or even differences in specific situations that require assignment of different values. Much more tricky is the task of replacing the hypotheses about conclusive states with more realistic assumptions. In principle, one ought to allow for the possibility that ingenuity can make a false theory continue to appear plausible, and the idea of simple opposition between victory for one theory or an inconclusive dispute should give way to study of the evolution of the probabilities assigned to the rival theories. So, in effect, we want an estimate of the probability that each theory will be assigned if there's a particular division of cognitive labor, and to use this to specify the probability of making a *correct* decision at the time of reckoning. These complications will be explored later in this section and in the following sections.

At this point, it's necessary to make some assumptions about the parameter m and the functions p_i^*. There are many ways to formulate the idea that both theories should be given a chance: we could require that each theory must be assigned sufficient workers to give it some high fraction of its maximal chance of overcoming its problems (on the supposition that it is true), or that the chances of each overcoming its problems be comparable (i.e., the probability that T_1 successfully responds, given that T_1, be roughly the same as the probability that T_2 successfully responds, given that T_2), or that m be simply some fixed relative frequency. I shall not try to choose among these various ways of proceeding, for the reason that my assumptions about the return functions p_i^* will allow for the existence of values of n that satisfy the constraint, given any of the plausible choices for m.

As with the Lucky Jim models, I shall consider two possible kinds of return functions: (a) $p_i^*(n) = p_i(1 - e^{-kn})$, (b) $p_i^*(n) = p_i(3n^2 - 2n^3/kN)/k^2N^2$ $(n < kN)$, $p_i^*(n) = p_i$, otherwise. We can simplify the discussion by supposing that $p_1 = p_2$, in both cases. For the return functions measure the probability that the theory, if true, will be able to overcome its problems if n workers are assigned to it, and it does not strain credulity to suppose that these probabilities, for rival theories, might sometimes be the same functions of n. In case (a), we may also assume that $p = 1$, that, given the assignment of an infinite work force, the probability that a true theory overcomes its problems is 1. In case (b), this supposition seems much more questionable.

In light of the ideas of the last paragraph, some of the apparent implausibility of my previous assumptions may be diminished. For, we can think of the problem I am exploring in the following way: We want to compare the merits of two community cognitive strategies, one that involves leaping with the balance of the evidence, however slight, and one that involves patiently waiting for a state that is genuinely epistemically conclusive before reaching a decision. My idealizing hypothesis that, in an epistemically conclusive state, the theory that has appeared to resolve all its problems is correct should appear less outlandish if you think of the set of demands imposed for solving *all* the problems as very rigorous. Under these circumstances, the probability of our ever achieving the desired conclusive state may be quite small (a fact that will be reflected in the choice of the parameter k that appears in the return functions), so that the apparent optimism of the supposition that choices made in conclusive states are correct is, in large measure, retracted by recognizing that it is very difficult to reach such states.[37]

---

37. Of course, the point made in note 36, that we should think in terms of the evolution of probabilities and of choices made at later stages as the probability of one theory comes to seem much greater than that of its rival, remains correct. On the new way of looking at things, there is an obvious class of strategies that are missing from the discussion: work on both theories and transfer workers from one theory to its rival as the probabilistic relations change. However, my depiction of the problem as a choice between two extreme types of strategies has the merit that, if I can show that there are situations under which one of the B_n's is preferable, it's clear that the optimal distribution involves a genuine

From the details of the algebra (see Technical Discussion 7) we can draw an obvious moral. As in the case of the Lucky Jim models, there are specifiable circumstances, albeit highly idealized, in which the distribution in an epistemically pure community diverges from the optimum and in which extra-epistemic incentives bring the community to the optimal distribution. As in the examples of Section 15, we can understand how it is possible that social structures within the scientific community can work to the advantage of the community epistemic projects by exploiting the personal motives of individuals.

This is a useful beginning. However, the highly artificial assumptions that have been made in my analysis foster the worry that my conclusions could easily be subverted were we to proceed in a more realistic way. Moreover, those assumptions also make it impossible to consider various factors and combinations of factors that ought to be recognized. In consequence, an extremely large part of the work of following sections will be devoted to showing how we can do better. I shall offer a more refined model of division of labor for theory choice, one that not only is more realistic but also provides opportunities for introducing features that are absent from the relatively simple Lucky Jim models.

Technical Discussion 7

Consider case (a), and make the simplifying assumption that $p = 1$. Then a strategy B_n is superior to the more promising of the A strategies if

$$[q_1(1 - e^{-kn}) + q_2(1 - e^{-k(N-n)})]u_2 > (q_1 - q_2)u_1. \qquad (16.3)$$

As with the Lucky Jim models, the optimal B_n is given when $n = (kN + \ln q_1 - \ln q_2)/2k$. By our assumption that q_1 and q_2 are close in value, n will be about $1/2$, unless k is very small—and, if n is close to $1/2$, both theories have been given a chance. A sufficient condition for the optimal B_n to be superior to the more promising of the A strategies is that

$$u_2(1 - e^{-kN/2}) > u_1(q_1 - q_2). \qquad (16.4)$$

Intuitively, unless k is very small indeed, so that assigning half the available workers to each rival would not provide a significant chance of achieving a resolution of the issue, or if the utility of immediate action on the basis of the true theory is very high, we can expect that an optimal distribution will involve a genuine division of labor.

Turn now to case (b). Imagine, as with the Lucky Jim models, that $k < 1/2$, so that it would be possible to assign scientists to each theory in a way that would give each a maximal run for its money. If n lies in the interval

---

division of labor—for that distribution is either given by the best of the B_n's or by one of the missing strategies, all of which begin by dividing the labor.

$[kN, (1 - k)N]$, then the condition that both theories have been given a chance is surely satisfied. So the crucial inequality for the preferability of one of the B_n is

$$pu_2 > (q_1 - q_2)u_1. \tag{16.5}$$

Thus, unless the maximal chances of a true theory's overcoming its problems are low (p is small) or the utility of immediate action is high (u_1 is large relative to u_2), the optimal distribution again involves a genuine division.[38]

Let us suppose that the important motive is each scientist's desire to be singled out by posterity as an early champion of the accepted theory. If the community is initially divided, with distribution $\langle n, N - n \rangle$, if the return functions take the form (b), and if $kN < n < (1 - k)N$, then there is a stable attainable distribution, $\langle n^*, N - n^* \rangle$, where

$$n^* = q_1 N/(q_1 + q_2). \tag{16.6}$$

Provided only that $k < q_2$ (a very weak assumption, given that $k < 1/2$ and q_2 is close to $q_1 (= 1 - q_2)$), this will be optimal.

---

17. Theory Choice: Community Optimum

Attempts to tackle the evolution of the probabilities of the rival theories directly lead to enormous complications, and I shall continue to make some simplifying assumptions. We can think of the community decision problem in the following way: First, suppose that the strategies considered assign n workers to T_1, $N - n$ to T_2, and that these assignments are maintained throughout the subsequent changes in the fortunes of the rival theories, until a point at which the situation becomes determinate. Second, we may assume that there are three possible outcomes from this assignment during the next temporal stage: T_1 may appear to overcome its problems, T_2 may appear to overcome its problems, or the situation may remain indeterminate. We do not allow for a case in which both rivals appear to be successful. Third, if the situation remains indeterminate, then we imagine that the same distribution of scientists continues to pursue the rivals. Finally, we suppose that an apparently successful theory will become part of consensus practice, and that, if this is the case, then there are two possibilities: that the theory is correct, or that it is incorrect. Using these ideas, we can represent the community decision problem as in Figure 8.9.

For the present, I shall suppose that T_1 and T_2 are exclusive and exhaustive, and that the initial probabilities assigned to them, given the available evidence, are p, $1 - p$. $F_1(n)$ is the probability that T_1 will appear to solve its problems, given the assignment of n workers, and this will surely increase with n. Without

---

38. I note explicitly that the distribution in an epistemically pure community will presumably be $\langle N, 0 \rangle$. Since $q_1 > q_2$, pure scientists will choose T_1 over T_2.

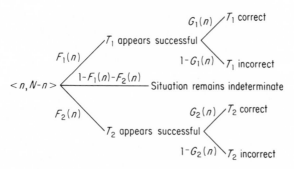

Figure 8.9

too much loss of realism, we can keep matters simple by setting $F_1(n) = pn/N$. Using similar reasoning, $F_2(n)$, the probability that T_2 will appear to solve its problems given an assignment of n to T_1, can be taken as $(1 - p)(N - n)/N$. $G_1(n)$ is the probability that if T_1 solves its problems given an assignment of n workers to T_1 then T_1 is correct. Now if T_1 has had all the available resources lavished upon it, then its appearance of success is far less impressive than if it has done a lot with a little. We can use this intuitive idea to refine the "give both a chance" constraint. Suppose that the status of T_1 will be effectively unchanged if all the available resources have been devoted to articulating and defending it: that is, $G_1(N) = p$. On the other hand, if T_1 succeeds with *no* resources devoted to it, that is a considerable accomplishment, and we can then take T_1 to be maximally probable: thus, let $G_1(0) = 1 - \alpha$ where α is small. If we suppose that the probability is a linear function of n then we should set

$$G_1(n) = ((1 - \alpha)(N - n) + pn)/N. \qquad (17.1)$$

By parity of reasoning:

$$G_2(n) = ((1 - \alpha)n + (1 - p)(N - n))/N. \qquad (17.2)$$

Now consider the rewards. I shall continue to assume symmetry between being right and being wrong, letting the payoff for a correct decision be u and the payoff for an incorrect decision be $-u$. However, it will now be assumed that the rewards decay at a rate β. We imagine that the period following community decision is divided into a series of "rounds," after each of which the situation is assessed. Postponing a decision from the end of one round to the next involves decreasing the value from V to βV.

Besides the various strategies for dividing the work force, there is an alternative strategy of amending consensus in the initial situation: "jumping at once." The expected utility of jumping at once is readily computed as

$$U = pu - (1 - p)u = u(2p - 1). \qquad (17.3)$$

Here we assume that the decision is made in favor of T_1, the initially more probable theory, and, since $p > 1/2$, U is positive. The expected utility of the strategy that assigns n scientists to T_1, $N - n$ to T_2, $\langle n, N - n \rangle$, is V_n where

$$V_n = \{F_1(n)G_1(n) + F_2(n)G_2(n)\}\beta u - \{F_1(n)(1 - G_1(n)) + \quad (17.4)$$
$$F_2(n)(1 - G_2(n))\}\beta u + (1 - F_1(n) - F_2(n))\ \beta V_n.$$

That is,

$$V_n = u\beta \sum_{i=1,2} (2F_i(n)G_i(n) - F_i(n))/\{1 - \beta + \beta(F_1(n) + F_2(n))\}(17.5)$$

We can now formulate the community decision problem precisely. Our task is to discover the conditions under which there is a value of n such that $V_n > U$, the conditions under which dividing the labor is preferable to jumping at once. Subject to these conditions, we want to identify the value of n for which V_n is maximal. Technical Discussion 8 takes up these tasks in order.

The introduction of the parameters α and β enables us to refine the simple treatment of the last section by recognizing that even those theories that solve their problems may not attain probability one, and that there are costs involved in delaying. Thus the new analysis permits us to appreciate the existence of situations in which dividing the labor is unprofitable because our epistemic position cannot be expected to improve in any significant fashion. We can also see that when a theoretical decision is urgently needed—perhaps for the resolution of a practical question—the utilities may decay at a fairly rapid rate (β is significantly less than one), so that jumping at once is preferable. Moreover, the specification of the probability functions G_1 and G_2 enables us to formulate the notion that both theories must be given a chance in a more sophisticated fashion.

Technical Discussion 8 reveals that, when α and β are close to 0, 1, respectively, the model of community decision I have constructed favors diversity. Even when p is relatively high (T_1 is antecedently quite probable) the community should still devote a sizable portion of its effort to T_2. Thus, if the assumptions that I have made are (approximately) correct, the *optimal* decision for the community involves a much more pronounced division of labor than we might have anticipated (or perhaps than is found in most scientific communities). Set against this finding is the result that, within limits, the *amount* of diversity does not matter much. Provided p is fairly high, almost any value of r (the relative frequency of workers assigned to T_1) between 0.5 and 1 will yield an acceptable result. However, as p decreases, diversity becomes more significant, and, when $p < 0.7$, it is clear that a roughly equal division of the labor is a reasonably good all-purpose strategy. Indeed, given the uncertainties that attend the determination of precise probabilities for rival theories, the table shows that a tolerable rule of thumb is to divide the community labor roughly equally between two theories if both are seen as serious candidates. Moreover, even when α and β depart from 0 and 1, in a wide range of instances, division of labor is profitable for the community. Pronounced division of labor (setting $r < p$) is favored if the costs of delay are small and if it is possible to achieve a very high probability for the successful theory. Furthermore considerable deviations from the optimal value of r can be tolerated without severe epistemic loss, and the value of r should increase as we increase the costs of delay (lower β) and the irresoluble error (higher α).

Technical Discussion 8

The expected utility of dividing the community $\langle rN, (1 - r)N \rangle$ is V_{rN}, where

$$V_{rN} = \beta u \{2pr((1 - \alpha)(1 - r) + pr) - pr + 2(1 - p)(1 - r) \quad (17.1)$$
$$((1 - \alpha)r + (1 - p)(1 - r)) - (1 - p)(1 - r)\}/$$
$$((1 - \beta) + \beta(pr + (1 - p)(1 - r))).$$

After some algebraic simplification, this reduces to:

$$V_{rN} = \beta u \{r^2(2\alpha - 4p + 4p^2) + r(6p - 1 - 2\alpha - 4p^2) + \quad (17.2)$$
$$(1 - 3p + 2p^2)\}/(1 + 2\beta pr - \beta p - \beta r)$$

Dividing the community is preferable to jumping at once provided that (for some value of r) $V_{rN} > U$, where

$$U = u(2p - 1). \quad (17.3)$$

Working out the algebra yields the crucial inequality:

$$r^2(2\alpha\beta + 4\beta p^2 - 4\beta p) + r(10\beta p - 2\beta - 2\alpha\beta - 8\beta p^2) + \quad (17.4)$$
$$(1 + \beta - 2p - 4\beta p + 4\beta p^2) > 0$$

which I shall sometimes abbreviate as

$$T(\alpha, \beta, p, r) > 0. \quad (17.5)$$

Intuitively, when $p = 1$, jumping at once ought to be preferable, for, under these circumstances, dividing the labor incurs the costs of delay and only promises a maximal probability for T_1 of $1 - \alpha$. When $p = 1$, (17.4) becomes

$$-2\alpha\beta r^2 - 2\alpha\beta r - 1 + \beta > 0 \quad (17.6)$$

and this is clearly unsatisfiable.

The conditions under which dividing the labor is preferable to jumping at once plainly depend on the values of p, α, and β. We can investigate these conditions by looking at the behavior of T as the crucial parameters are varied. The first point to note is that

$$\partial T/\partial \alpha = 2\beta r(r - 1) < 0 \quad (17.8)$$

so that (as we should expect) T increases as α decreases. Since $(1 - \alpha)$ represents the maximal probability one can expect to confer upon either of the rivals by dividing the labor, smaller values of α correspond to more valuable potential final states, and thus make it more worthwhile to delay.

Provided that α is small, $V_{rN} > 0$. Now if $V_{rN} > 0$, we can write

$$V_{rN} = \beta H/(1 + 2\beta pr - \beta p - \beta r) \quad (17.9)$$

where H does not contain β, and $H > 0$. Hence

$$T = \beta H - (1 + 2\beta pr - \beta p - \beta r)(2p - 1) \qquad (17.10)$$

$$\partial T/\partial \beta = H + (2p - 1)(p(1 - r) + r(1 - p)) > 0. \qquad (17.11)$$

Thus, if α is small, T increases as β increases. Again, this should be unsurprising, since higher values of β correspond to situations in which there is less cost involved in delaying.

Let us now consider the behavior of T when p and r are allowed to vary. Because a general treatment is algebraically troublesome, I shall consider some special cases.

First, let us examine instances in which α and β are significantly different from 0 and 1, respectively. If $\alpha = 1/4$, $\beta = 3/4$, $T(\alpha, \beta, p, r)$ becomes

$$T(1/4, 3/4, p, r) = r^2 3p(p - 1) + 3r(r - 1)/8 + \qquad (17.12)$$
$$3r(5p - 4p^2 - 1)/2 + (7/4 - 5p + 3p^2)$$

$T(1/4, 3/4, p, r)$ decreases with p, so that, the closer the probability values of the rivals, the more profitable it is to divide the labor rather than jumping at once. Combining our results so far, we know that

For all α, β, p, if $T(\alpha_0, \beta_0, p_0, r_0) > 0$, and if $\qquad (17.13)$
$\alpha_0 < 1/4$, $\beta_0 > 3/4$, $\alpha < \alpha_0$, $\beta > \beta_0$, $p < p_0$, then
$T(\alpha, \beta, p, r_0) > 0$.

Our simplified treatment of the last section suggests that setting $r = p$ might be close to a maximum for V_{rN}, so it is worth investigating the special case $r = p$ to discover how close the probabilities of T_1 and T_2 need to be for dividing the labor to be preferable to jumping at once. Tedious but straightforward calculations show:

When $p < 0.58$, $T(1/4, 3/4, p, p) > 0$ $\qquad (17.14)$

When $p > 0.59$, $T(1/4, 3/4, p, p) < 0$. $\qquad (17.15)$

Assuming, then, that $r = p$ comes close to maximizing V_{rN}, we can conclude that, when the maximum probability attainable from dividing the labor is relatively low $(1 - \alpha = 3/4)$ and the decay in utility relatively rapid $(\beta = 3/4)$, the probabilities of T_1 and T_2 have to be rather close for dividing the labor to be worthwhile.

Contrast this with the case in which $\alpha = 0.1$, $\beta = 0.95$, in which the maximal probability obtaining from working further on the rivals is 0.9, and the rate of decay in value is much slower. Calculations of the values of $T(0.1, 0.95, p, p)$ show that

When $p < 0.77$ $\qquad T(0.1, 0.95, p, p) > 0$ $\qquad (17.16)$

When $p > 0.78$ $\qquad T(0.1, 0.95, p, p) < 0$. $\qquad (17.17)$

Under these circumstances, it would be worth dividing the labor even when T_1 is three times as initially probable as T_2.

Finally, if $\alpha = 0$, $\beta = 1$, it is possible for work on the rivals to boost the probability of one of them to 1, and there are no costs in delaying. As we have already seen, if $p = 1$, there is no gain in dividing the labor. However, if $p < 1$, we can always find r so that $T(0, 1, p, r) > 0$. Hence, if probability 1 is attainable, if delay costs nothing, and if the current value of the probability of T_1 is less than 1, dividing the labor is preferable to jumping at once.

Let us now turn to the second part of the task set at the beginning of this section. How do we choose r so as to maximize V_{rN}? I shall start with the case in which $\alpha = 0$, $\beta = 1$, and subsequently take a look at the effects of increasing α, and decreasing β.

When $\alpha = 0$, $\beta = 1$,

$$V_{rN} = \{r^2(4p^2 - 4p) + r(6p - 4p^2 - 1) + (1 - 3p + 2p^2)\}/ \qquad (17.18)$$
$$(1 + 2pr - p - r)$$

If $\partial V_{rN}/\partial r = 0$, then

$$2r^2(2p - 1) + 4r(1 - p) - 1 = 0. \qquad (17.19)$$

That is,

$$r = ((p^2 + (1 - p)^2/2)^{1/2} + (1 - p))/(2p - 1)$$

Let r^* be the optimal value of r, as given by this formula. Then we can illustrate the effects of decisions to divide the labor in various ways by the following table[39]:

p	r^*	$V_{0.5N}$	V_{r^*N}	V_N
0.6	0.55	0.52	0.53	0.2
0.7	0.6	0.58	0.6	0.4
0.75	0.62	0.625	0.65	0.5
0.8	0.64	0.68	0.71	0.6
0.9	0.68	0.82	0.84	0.8
0.95	0.69	0.905	0.92	0.9

I shall conclude by considering the effect of letting α depart from 0, β depart from 1. The general case is too complicated to provide a revealing treatment, so I shall focus on the special case in which $\alpha = 0.1$, $\beta = 0.9$. Again, let r^* be the optimal value of r (that which maximizes V_{rN}). Substituting values of p and using elementary calculus to find the associated maxima of V_{rN}, we obtain the following results[40]:

---

39. Here I normalize the values by setting $u = 1$.

40. Once again $u = 1$. Plainly, in this instance, it is not worth considering values of p larger than 0.9, since the maximum probability that can now be attained for either theory is 0.9! It should be noted that, when $p = 0.8$, the value of r^* is 0.605 (to three significant figures); the associated value of U (the utility of jumping at once) is $(2p - 1) = 0.6$. Hence, in this instance, dividing the labor *in an optimal fashion* is superior to jumping at once, but, because the payoffs are so close, most division strategies will be inferior.

p	r^*	$V_{0.5N}$	V_{r^*N}	V_N
0.6	0.59	0.38	0.39	0.19
0.7	0.68	0.44	0.47	0.38
0.75	0.74	0.48	0.53	0.48
0.8	0.81	0.53	0.6	0.59

The effect of increasing α and decreasing β has been to increase the optimal value of r. Thus, when the maximal probability we can hope to obtain for a successful theory is lower, and when the costs of delay are higher, the community division of labor should be less pronounced. We should note that, for the special case under consideration, the strategy of setting $r = p$ is a good heuristic, and that, as in the earlier example, a fairly wide range of values of r will yield acceptable results.

18. Theory Choice: Individual Responses

Let us now consider the ways in which various types of scientists would distribute themselves. As in Section 16, I shall suppose that the community contains epistemically sullied agents who are motivated by the desire to be hailed as early champions of an ultimately accepted theory. Thus, for each member of the community, we can represent the utility of defending T_1 in a situation in which there are $n - 1$ other champions of T_1 as proportional to

$$\text{Prob}(T_1 \text{ is ultimately accepted})/n.$$

The crucial question at this point is, How is the pertinent probability computed?

One possibility is that the scientists are thoroughly cynical. They do not concern themselves with questions about whether T_1 is accepted *and true*. Apparent success is enough—for they believe that the community will adopt T_1 into consensus practice if T_1 manages to overcome its problems. The cynical computation of $\text{Prob}(T_1$ is ultimately accepted) goes as follows:

$$\begin{aligned}
&\text{Prob}(T_1 \text{ is ultimately accepted}) \qquad\qquad\qquad\qquad (18.1)\\
&= \text{Prob}(T_1 \text{ accepted on the first round}) + \text{Prob}(T_1 \text{ accepted on the}\\
&\quad \text{second round}) + \ldots\\
&= F_1(n) + (1 - F_1(n) - F_2(n))F_1(n) + (1 - F_1(n) - F_2(n))^2 F_1(n)\\
&= F_1(n)/(F_1(n) + F_2(n)).
\end{aligned}$$

It is now evident that the expected utilities of defending T_1 and of defending T_2 are proportional to

$$\begin{aligned}
&pr/rN(pr + (1 - p)(1 - r)) \quad \text{and}\\
&(1 - p)(1 - r)/(1 - r) \, N(pr + (1 - p)(1 - r))
\end{aligned}$$

where $n = rN$. Since $p > 1 - p$, the expected utility of defending T_1 is greater than the expected utility of defending T_2 *whatever* the value of r. So we can expect a bandwagon to form in favor of T_1.

This conclusion is sobering. Even a community of sullied agents will not succeed in dividing the labor unless its members refrain from adopting a thoroughly cynical attitude to the way in which community decisions will be made. To put the point dramatically (but somewhat inaccurately) even if personal motivations (such as the desire for credit or enduring fame) play a major role in the decisions of individuals it is important that those individuals believe that they participate in an enterprise that is governed by devotion to truth. *Mauvaise foi* may be an essential part of *la condition scientifique*.

This result was prevented in Section 16 by supposing that the members of the sullied community calculated the probability that T_1 will ultimately be accepted in a more idealistic fashion. The idealistic computation simply assumes that the chance that a theory will be accepted is the probability that it is true and uses current estimates of that probability. Hence, if the agents perform the idealistic computation, the expected utilities of defending T_1 and T_2 will be proportional to

$$p/rN \quad \text{and} \quad (1 - p)/(1 - r)N.$$

Thus the scientists can be expected to achieve the distribution $\langle pN, (1 - p)N \rangle$. As we saw in the last section, this is a good approximate solution to the community decision problem—provided, of course, that the optimal strategy is to have *some* type of division of labor (rather than jumping at once).

Technical Discussion 9 reveals that even a small dose of idealism will yield a distribution whose expected utility is not very far from the optimum value, provided that the difference in probability of the two theories is not too great. Thus, in relatively uncertain situations, which are (as is apparent from the earlier tables) just those situations in which dividing the labor well yields the greatest increase in epistemic benefits, epistemically sullied agents who are relatively cynical (attaching three times as much weight to mere acceptance as they do to considerations of truth) achieve a reasonably good distribution. Thorough cynicism is a disaster, but starry-eyed idealism is by no means necessary to serve the community well.

Technical Discussion 9

It is not difficult to show that the division generated by the idealistic computation (henceforth the *idealistic division*) is preferable to jumping at once in the special case in which $\alpha = 0$, $\beta = 1$. For, setting $\alpha = 0$, $\beta = 1$, $r = p$, the inequality of the last section (17.4) becomes

$$2p^4 - 6p^3 + 7p^2 - 4p + 1 > 0 \tag{18.2}$$

which holds for all values of p. As noted in discussion of the other two special cases $((\alpha = 1/4, \beta = 3/4), (\alpha = 0.1, \beta = 0.95))$, there will be occasions on which jumping at once is preferable to dividing the labor, and, on these occasions, performing the idealistic computation will lead the community to a suboptimal strategy. (Thus, for example, if $\alpha = 1/4$, $\beta = 3/4$, $p = 0.6$, the idealistic division will be inferior to jumping at once.)

Perhaps actual scientists are neither thoroughly idealistic, believing that truth will win out in the long run, nor thoroughly cynical, thinking that apparent success, even that manufactured by the heavy investment of resources, will determine community acceptance. Consider the possibility that the estimation of the crucial probability, Prob(T_1 will ultimately be accepted), is a result of a mixture of the two polar attitudes. Suppose then that the members of the sullied community effectively compute the probability as

$$\tau \text{Prob}_C + (1 - \tau)\text{Prob}_I \quad \text{where}$$
$$\text{Prob}_C = F_1(rN)/(F_1(rN) + F_2(rN)) \quad \text{and} \quad \text{Prob}_I = p.$$

The resultant distribution will choose that value of r for which

$$\{\tau pr/(1 + 2pr - p - r) + (1 - \tau)p\}/r = \tag{18.3}$$
$$\{\tau(1 - p)(1 - r)/(1 + 2pr - p - r) + (1 - \tau)(1 - p)\}/(1 - r).$$

To appreciate the effects of a mixture of idealism and cynicism, let us compare the expected utilities of the distributions obtained for different values of τ. As in one of the numerical examples considered in the last section, I shall suppose that $\alpha = 0$, $\beta = 1$. Substituting for τ and p we can solve the preceding equation to obtain the value of r, compute the value of V_{rN}, and compare the result with $V_{r}^{*}{}_{N} = V_{max}$ (obtained in the last section on p. 363 above). Setting $\tau = 1/2$—equal weighting of idealism and cynicism—we find the following results:

p	r	V_{rN}	V_{max}
0.6	0.68	0.49	0.53
0.7	0.78	0.54	0.6
0.8	0.88	0.65	0.71
0.9	0.95	0.81	0.84

If the members of the community are more cynical, setting $\tau = 3/4$, then the analogous results are

p	r	V_{rN}	V_{max}
0.6	0.79	0.43	0.53
0.7	0.89	0.49	0.6
0.8	0.94	0.63	0.71
0.9	0.97	0.81	0.84

In a completely idealistic community, $\tau = 0$, we have

p	V_{pN}	V_{max}
0.6	0.52	0.53
0.7	0.58	0.6

0.8	0.68	0.71
0.9	0.82	0.84

A useful way to summarize the significance of these results is to introduce a normalized measure of the divergence from the maximum value. For a distribution $\langle rN, (1 - r)N \rangle$ let the *mismatch*, $\mu(r)$, be defined as

$$\mu(r) = (V_{max} - V_r)/(V_{max} - V_N). \tag{18.4}$$

Since, in the cases that interest us, V_N is typically the minimum of V_r ($1/2 < r < 1$), our mismatch measures the proportion of the maximal deviation from the optimum that a particular distribution generates.[41] It also has the advantage that V_N is the value achieved by the members of a completely cynical community. The following table shows the values of the mismatch for various combinations of p and τ.

	$\tau = 0$	0.5	0.75	1
$p = 0.6$	0.03	0.12	0.3	1
$p = 0.7$	0.1	0.3	0.55	1
$p = 0.8$	0.27	0.55	0.72	1
$p = 0.9$	0.5	0.75	0.75	1

19. The Influence of Tradition

It is now time to complicate our basic picture of individual decision making. I have supposed so far that there are no prior commitments or allegiances that would affect the credit that a scientist is likely to receive from defending either T_1 or T_2. What happens if we introduce some asymmetries here?

Imagine that, as a consequence of training, nationality, personal loyalties, intellectual affiliations, or whatever, the scientific community is divided into three subgroups: a group of traditionalists who are associated with the dominant theory T_1, a group of rebels who are associated with the underdog theory T_2, and a group of neophytes who are, as yet, uncommitted. The utilities for members of these groups are different through a straightforward discounting effect: those who do not belong to the group associated with a theory are less likely to be assigned credit for championing it or for doing work that contributes to its eventual acceptance.

Suppose that the neophytes perform the idealistic computation. Then they effectively distribute themselves so as to remedy imbalances caused by existing allegiances. Provided that there are enough of them, there will be an internal equilibrium in the class of neophytes. However, there may be too few neophytes to rectify an unbalanced distribution.

But what if everyone, neophytes included, is thoroughly cynical? As we

---

41. As the tables in Technical Discussion 8 reveal, this is not generally correct. Nevertheless, the mismatch *approximates* the proportion of the maximal deviation and is more easily applicable than computing $(V_{max} - V_r)/(V_{max} - V_{min})$.

saw in the last section, cynical computations lead to disaster, producing a bandwagon in favor of T_1. Can tradition effects break this monopoly? Can they allow for the introduction of maverick ideas? Provided that the expected credit from contributing to the traditional theory is heavily discounted for neophytes then a lone neophyte can be expected to begin a defense of T_2, even when the value of p is very high. *Moral:* discrimination against the work of newcomers can serve as inspiration to develop underdog theories, even in sullied communities whose members are thoroughly cynical. Tradition effects can thus undo some of the bad consequences of cynicism.

Technical Discussion 10

Tradition Effects

Let the community of scientists be of size N (as in our previous examples), divided into n_t traditionalists, n_r rebels, and $N - (n_t + n_r)$ neophytes. For a traditionalist, the expected utility of defending T_1 is

Prob(T_1 is ultimately accepted)$/(n_t + \Theta_r n_r + \Theta_u(N - n_t - n_r))$.

The expected utility of defending T_2 is

Θ_tProb(T_2 is ultimately accepted)$/(n_t + \Theta_r n_r + \Theta_u(N - n_t - n_r))$.

For a rebel, on the other hand, the expected utilities are

Θ_rProb(T_1 is ultimately accepted)$/(n_t + \Theta_r n_r + \Theta_u(N - n_t - n_r))$ and

Prob(T_2 is ultimately accepted)$/(n_t + \Theta_r n_r + \Theta_u(N - n_t - n_r))$.

   In the extreme case, members of a group associated with a rival theory have no hope of receiving credit for work that leads to the establishment of a theory. If this is so, then $\Theta_t = \Theta_r = 0$. Neophytes are not treated so harshly, and we can suppose that $\Theta_u = \Theta$ where $0 < \Theta < 1$. Under these circumstances, the decision problem for traditionalists and rebels is easy: traditionalists choose T_1 and rebels choose T_2. Let us suppose that the neophytes distribute themselves by using the idealistic computation, so that Prob(T_1 is ultimately accepted) $=$ Prob$_I$. Then neophytes choose T_1 if

$$p/(n_t + \Theta n) > (1 - p)/(n_r + \Theta(N - n_t - n_r - n))   \qquad (19.1)$$

where n is the number of neophytes who commit themselves to T_1.

   If there are enough neophytes, then there will be an internal equilibrium within the class of neophytes, with n^* defending T_1, where

$$n^* = p(N - n_t - n_r) + \{pn_r - (1 - p)n_t\}/\Theta.   \qquad (19.2)$$

The first term here represents the standard frequency of T_1 defenders within the class of neophytes: if there were no division into rebels and traditionalists in the community, then this would be the idealistic distribution. The second term is a correction factor, representing the possibility that existing allegiances make for an unbalanced distribution. From the view of the idealistic neophytes, pn_r ought to be $(1 - p)n_t$. If n_r is too large, then the neophytes will correct by adding extra supporters to T_1 (the correction term will be positive). If n_r is too small, then the complementary effect will occur, the correction term will be negative, and the neophytes will attempt to rectify the situation by disproportionately supporting T_2.

But there may be too few neophytes. Consider a simple case, that in which $\Theta = 1/2$. Now $n^* = pN + pn_t + pn_r - 2n_t$. This reaches its maximum value of $N - n_t - n_r$ when

$$p = (N + n_t - n_r)/(N + n_t + n_r). \tag{19.3}$$

If p exceeds this value then, do what they may, the neophytes will be unable to restore the idealistic distribution across the community.

Let us now try to assess the effects of tradition on sullied communities whose members use the idealistic computation. Without tradition, the number of T_1 supporters would be pN. With tradition, the number is

$$n_t + p(N - n_t - n_r) + (pn_r - (1 - p)n_t)/\Theta \quad \text{or} \quad N - n_r$$

whichever is less. Assuming that the number of T_1 defenders is given by the former equation, we can write the change in support due to tradition as

$$\Delta n^* = (1/\Theta - 1)\{p(n_t + n_r) - n_t\}. \tag{19.4}$$

Plainly this is an increasing function of p, and it increases as Θ decreases. Thus departures from the idealistic distribution are to be expected when T_1 is antecedently more probable, and when neophytes are treated as second-class citizens.

As in previous sections, a numerical example helps to illustrate the general point. Suppose that $n_r = n_t = N/4$, and that $\Theta = 1/2$. Then we can compare the support for T_1 (the number of community members who defend T_1) and the expected value for the community from that distribution, as follows:

	No Tradition		Tradition	
	Support	*Value*	*Support*	*Value*
$p = 0.6$	$0.6N$	0.52	$0.65N$	0.47
$p = 0.7$	$0.7N$	0.58	$0.75N$	0.56
$p = 0.8$	$0.8N$	0.68	$0.75N$	0.69
$p = 0.9$	$0.9N$	0.82	$0.75N$	0.83

One interesting feature of this table is that it reveals how an overload of rebels can work *favorably* with tradition in the last two cases.

Resisting Bandwagons. Suppose, as before, that $\Theta_t = \Theta_r = 0$, and that $\Theta_u = \Theta$. Imagine that we begin from a situation in which all members of the community support T_1. Then, for a neophyte, the expected utility of defending T_1 is

$$\Theta F_1(N)/(n_t + \Theta(N - n_t))(F_1(N) + 0) = \Theta/(n_t + \Theta(N - n_t)).$$

The expected utility of defending T_2 is

$$\Theta F_2(N-1)/(0+\Theta)(F_1(N-1)+F_2(N-1)) = (1-p)/((N-2)p+1).$$

So defending T_2 is preferable if

$$(1-p)(n_t + \Theta(N-n_t)) > \Theta((N-2)p+1). \tag{19.5}$$

That is:

$$p < (n_t(1-\Theta) + \Theta(N-1))/(n_t(1-\Theta) + 2\Theta(N-1)). \tag{19.6}$$

As we should expect, this inequality is unsatisfiable if $n_t = 0$ or if $\Theta = 1$. However, if $n_t = N-1$, the inequality requires only that

$$p < 1/(1+\Theta). \tag{19.7}$$

Tradition effects thus enable neophytes to resist the bandwagon generated from cynical computations. This effect can easily be amplified if rebels are much more hospitable to neophytes than are traditionalists ($\Theta_r = 1$, $\Theta_t < 1$).

We can approach the possibility of asymmetries in the awarding of credit in a general way by supposing that the payoffs to neophytes for defending novel (underdog, maverick) theories are greater than those for defending well-articulated (mainstream) theories. Thus the payoff matrix for a neophyte, honored as the champion of a theory, might be

	T_1 Succeeds	*T_2 Succeeds*
Defend T_1	*tu*	*0*
Defend T_2	*0*	*u*

where $0 < t < 1$. When the neophytes use the cynical computation of the probability that T_1 will ultimately succeed, the expected utilities, if n defend T_1 and $N-n$ defend T_2 are

Defend T_1: $tupn/n(pn + (1-p)(N-n))$

Defend T_2: $(1-p)u(N-n)/(N-n)(pn + (1-p)(N-n))$.

The support for T_2 will thus increase if

$$p < 1/(1+t) \tag{19.8}$$

reinforcing our previous conclusion that asymmetries in the awarding of credit can block the potential bandwagon for T_2, even when the members of the community are cynical.

Finally, I want to note an interesting relationship. Suppose that we ignore tradition effects, including asymmetries in the awarding of credit, and consider a sullied community whose members use the idealistic computation. Then the condition that there be some support for T_2 within the community is

$$p < N/(N+1). \tag{19.9}$$

Compare this with a sullied community whose members use the cynical computation, and in which there is a marked asymmetry in the awarding of credit, $t = 1/N$. Here the condition that there should be some support for T_2 is

$$p < N/(N + 1). \tag{19.10}$$

The equivalence of the two cases underscores a general moral: diversity within a scientific community can be fostered in a number of different ways; although, as we have seen, cynicism might lead to a condition of cognitive uniformity, there are countervailing factors that can halt the potential bandwagon for the dominant theory.

---

20. The Pure, the Sullied, and the Mixed-Up

I shall now offer a more systematic account of various types of epistemic agents. Let us start with the epistemically pure.

Those whose sole concern is to defend the truth can be understood as assigning values to outcomes in accordance with the following matrix:

	T_1 *Correct*	T_2 *Correct*
Defend T_1	v	$-v$
Defend T_2	$-v$	v

The expected utilities of the courses of action are $(2p - 1)v$, $(1 - 2p)v$, respectively. Since $2p > 1$, a community of pure agents would achieve the distribution $\langle N, 0 \rangle$.

So far, this simply recapitulates our earlier discussions. However, we can use our approach to represent the possibility that there is an epistemic asymmetry between the two theories. Perhaps T_2 would be epistemically more valuable than T_1 (opening up new directions for tackling questions hailed as significant, for example). We can represent this possibility with a different payoff matrix:

	T_1 *Correct*	T_2 *Correct*
Defend T_1	v	$-V$
Defend T_2	$-v$	V

where $V > v$. Now the expected utilities are $p(v + V) - V$, $V - p(v + V)$, and if $V/v > p/(1 - p)$ then an epistemically pure agent will prefer T_2 to T_1. But this is of little comfort when cognitive diversity is valued, since if *one* such agent prefers T_2 then *all* will do so. Depending on the sign of $V/v - p/(1 - p)$, the community heads either to $\langle 0, N \rangle$ or to $\langle N, 0 \rangle$.

The attitudes of sullied agents can be represented in similar fashion. We can either think in terms of the payoff in situations where the agent is hailed as the first champion of the accepted theory or we can think of the payoff for *defending* the accepted theory, where it is recognized that this will be an *expected* value that depends on the extent of the competition. Because it will be easier to make comparisons between the pure and the sullied by doing so, I shall follow the latter approach, writing the payoff matrix for the sullied as:

	T_1 *Accepted*	T_2 *Accepted*
Defend T_1	u/n	0
Defend T_2	0	$u/(N - n)$

where u is the payoff for being recognized as the first defender of the accepted theory, and n is the total number of those defending T_1. Given the idealistic computation of the probability that T is ultimately accepted, the community distribution will be $\langle pN, (1 - p)N \rangle$. Given the cynical computation it will be $\langle N, 0 \rangle$. So much is familiar.

Once we have these representations, there is an obvious way of extending the discussion. It is hardly likely that scientists are so pure that they care about nothing but defending the True. Equally, it is not very plausible to imagine that they care nothing about the True. We might expect them to be thoroughly mixed-up, motivated in part by considerations of receiving credit for the work, inspired also by the goal of replacing false belief, or agnosticism, with true belief. These mixed motivations can be represented by offering a payoff matrix for *hybrid* agents:

	T_1 *correct* T_1 *accepted*	T_2 *correct* T_1 *accepted*	T_1 *correct* T_2 *accepted*	T_2 *correct* T_2 *accepted*
Defend T_1	$u/n + v$	$u/n - v$	v	$-v$
Defend T_2	$-v$	v	$u/(N - n) - v$	$u/(N - n) + v$

Assume first that the hybrid agents compute the probability that T_1 will ultimately be accepted in the idealistic way. In this case, the probabilities for the four outcomes are $p^2, p(1 - p), p(1 - p), (1 - p)^2$. The expected utilities of defending T_1 and T_2 are, respectively,

$$p^2(u/n + v) + p(1 - p)u/n - (1 - p)^2v$$
$$-p^2v + p(1 - p)u/(N - n) + (1 - p)^2u/(N - n) + (1 - p)^2v$$

T_2 is preferable just in case

$$u/v > 2(2p - 1)n(N - n)/(n - Np). \tag{20.1}$$

Suppose that $u = mNv$, $n = rN$. Then T_2 is preferable if

$$m > 2(2p - 1)r(1 - r)/(r - p) \tag{20.2}$$

and the population would reach an equilibrium when

$$2(2p - 1)r^2 - (2(2p - 1) - m)r - mp = 0 \tag{20.3}$$

(assuming, of course, that this equation has a root in $[0,1]$).

We know that in communities of pure agents there is no cognitive diversity. Let us therefore ask how the values of u and v must be related to obtain some defense of T_2 in a community of hybrid agents. The condition for its being profitable for one member of a hybrid community to switch from defense of T_1 to defense of T_2 is

$$p^2(u/N + v) + p(1 - p)u/N - (1 - p)^2v < -p^2v + p(1 - p)u \tag{20.4}$$
$$+ (1 - p)^2u + (1 - p)^2v.$$

That is,

$$p < (mN + 2)/(m(N + 1) + 4) = p^o. \tag{20.5}$$

When $m = 0$, our "hybrid" agents are in fact epistemically pure, and the condition reduces to the (unsatisfiable) $p < 1/2$. As m goes to infinity, the agents are sullied, and the condition becomes $p < N/(N + 1)$, which is the condition for the representation of T_2 in a sullied community.

The advantage of our new perspective is that it enables us to handle intermediate cases. We can characterize these by the value of p^o, the maximum value of p for which the community achieves some representation of T_2. When $m = 1 (u/v = N)$, $p^o = (N + 2)/(N + 5)$; hence, even in a small community, $N = 10$, T_2 will achieve some representation unless p is close to 1 ($p > 0.8$). Suppose, however, that our hybrid agents assign roughly equal value to truth and credit for championing an accepted theory, $u = v$, $m = 1/N$. Then $p^o = 3N/(5N + 1)$; now, even in a large community, (N is infinite), p must be less than 0.6 for T_2 to be represented. Provided that their members are not too pure, and provided that those members use the idealistic computation of the probability of theory acceptance, communities of hybrid agents have enough in common with sullied communities to achieve a tolerable division of cognitive labor. However, if the scientists calculate the probability of theory acceptance by using the cynical computation, then, as before, in the absence of tradition effects (or other asymmetries), the result will be the uniform distribution $\langle N, 0 \rangle$.

Technical Discussion 11

If the community is very high-minded, and its members attach only very little value to social recognition, then, of course, T_1 and T_2 must be very close in perceived probability for T_2 to be represented. If $m = 1/2N$ then the maximum value of p^o (as N becomes large) is approximately 0.55. If $m = 1/10N$ (the community members value truth 10 times as much as they care about credit), then, even in a large community, the probabilities of the rivals must be very close if T_2 is to attract any representatives (the maximum value of p^o is about 0.512). Communities of more sullied scientists can obtain some representation of T_2 even when the probabilities are much further apart. If $m = 2/N$, the maximum value of p^o is 2/3; if $m = 4/N$, then the maximum value of p^o is 3/4; and if $m = 10/N$, then the maximum value of p^o is about 0.86.

How much good does it do the community if its members assign some weight to considerations of credit? Suppose that $u = v$, so that $m = 1/N$. As we have seen, the condition for *some* representation of T_2 is that $p < 3N/(5N + 1)$. Let us suppose that $N = 10$, $p = 0.55$. Then, solving the equilibrium equation for $r (= n/N)$, we obtain the value 0.83. The community optimum value of r is about 0.5, so that, initially, we might conclude that the hybrid community does rather poorly. However, when we compute values of V_{rN} for $r = 0.5$, 0.83, and 1, respectively, we discover that they are 0.5, 0.31, and 0.06. Thus, the expected utility of the distribution in the hybrid community, while lower than that obtained in a sullied community, is much closer to it than the value obtained by the pure community.

---

21. Cognitive Variation

Throughout the discussion of distributions of cognitive effort within scientific communities I have treated those communities as homogeneous in many important respects. Although Section 19 introduced the idea that members of different social groups might receive different rewards from defending different theories, I have supposed that all members of a community are motivated similarly: *all* are sullied in a sullied community, *all* are pure in a pure community, and *all* assign the same value to the tradeoff between credit and truth (the same m) in the hybrid communities of the last section. Equally, I have supposed that all assign the same value to the probability p. It is time to explore the consequences of relaxing both assumptions.

The problem of understanding distributions in communities containing some pure and some sullied members—or, more generally, in communities whose members assign different values to the parameter m—is formally equivalent to problems we have already explored. Suppose that we have a community with N_1 pure and N_2 sullied agents. The pure agents will all defend T_1. The sullied agents, on the other hand, will do the best they can to achieve a communitywide distribution $<p(N_1 + N_2),(1 - p)(N_1 + N_2)>$. If $pN_2 > (1 - p)N_1$, then they will be able to do so. Otherwise, the community will achieve the distribution $\langle N_1, N_2 \rangle$, with overrepresentation of the dominant theory. The situation is evidently analogous to that of the distribution of neophytes, discussed in Section 19.

Technical Discussion 12 shows that if we have a more finely divided community, containing homogeneous subgroups that make different tradeoffs between truth and credit, then we obtain a generalization of the same result. The underdog theory T_2 invades the most sullied subgroup and continues to spread through groups with increasing purity, until there is either an equilibrium within some group, or else there is an uninvadeable group.

We can also consider inhomogeneities that arise from differences in assessment of the rival theories. Here it can be shown that, under some conditions, pure and sullied communities reach the same distribution; under other circumstances, sullied communities reach that cognitive diversity that their pure cousins fail to attain.[42]

Finally, we can combine the idea that different subgroups assign different probabilities to the rival theories with asymmetries in utilities. Suppose, then, that the payoff matrix for the S_is is

---

42. Technical Discussion 12 does not show this, but application of the same methods it employs readily demonstrates the point.

	T_1 Accepted	T_2 Accepted
Defend T_1	$k_i u/n$	0
Defend T_2	0	$u/(N - n)$

where the k_i may be greater or less than 1. Let us begin with the case in which there are two subgroups, S_1 and S_2, whose members agree in assigning a probability p to T_1. Let $k_1 > 1$, $k_2 < 1$, so that S_1's obtain greater rewards from T_1, S_2's greater rewards from T_2.

Suppose that the initial distribution is $\langle\langle N_1, 0\rangle, \langle 0, N_2\rangle\rangle$. T_2 invades S_1 just in case

$$k_1 p/N_1 > (1 - p)/(N_2 + 1) \quad \text{i.e.,} \quad k_1 < N_1(1 - p)/(N_2 + 1)p. \quad (21.1)$$

T_1 invades S_2 just in case

$$k_2 p/(N_1 + 1) > (1 - p)/N_2 \quad \text{i.e.,} \quad k_2 > (N_1 + 1)(1 - p)/N_2 p. \quad (21.2)$$

Satisfaction of both conditions is impossible, for that would contravene our supposition that $k_1 > k_2$.

Now let us consider a more interesting situation in which the community contains subgroups for whom both the incentives and the probabilities are different. Specifically, let there be four subgroups, S_{11}, S_{12}, S_{21}, and S_{22}, of sizes N_{11}, N_{12}, N_{21}, N_{22}, respectively. The payoffs are given by the utility matrix introduced in the previous example, where the parameter k_i and the probability assigned to T_1 are given in the following way:

$$S_{11}\text{—}k_1 > 1, \quad p > 1/2 \qquad S_{21}\text{—}k_2 < 1, \quad p > 1/2$$
$$S_{12}\text{—}k_1 > 1, \quad q < 1/2 \qquad S_{22}\text{—}k_2 < 1, \quad q < 1/2$$

S_{11} and S_{22} are in cognitive harmony, in the sense that their views about which theory is more probable accord with their ideas about which theory offers greater incentives. By the same token, S_{12} and S_{21} suffer the correlative form of cognitive dissonance.

Technical Discussion 12 shows that moving members of the extremal groups, S_{11}, S_{22}, away from the theories to which both assignments of probability and incentives incline them is very difficult, so that support for both theories will almost always obtain in the community. It also brings out the possibility of an interaction effect between considerations of incentive and considerations of plausibility that can diminish cognitive diversity. Differential incentives and distinct assignments of probability appear to be independent solutions to the problem of maintaining cognitive diversity, but it is possible that, when they occur together, the beneficial effects of each can be cancelled.

Technical Discussion 12

Variations in Purity

Imagine that we have a community whose members are all hybrid agents but who assign different values to m. Assort them into groups, $S_1, \ldots, S_k$ with increasing values of m. Let the group sizes be $N_1, \ldots, N_k$, and the hybridization parameters $m_1, \ldots, m_k$. Assume that, initially, all members of the community defend T_1. It will be easiest for T_2 to gain representation in the most sullied group, S_k. The condition for T_2 to "invade" this group, given in the last section, is

$$p < (m_k N + 2)/(m_k(N + 1) + 4) \tag{21.3}$$

where $N = N_i$. Suppose that this condition is met. Then we can use the equilibrium equation of the last section

$$2(2p - 1)r^2 - (2(2p - 1) - m_k)r - m_k p = 0 \tag{21.4}$$

to compute the value of r ($=$ number of T_1 defenders/N) that members of S_k would prefer to see in the general population. If $N_k > (1 - r)N$, then the S_k members can have their way and the community distribution is $\langle r_k N, (1 - r_k)N \rangle$, where r_k is the value of r in question. If $N_k < (1 - r)N$, then all members of S_k support T_2. There are now two possibilities. Either T_2 is still underrepresented from the point of view of members of S_{k-1} or it is not. In the latter case, the community distribution becomes $\langle N - N_k, N_k \rangle$. In the former, there are two subcases. From the perspective of members of S_{k-1}, the preferred distribution in the total population is $\langle r_{k-1}N, (1 - r_{k-1})N \rangle$ where r_{k-1} is given by

$$2(2p - 1)r^2 - (2(2p - 1) - m_{k-1})r - m_{k-1}p = 0. \tag{21.5}$$

If $N_{k-1} + N_k > (1 - r_{k-1})N$, then there are enough members of S_{k-1} to enforce their way, and the community distribution is the preferred distribution for S_{k-1}. If not, then all members of S_{k-1} defend T_2. We now consider the possibility that T_2 can invade S_{k-2}. Iterating the analysis until either some group achieves its preferred distribution or invasion of T_2 is blocked, we compute the community distribution.

Interactions between Motivational and Cognitive Diversity

Consider the community of four subgroups, the cognitively harmonious S_{11} and S_{22} and the cognitively dissonant S_{12} and S_{21}. Suppose that the community starts from the distribution $\langle\langle N_{11}, 0\rangle, \langle N_{12}, 0\rangle, \langle N_{21}, 0\rangle, \langle 0, N_{22}\rangle\rangle$. T_1 invades S_{22} only if

$$k_2 q u/(N_{11} + N_{12} + N_{21} + 1) > (1 - q)u/N_{22}. \tag{21.6}$$

That is,

$$N_{22} > (N + 1)(1 - q)/(1 + (k_2 - 1)q). \tag{21.7}$$

Plainly, this condition is satisfiable only if N_{22} is very large. Similarly, if the community reaches the distribution $\langle\langle N_{11}, 0\rangle, \langle N_{12}, 0\rangle, \langle N_{21}, 0\rangle, \langle 0, N_{22}\rangle\rangle$, then T_2 will invade S_{11} only if

$$(1 - p)u/(N + 1 - N_{11}) > p k_1 u/N_{11}. \tag{21.8}$$

That is,

$$N_{11} > (N + 1)pk_1/(1 + (k_1 - 1)p) \qquad (21.9)$$

which is satisfiable only if N_{11} is very large.

Now let us consider the behavior of the "cognitively dissonant" groups. There are two obvious special distributions that we might expect to be favored:

(a) $\langle\langle N_{11}, 0\rangle, \langle N_{12}, 0\rangle, \langle 0, N_{21}\rangle, \langle 0, N_{22}\rangle\rangle$

(b) $\langle\langle N_{11}, 0\rangle, \langle 0, N_{12}\rangle, \langle N_{21}, 0\rangle, \langle 0, N_{22}\rangle\rangle$.

Under (a) members of the community follow their preferences, defending the theory that provides greater incentives for their work. Under (b), they follow the probabilities, supporting the theory that they regard as more probable.

Suppose that the community begins from (a). T_2 can invade S_{12} just in case

$$k_1q/(N_{11} + N_{12}) < (1 - q)/(N_{21} + N_{22} + 1). \qquad (21.10)$$

That is,

$$q < (N_{11} + N_{12})/(N_{11} + N_{12} + k_1(N_{21} + N_{22} + 1)). \qquad (21.11)$$

Similarly, T_1 can invade S_{21} just in case

$$k_2p/(N_{11} + N_{12} + 1) > (1 - p)/(N_{21} + N_{22}). \qquad (21.12)$$

That is,

$$p > (N_{11} + N_{12} + 1)/(N_{11} + N_{12} + k_2(N_{21} + N_{22}) + 1). \qquad (21.13)$$

Consonant with our initial assumptions, (21.11) and (21.13) are simultaneously satisfiable. It follows that it is possible for the distribution (a) to collapse into (b): if (21.11) and (21.13) are both met, then it will pay one member of S_{12} to switch to T_2 and one member of S_{21} to switch to T_1. The conditions for other members of these groups to change allegiances will then be just (21.11) and (21.13); so we should have sequential switching in each group; if both groups have the same size ($N_{12} = N_{21}$) then, plainly, this can lead to the replacement of (a) with (b), but, even if the sizes are different, it is possible that the process should continue so that the changeover is complete.

Let us now consider (b). Starting from (b), T_1 would invade S_{12} just in case

$$q > (N_{11} + N_{21} + 1)/(N_{11} + N_{21} + k_1(N_{12} + N_{22}) + 1). \qquad (21.14)$$

Similarly, T_2 invades S_{21} just in case

$$p < (N_{11} + N_{21})/(N_{11} + N_{21} + k_2(N_{12} + N_{22} + 1)). \qquad (21.15)$$

For these to be jointly satisfiable, consonant with our initial assumptions, it must be the case that the right-hand side of (21.14) is less than 1/2 and the right-hand side of (21.15) is greater than 1/2. So (21.14) and (21.15) are simultaneously satisfiable only if

$$N_{11} + N_{21} < k_1(N_{12} + N_{22}) \quad \text{and} \tag{21.16}$$

$$N_{11} + N_{21} > k_2(N_{12} + N_{22} + 1). \tag{21.17}$$

This can easily be achieved provided that k_1 is large and k_2 is small. Hence there are cases in which (b) collapses to (a).

Now consider that we begin from (b), the distribution that would be favored by the epistemically pure. Is it possible that S_{12} should be invaded and that S_{21} should resist invasion? To achieve this, we would require

$$k_1 > (N_{11} + N_{21} + 1)(1 - q)/(N_{12} + N_{22})q \tag{21.18}$$

$$k_2 > (N_{11} + N_{21})(1 - p)/(N_{12} + N_{22} + 1)p. \tag{21.19}$$

There is no difficulty in satisfying (21.18) if k_1 is sufficiently large. For (21.19) to be satisfiable, it must be compatible with the stipulation that $k_2 < 1$. But that simply requires that

$$(N_{11} + N_{21})(1 - p) < p(N_{12} + N_{22} + 1) \quad \text{or} \tag{21.20}$$

$$p > (N_{11} + N_{21})/(N + 1). \tag{21.21}$$

Thus if k_1 is sufficiently large (the incentives for S_{12}'s to defend T_1 are very strong) and if p is sufficiently large, the representation of T_2 in the sullied community is *lower* than the representation in the pure community. Here we see an interaction between two types of forces that independently favor cognitive diversity.

---

22. Nonexhaustive Alternatives

I have lavished considerable attention on a simple problem situation involving choice between rival theories. By simplifying instances of actual theory choice, it has been possible to explore optimal strategies for the community and a number of types of factors that might influence the decision making of individuals. Yet we should wonder whether similar results obtain when the situation is made more realistic. In this section, I shall respond, far too briefly, to worries of this kind, by considering what happens if we relax the assumption that our two theories are exclusive and exhaustive. Suppose that the sum of the probabilities of T_1 and T_2 is strictly less than 1. What is the optimal strategy for the community? How would we expect communities of various types to respond?

Plainly, the structure of the community decision problem with which we have been working since Section 16 must now be changed. In particular, we should consider the devotion of resources in the community to explore as yet unarticulated alternatives. Suppose, then, that the work force, of size N, is divided into groups of size r_1N, r_2N, and r_3N ($r_1 + r_2 + r_3 = 1$) that champion

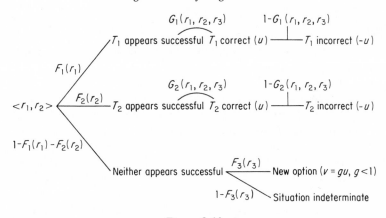

Figure 8.10

T_1, champion T_2, and explore further possibilities, respectively. I shall represent the possible outcomes by the tree in Figure 8.10.

In accordance with the discussion of Section 16, I shall suppose that $F_1(r_1) = p_1 r_1$, $F_2(r_2) = p_2 r_2$, $F_3(r_3) = p_3 r_3$. Relying on the treatment of the value of exploring alternatives to the explicitly formulated hypotheses (see Chapter 7), I shall assume that $G_i(r_1, r_2, r_3) = p_i$ (for $i = 1, 2$) if $r_3 < r^*$. If too little attention is given to seeking rivals, then apparent success in solving problems does nothing to boost the probability that a theory is true. However, the value of r^* ought to depend on the probabilities assigned to the extant rivals, and I shall take $r^* = h(1 - \{p_1 + p_2\})$, where h is a constant, less than one. Appealing to the ideas of Section 16, let $G_1(r_1, r_2, r_3) = ((1 - \alpha)r_2 + p_1 r_1)/(r_1 + r_2)$, $G_2(r_1, r_2, r_3) = ((1 - \alpha)r_1 + p_2 r_2)/(r_1 + r_2)$, when $r_3 > r^*$.

Without loss of generality, we can take T_1 to be the more probable of the two formulated rivals, $p_1 > p_2$. The utility of jumping at once by accepting T_1 is then

$$U = u(2p - 1). \tag{22.1}$$

When $r_3 > r^*$, the expected utility of dividing the labor $\langle r_1, r_2 \rangle$ is

$$V_{r1,r2} = \beta p_1 r_1 u(2\{(1 - \alpha)r_2 + p_1 r_1\}/(r_1 + r_2) - 1) + \tag{22.2}$$
$$\beta p_2 r_2 u(2\{(1 - \alpha)r_1 + p_2 r_2\}/(r_1 + r_2) - 1) +$$
$$\beta\{1 - (p_1 r_1 + p_2 r_2)\}(1 - (p_1 + p_2))(1 - (r_1 + r_2))gu +$$
$$\beta V_{r1,r2}(1 - (p_1 r_1 + p_2 r_2))(1 - (1 - \{p_1 + p_2\})(1 - \{r_1 + r_2\})).$$

Whence

$$V_{r1,r2}\{1 - \beta(1 - (p_1 r_1 + p_2 r_2))\}\{p_1 + p_2 + r_1 + r_2 - (p_1 + p_2)(r_1 + r_2)\}(22.3)$$
$$= \beta u\{p_1 r_1(2\{(1 - \alpha)r_2 + p_1 r_1\}/(r_1 + r_2) - 1) +$$
$$p_2 r_2(2\{(1 - \alpha)r_1 + p_2 r_2\}/(r_1 + r_2) - 1) +$$
$$g(1 - (p_1 r_1 + p_2 r_2))(1 - (p_1 + p_2))(1 - (r_1 + r_2))\}.$$

Quite evidently, the task of comparing the expected utilities of various divisions with the utility of jumping at once is far more complex in the present case. Technical Discussion 13 explores the behavior of $V_{r1,r2}$ by considering a numerical example.

The numerical example only illustrates a possibility, revealing that, when there are no significant costs of delay and when the probability of a theory can be boosted close to one, it will sometimes be better for the community not to jump to an immediate decision. How the resources are best assigned to defense of each of the articulated rivals and to exploration of new options varies with the parameter values, but, as noted, if a very high value is placed on the emergence of new ideas, then it may be optimal to devote all the resources to exploration.

Let us now turn to the decision problem for an individual scientist. Our scientist may defend T_1, defend T_2, or explore new approaches. As in our earlier discussions, we may suppose that defense of T_1 would yield a return of u if T_1 were accepted and if the scientist were hailed as an early champion of this theory. So we can take the expected utility of supporting T_1 to be p_1u/n_1 where n_1 is the number of those in the community who devote themselves to trying to show that T_1 should be accepted. Similarly, we can take the expected utility of defending T_2 to be p_2u/n_2. Let q be the probability that no accepted theory emerges, given that neither T_1 nor T_2 is accepted. Then, by exploring, the scientist effectively buys into a lottery which will pay up only if two conditions are met: first, neither T_1 nor T_2 must win acceptance; second, some accepted theory must emerge from the process of exploration. Given that these conditions are met, the scientist's chances of receiving the payoff are inversely proportional to the number of explorers. The payoff for first articulating a new option, ultimately accepted by the community, is, I assume, large. Let the value be w, where $w > u$. The expected utility of exploring is then

$$(1 - (p_1 + p_2))(1 - q)w/(N - n_1 - n_2).$$

I am especially interested in the pressure on individuals to explore new options, for, as my discussion of the community optimum shows, some exploration is always welcome and a large amount of exploration may even be desirable. Assume, then, that we begin with a community in which everyone defends one of the two articulated theories. The defenders are balanced so that the expected utility of defending T_1 is that of defending T_2. So the initial distribution is $\langle p_1N/(p_1 + p_2), p_2N/(p_1 + p_2), 0 \rangle$. Under these circumstances, the expected payoff if one person switches to exploration is $(1 - (p_1 + p_2))(1 - q)w$. Some exploration will occur if

$$(1 - (p_1 + p_2))(1 - q)w > u(p_1 + p_2)/N. \qquad (22.4)$$

That is,

$$w/u > (p_1 + p_2)/N(1 - q)(1 - (p_1 + p_2)). \qquad (22.5)$$

In the least favorable case, $w = u$, and the condition requires that

$$(1 - q) > (p_1 + p_2)/N(1 - (p_1 + p_2)). \tag{22.6}$$

The left-hand side represents the chance that an accepted theory will emerge from the exploratory process, given that neither T_1 nor T_2 is accepted. The more confident the scientist is that failure of T_1 and T_2 will correspond to success of one of the new ideas generated by the explorers, the more easy it is to satisfy the inequality. But, given a fairly large community, exploration is often worthwhile even if the scientist thinks that it yields only a small chance of turning up an ultimately accepted theory. For example, if $p_1 + p_2 < 1/2$, $(1 - q)$ is required only to be larger than $1/N$; if $p_1 + p_2 < 3/4$, $(1 - q)$ must be bigger than $3/N$. As Technical Discussion 13 shows, when it is overwhelmingly likely that exploration will deliver a successful theory, provided that the articulated candidates fail ($q = 0$), the equilibrium number of explorers will be $(1 - (p_1 + p_2))$. Even in the worst case, when $w = u$, if $q = 0$, the pressures on individual scientists yield sufficiently many explorers to give an acceptable distribution. We can have some confidence that moving from the highly idealistic problem of §17 to the slightly more realistic version considered here does not undermine our main conclusions.

Technical Discussion 13

Numerical Illustration

Let us suppose that $\alpha = 0$, $\beta = 1$ (conditions, as we have seen, favorable to dividing the labor). Let $p_1 = 1/2$, $p_2 = 1/4$, $r_1 = 1/2$, $r_2 = 1/4$, $g = 1/2$ (finding a new option is half as valuable as discovering the truth), and let $h < 1$. Under these circumstances $r_3 (= 1/4) > r^* (= h/4)$ so that our expression for V_{r_1,r_2} applies.

Straightforward substitution shows that

$$V_{1/2,1/4} = 209u/546 \qquad U = 0$$

so that the distribution $\langle 1/2, 1/4, 1/4 \rangle$ is preferable to jumping at once. However, when the value of generating a new option is set so high, the optimal strategy is to devote *all* of the work force to exploration

$$V_{0,0} = u/2.$$

Suppose, instead, that the value of articulating a new option is lower, $g = 1/4$. Under these conditions, we obtain the following values

$$V_{1/2,1/4} = 385u/1092 \qquad V_{3/4,0} = 5u/212$$

$$V_{0,0} = u/4 \qquad U = 0.$$

Hence, under these circumstances, an internal division of the community, with some scientists assigned to each of the three strategies, is preferable.

Amount of Exploration. Imagine that there are n explorers. The remaining $N - n$ will be distributed so that

$$n_1 = p_1(N - n)/(p_1 + p_2), \; n_2 = p_2(N - n)/(p_1 + p_2). \qquad (22.7)$$

The expected utility of defending either T_1 or T_2 will be

$$(p_1 + p_2)u/(N - n).$$

The equilibrium number of explorers is $n^\$$ where

$$n^\$\{(p_1 + p_2)u + (1 - (p_1 + p_2))(1 - q)w\} \qquad (22.8)$$
$$= (1 - (p_1 + p_2))(1 - q)w.$$

In the worst case for encouraging exploration, when $u = w$,

$$n^\$ = (1 - (p_1 + p_2))(1 - q)N/(1 - q\{1 - (p_1 + p_2)\}). \qquad (22.9)$$

If q is high—that is, if there is a strong chance that exploring will not result in any successful new theory, even if T_1 and T_2 fail—then the equilibrium number of explorers may be fairly small.

However, suppose that $q = 0$. Recall that, for the successes of T_1 to increase its probability, G_1 must take a value greater than p_1, and this requires that r_3 be larger than r^* ($< 1 - (p_1 + p_2)$). If scientists are sure that, if T_1 and T_2 do not succeed, then something that emerges from exploration will ultimately be accepted, then the community will contain enough explorers to allow for serious increases in the probability of T_1 or T_2. At equilibrium, $r_3 = 1 - (p_1 + p_2)$.

---

23. Consensus Formation

The last issue that I shall discuss also relates to an idealizing assumption that I have adopted in previous sections. Several of my analyses depend on viewing consensus as being formed in a particular way: in studying replication I assumed that a finding would be accepted if it was replicated once (later, if it was replicated r times); in looking at the division of labor problem, I have supposed that a theory becomes part of consensus by being accepted by all members of the community. I now want to look, rather quickly, at the question of which consensus-forming mechanisms would be best for a scientific community.

At least three kinds of considerations affect the costs and benefits of consensus-forming mechanisms. First, the security of the consensus depends on the number of independent agents who reach the consensus decision. In a community in which there is a scientific dictator whose decisions are always embodied in consensus practice, there is obviously a significant risk that the community will go astray. At the opposite extreme, we can envisage a complete consensual democracy, in which independent decisions are required of all community members if consensus practice is to be modified in a particular way. However we locate the community on the continuum whose extremes are dictatorship and consensual democracy, I shall suppose that the decisions with which we are concerned sometimes involve the acceptance (or rejection) of an experimental result, sometimes the evaluation of an elaborate argument, of the type that I have viewed as being central to major changes in science (see Chapters 6 and 7). In either case, I suppose that the probability of correct modification of consensus practice can be written as

$$1 - \prod_i (1 - a^*(X_i))$$

where a^* is a measure of the authority attributed within the community, and the product is taken over all those who make *independent* decisions in fashioning the consensus.[43]

Second, there are correlative losses involved in the proliferation of decision makers. In reaching independent decisions about the merits of reforming consensus practice, each of the agents temporarily abandons a research project that might have been of value to the community. The costs of losing this work can be written as Σd_i, where, again, the summation is over the set of independent decision makers.

Finally, the imposition of consensus by a group that is smaller than the entire community runs the risk of alienating (or banishing) some members of the community. Those who do not agree may be forced out or deprived of resources that they need to continue their research. Let us suppose that, for each scientist *not* in the decision-making group, there is a probability P_i of being at odds with the consensus in such a way that would lead to a loss of magnitude e_i. The simplest instances will be those in which outsiders rebel and quit the field altogether, but we can allow for less extreme forms of behavior to be covered by the formalism.

Technical Discussion 14 explores the problem for some simple stratified societies containing both high-status scientists (*aristos*) and low-status scientists (*proles*). If the community is to have a consensus-forming clique of a particular size, then it is preferable that that clique include all aristos (if there are enough aristos to reach that size). Under some circumstances, pronounced elitism can be favored, and there are situations in which the clique should include just the aristos. Whether the parameter values that favor elitist systems of consensus formation are those found in actual scientific communities is, of course, a matter for empirical investigation.

Technical Discussion 14

Let the value of adopting a correct modification of practice be u, that of adopting an incorrect modification $-u$. Then the expected utility of a consensus-forming mechanism can be written as

---

43. I should note explicitly that there may be a difference between those who have the right to contribute to the formation of consensus and those who actually do so. Some of those whose decisions would be viewed as crucial to the formation of communitywide decision may forgo independent decision, so that the group that actually forms consensus may be smaller than the official elite.

$$(1 - 2\Pi_{i \in C}(1 - a^*(X_i)))u - \sum_{i \in C} d_i - \sum_{j \in C} P_j e_j$$

where C (the *clique*) consists of the group of independent decision makers. The community's problem is to choose the clique so that the preceeding quantity is maximized.

I shall explore the problem for the case in which community members divide into two subgroups, aristos and proles. I shall suppose that aristos are all alike and proles are all alike. Let the parameters be related as follows:

$$a^*(\text{aristo}) = Q \text{ (close to 1)}; \quad a^*(\text{prole}) = 1/2$$
$$d_{\text{aristo}} = kd_{\text{prole}} = kd \quad k > 1$$
$$e_{\text{aristo}} = le_{\text{prole}} = le \quad l > 1$$
$$P_{\text{aristo}} = P_{\text{prole}} = P.$$

Suppose that there are N_1 aristos and N_2 proles in the community ($N_1 + N_2 = N$). Then, if there are n ($<N_1$) aristos in the clique, the expected payoff is

$$(1 - 2(1 - Q)^n)u - nkd - ((N_1 - n)l + N_2)Pe.$$

If the clique consists of n ($<N_2$) proles, the expected payoff is

$$(1 - 2^{-(n-1)})u - nd - (N_1 l + (N_2 - n))Pe.$$

If the community is going to have a clique of size n is it preferable to choose aristos or proles (given that $n < N_1$ $n < N_2$)? Choosing the aristos is better if

$$2u\{1/2^n - (1 - Q)^n\} - nd(k - 1) + nPe(l - 1) > 0. \tag{23.1}$$

Suppose $k = l$.[44] Let $Q > 0.9$. Then sufficient conditions for the community to prefer a clique of aristos are:

$$Pe > d \tag{23.2}$$

$$u > 2^{n-1}n(k - 1)d/(1 - 5^{-n}). \tag{23.3}$$

Plainly, u (the value of choosing correctly) is likely to be very large in comparison with d (the cost due to delay). It is thus not hard to appreciate the fact that, in elitist communities (small n), aristos are preferred as clique members.

Imagine that we are interested in the optimum size of a clique of aristos. Let

$$H = (1 - 2(1 - Q)^n)u - nkd - ((N_1 - n)l + N_2)Pe \tag{23.4}$$

---

44. This may be unrealistic if some proles are young members of the community who may become aristos in time. For then the losses of their work due to delay are only a fraction of the losses of the work of an aristo. But, comparing the *total* values of a lifetime's work, the promising prole may be very close to the aristo. In that case, k might be significantly greater than 1, l approximately 1. I shall ignore this complication and other intricacies that result from the age structuring of populations.

$$\partial H/\partial n = -2(1 - Q)^n(\ln\{2(1 - Q)\}u - kd + lPe. \tag{23.5}$$

Since $\ln\{2(1 - Q)\} < 0$, $\partial H/\partial n > 0$ if $lPe > kd$. Thus, if the probability of rebellion is high enough, or the costs of delay small enough, the community would prefer to involve *all* the aristos in independent decision making. If, however, $lPe < kd$, then the optimum number of aristos is given by

$$2(1 - Q)^n = (lPe - kd)/u\ln\{2(1 - Q)\}. \tag{23.6}$$

In sheepish communities, where there is little danger of rebellion, quite a restricted oligarchy can be preferable. Suppose $P = 0$, $Q = 0.9$, $k = 3.2$. Then the expression for the optimal value of n reduces to

$$(0.1)^n = d/u. \tag{23.7}$$

Even if d is tiny in comparison with u ($u = 10^{10}d$, say), this will only give a value of 10 for n, which, in a large community, may be a small proportion of the aristos.

Elitism can become even more pronounced if we suppose that the costs of delay increase with the number of independent decision makers. In more democratic communities, even more time must be spent in articulating arguments for others to appraise, and this will lengthen the period through which the work of clique members is lost to the community. If, for example, the delay cost for a prole is dn^2, then, with the numerical values of the last paragraph, the value of u/d must be $3 \cdot 10^{12}$ for n to be as large as 10.

Another obvious amendment of my analysis is to suppose that the probability of rebellion varies inversely as the size of the clique. The decisions of small, autocratic consensus-forming groups are likely to arouse resistance among nonmembers. Suppose, then, that $P = h/n$. Under these conditions, the expected utility of having a clique of n aristos is

$$H = (1 - 2(1 - Q)^n)u - nkd - ((N_1 - n)l + N_2)he/n \tag{23.8}$$

$$\partial H/\partial n = -2(1 - q)^n u\ln\{2(1 - Q)\} - kd + he((N_1 - n)l + N_2)/n^2 \tag{23.9}$$
$$+ hel/n.$$

The optimal value of n is given by

$$2u\ln\{2(1 - Q)\}(1 - Q)^n = he(lN_1 + N_2)/n^2 - kd. \tag{23.10}$$

We only obtain an elitist solution (that is: $n < N_1$) if the right-hand side is negative. If

$$he(lN_1 + N_2)/N_1^2 > kd \tag{23.11}$$

then $\partial H/\partial n$ is always positive, and complete democracy for aristos is preferred.

To see that democracy for aristos is often preferable, consider the situation when $Q = 0.9$, $e = 100d$, $l = 2$, $k = 3.2$, $h = 0.032$. Under these conditions, the optimal value of n is given by

$$(0.1)^n = d(n^2 - (2N_1 + N_2))/un^2. \tag{23.12}$$

If $N_1^2 < 2N_1 + N_2$, then this equation is unsatisfiable and the community's best strategy is to demand that all the aristos belong to the clique. *Moral:* if there are sufficiently many proles in proportion to aristos. then all the aristos should be included in the group of people who make independent decisions.

Let us now suppose that the community is in a situation of complete consensual democracy for aristos and examine the conditions under which it is also worth involving proles in the decision making. The condition for including n proles in the clique is that

$$K = (1 - 2(1 - Q)^{N_1} \cdot 2^{-n})u - (N_1 k + n)d - (N_2 - n)Pe > 0. \qquad (23.13)$$

Suppose, first, that P is constant. Then

$$\partial K/\partial n = 2(1 - Q)^{N_1} \cdot 2^{-n} \cdot \ln(2)u - d + Pe. \qquad (23.14)$$

Plainly, if $Pe > d$, K increases with n, so that complete democracy is preferable. If, however, $P = 0$, $Q = 0.9$, $N_1 = 10$, then, for even a single prole to be included in the clique, u must be larger than $10^{10}d$.

If we amend the assumptions to suppose that $P = h/n$ then

$$K = (1 - 2(1 - Q)^{N_1} \cdot 2^{-n})u - (N_1 k + n)d - (N_2 - n)he/n \qquad (23.15)$$

which is approximated by

$$u - (N_1 k + n)d - N_2 he/n + he \qquad (23.16)$$
$$\partial K/\partial n = 0 \quad \text{only if} \quad n^2 = he N_2/d.$$

Some representation of proles can be achieved if hN_2 is sufficiently large and e/d is not too small. For example, if $e/d = 100$, $h = 0.033$, $N_2 = 30$, the preferred value of n is 10, so that one third of the proles would be included in the clique.

The probability function of the last paragraph supposes that probability of rebellion is inversely proportional to the number *of proles* in the clique—that is, it assumes that the resentment among the proles depends on their identification of themselves as an excluded group. Perhaps it is more realistic to suppose that the probability of rebellion is inversely proportional to the number of clique members: $P = h/(N_1 + n)$. Under these circumstances, we can approximate K by

$$u - (N_1 k + n)d - (N_2 - n)he/(n + N_1) \qquad (23.17)$$

$$\partial K/\partial n = -d + heN/(n + N_1)^2 \qquad (23.18)$$

so that the optimal value of n is given when

$$(n + N_1)^2 = heN/d. \qquad (23.19)$$

To obtain complete democracy would now require

$$h > Nd/e \qquad (23.20)$$

which can only occur if the members of the community are antiauthoritarian (high h), or the community is small (low N), or the costs of delay are very small in comparison to the value of a lifetime's work (d/e is very small).

Finally, it is easy to envisage circumstances under which no proles should be included in the consensus-forming clique. The condition for this to occur is

$$N_1^2 > heN/d \qquad (23.21)$$

When $N_1 = 10$, $N = 40$, $e/d = 100$, $h = 0.025$, this condition is satisfied, and, as we saw earlier, this kind of case allows for complete democracy for aristos.

---

24. Conclusions

As I noted in section 1, there are two main results of the foregoing analyses. First, motives often dismissed as beyond the pale of scientific decision making can, under a wide range of conditions, play a constructive role in the community's epistemic enterprise. Second, the details matter. The effect of various types of factors (authority, elitism), which we might have thought of as acting in a single fashion across scientific contexts, depends on features of the social situation and of the decision problem.

These results flow from a formal—but highly idealized—treatment of communities and of individual decision-making. What can we achieve by modeling scientific communities in so artificial a way? I have begun by employing an unrealistic model of human decision making (who can give precise probabilities? who has definite utility functions? who knows enough about others to be able to carry out, even in principle, the computations I have labored through?). I have proceeded by making numerous further assumptions in the interests of algebraic simplification (think of the many occasions on which I have regarded particular groups as homogeneous in some important respect). These are defects that I would be happy to overcome. It is possible that a more realistic approach to cognition could be formulated with sufficient precision to enable us to achieve clear results about the outcomes of decisions made by the members of a community. It is possible that a more elegant general approach would expose the crudities of my analyses and offer us a more global perspective on the phenomena I have tried to study. Lacking these potential advances, the discussions of the previous sections should be viewed as first efforts, attempts to raise questions that I take to have been slighted within epistemology and to devise concepts and tools for solving them. I would be delighted were others to improve those concepts, or to transform the tools, rendering my treatment of the problems—but not the problems themselves!—obsolete.

Finally, I want to address the puzzled reader who regards this chapter as part of a different book, one that does not contain its predecessors. Recall some major claims of earlier chapters. Fields of science, I have suggested, make progress in broadly cumulative fashion. As those fields progress, scientists come to employ cognitive strategies that are superior, when judged by versions of the external standard. Despite the existence of periods in which alternative ways of revising practice can be defended by equally good arguments, scientific communities typically resolve these indeterminacies by articulating a superior form of reasoning that can be used to support one of the rivals. In Chapter 7, I have tried to outline a picture of the reasoning of individual scientists, at times both of large upheaval and of small modifications, a picture that could be applied to show how apparently difficult indeterminacies are resolved.

All of these theses presuppose, more or less directly, that scientists' social involvement with one another does not interfere with the employment of epistemically virtuous individual reasoning. Defenders of Legend, holding a highly idealized picture of scientists, took it for granted that there could be no such interference. But we know that their vision of the scientist as pure seeker after truth is a myth. What are the consequences? Should we conclude that the growth of science is a process in which various kinds of social forces have shaped the doctrines that are accepted and the styles of reasoning that are prized?

The minimal contribution of this last chapter is to rebut the notion that one can infer directly from the existence of social pressures and nonepistemic motivations the conclusion that science does not advance in the fashion described earlier in this book. Philosophers who have studied Legend's critics have usually been haunted by the idea that the impotence of appeals to reason and evidence to resolve scientific controversies would create a vacuum in which unpleasant "social factors" would move the community in arbitrary directions. Earlier chapters tried to show that scientific arguments are more intricate and more powerful than pessimists have sometimes supposed. This chapter reveals that the operation of social systems in ways that we might initially view as opposed to the growth of knowledge can be dependent on the use of complicated reasoning and can contribute to the community's attainment of its epistemic ends. The worry that Legend's heroes have feet of clay *and that, in consequence, science cannot have the progressive characteristics often attributed to it,* turns out to rest on a fallacy.

Imagine that we had started with a different image. From the beginning, let us suppose, we had conceived of scientists as ordinary people, subject to complex combinations of social pressures. Controversies, we might have believed, are typically settled by enrolling allies, and those who are most persuasive and recruit the most powerful followers win the day.[45] The present chapter attempts to show that, when the dynamics of communities who operate in this way is analyzed, the agents must be viewed as making complex epistemic evaluations. To turn the philosophers' fear inside out, the processes of enrolling allies depend critically on the potential allies' assessments of the consequences of the options open to them. Competitive and cooperative situations, as we have seen again and again, call for refined judgments about

---

45. Here I deliberately use the language of (Latour 1987). However, it should be noted that Latour's position is far more subtle than the view I sketch, and he would be as unsatisfied as I am with my imagined polar opposite to Legend.

the merits of methods or of proposals for modifying practice. Thus we can say that a simple view of science as driven by "external" or "social" factors would create a vacuum *into which epistemic considerations would have to be introduced to explain how the social factors obtain their purchase on the individual actors.* The investigations of the past two chapters are thus complementary, the one showing how individuals reason and the other showing how their efforts, in admittedly artificial social contexts, combine to yield distributions of cognitive effort.

Yet, in the end, I want to claim more. *Part* of the epistemological task consists in responding to important skeptical concerns. Beyond that, epistemology should strive to formulate (fallible) claims about good reasoning, thus attempting to improve the ways in which we revise our practices. A rightly respected tradition has contributed much to one side of the meliorative epistemological project: thanks to the efforts of Locke and Hume, Kant, Whewell and Mill, Frege, Russell, and Carnap, we have a far clearer vision of good individual reasoning. The other facet of the meliorative project has been, as I have noted, almost completely neglected. Yet, just as it is important to uncover rules for the right direction of the individual mind, so too, it is necessary to understand how *community* strategies for advancing knowledge might be well or ill designed. So I conceive this chapter not simply as an attempt to complete my discussion of Legend and its critics, my case for objectivity without illusion, but as a first foray into epistemological *terra incognita.* I doubt that I have done more than scratch the surface of unfamiliar terrain, but I am confident that epistemological rewards await those who are prepared to dig more deeply.

Envoi

Intellectual celebration often attends the demise of Legend, for, if Legend is dead, then, it seems, everything is permitted. "Modernity," it is sometimes claimed, "is bewitched by science. When we recognize that science has no claim to 'objective truth,' then we shall understand and appreciate the value of humanistic and artistic perspectives on our world, we shall recognize the important contributions of diverse ethnic groups." We shall inhabit, it seems, a richer, braver new world.

The central argument of this book is that the celebration is both premature and misguided. No one should question the value of humanistic and artistic perspectives or dismiss, without scrutiny, the ideas of other cultures. But, for all its difficulties, Legend was broadly right about the characteristics of science. Flawed people, working in complex social environments, moved by all kinds of interests, have collectively achieved a vision of parts of nature that is broadly progressive and that rests on arguments meeting standards that have been refined and improved over centuries. Legend does not require burial but metamorphosis.

I have endeavored to provide a philosophical framework for the study of science which combines the insights of Legend with the insights of its critics. Philosophy of science does not die with Legend. Instead, as I must admit, apologetically, at the end of a long book, it needs to pursue many of the loose ends that I have left in the preceding pages. Although I have considered many historical examples, a broader range of instances and greater depth of analysis are both needed. Although I have gestured toward connections between my framework and ideas in the cognitive sciences, there is no doubt that the view of science I provide could be enriched by more systematic and more detailed use of those ideas. The treatment of individual scientific reasoning should be made more extensive and more precise—although I trust that my discussions provide a bridge between traditional philosophical discussions of scientific reasoning and the complex practice of science. Perhaps most importantly, my explorations of social epistemology in Chapter 8 are, as frequently acknowledged, preliminary efforts.

Any philosophy of science worth its salt ought to be able to shed light on the discussions and controversies in which scientists are actively engaged. In recent years, despite the success of philosophical interventions in particular scientific debates, there has been no general account on which philosophy of the special sciences could draw. I hope that some of the concepts introduced will prove useful in illuminating parts of scientific activity—as, for example, a predecessor of the notion of a practice was valuable to me in discussing the growth of mathematics (Kitcher 1983a) and the structure of sociobiology (Kitcher 1985b). In particular, I believe that the project of trying to map the significant questions of a field can bring into focus areas of science in which there are many rival ideas about the proper development of the field.

Finally, as I have remarked several times, the foregoing chapters leave untouched some of the largest questions about science. Much of the opposition to "scientism"—the trendy castigation of "modernity" as bewitched by science—rests on faulty views about scientific knowledge. Yet, even if the metamorphosis of Legend attempted here clears away those errors, it does not address the issue of the value of science. To claim, as I have done, that the sciences achieve certain epistemic goals that we rightly prize is not enough—for the practice of science might be disadvantageous to human well-being in more direct, practical ways. A convincing account of practical progress will depend ultimately on articulating an ideal of human flourishing against which we can appraise various strategies for doing science. The extreme positions are clear. At one pole, it is suggested that science, as practiced, is a terrible thing, and that human beings should want none of it; at the other, that science, as we have fashioned it, is already perfect. Neither extreme is likely to be right.

In accordance with the approach taken in Chapter 8, we can envisage a very general problem of optimization. Given an ideal of human flourishing, how should we pursue our collective investigation of nature? Beyond my attempt to understand the *epistemic* features of the scientific enterprise lies this far broader question about science, a question that a critical philosophy of science ought to address. It would, I believe, be highly surprising if any of our institutions, evolved, as they have, by serendipitous routes, were to be vindicated by a careful optimality analysis. Given a clear view of the epistemic achievements and prospects of science, how should we modify the institution so as to enhance human well-being? Reflective understanding and constructive critique should, I believe, replace both sleepy complacency and Luddite rage. The philosophers have ignored the social context of science. The point, however, is to change it.

Bibliography

Allen, Garland. (1978a). *Life Science in the Twentieth Century*. Cambridge: Cambridge University Press.

———. (1978b). *Thomas Hunt Morgan*. Princeton: Princeton University Press.

Anderson, John. (1983). *The Architecture of Cognition*. Cambridge, MA: Harvard University Press.

Appel, Toby. (1987). *The Geoffroy-Cuvier Debate*. New York: Oxford University Press.

Aristotle. *Physics* in Sir David Ross (ed), (1966). *The Works of Aristotle,* volume 2. Oxford: Oxford University Press.

———. *Posterior Analytics* in Richard McKeon (ed), *The Basic Works of Aristotle*. New York: Random House.

Barber, Bernard. (1952). *Science and the Social Order*. Glencoe, IL: Free Press.

Barnes, Barry. (1974). *Scientific Knowledge and Sociological Theory*. London: Routledge.

———. (1992). "Realism, Relativism, and Finitism," in Diederick Raven et al. (eds), *Cognitive Relativism and Social Science*. New Brunswick, NJ: Transaction Books, 131–147.

———, and Bloor, David. (1982). "Relativism, Rationalism and the Sociology of Knowledge," in M. Hollis and S. Lukes (eds), *Rationality and Relativism*. Cambridge, MA: MIT Press, 21–47.

Beatty, John. (1985). "Speaking of Species: Darwin's Strategy," in (Kohn 1985), 265–281.

———. (1986). "Pluralism and Panselectionism," in P. Asquith and P. Kitcher (eds), *PSA 1984* Volume 2, East Lansing, MI: PSA, 113–128.

Benacerraf, Paul. (1973). "Mathematical Truth," *Journal of Philosophy,* 70, 661–679.

Bennett, Jonathan. (1966). *Kant's Analytic*. Cambridge: Cambridge University Press.

Bernstein, Richard. (1988). "The Rage against Reason," in (McMullin 1988) 189–221.

Biagioli, Mario. (1990). "Galileo, the Emblem Maker," *Isis,* 81, 230–258.

Bishop, Michael. (forthcoming). "The Theory-Ladenness of Perception," manuscript.

Block, Ned. (ed) (1981). *The Imagery Debate*. Cambridge, MA: MIT Press.

Bloor, David. (1974). *Knowledge and Social Imagery*. London: Routledge.

Bogen, James, and Woodward, James. (1988). "Saving the Phenomena," *Philosophical Review,* 18, 303–352.

Born, Max, and Wolf, Emil. (1980). *Principles of Optics*. Oxford: Oxford University Press.

Boscovich, Roger. (1763/1966). *A Theory of Natural Philosophy*. Cambridge, MA: MIT Press.

392

Bowler, Peter. (1983). *The Eclipse of Darwinism*. Baltimore: Johns Hopkins University Press.

Boyd, Richard. (1973). "Realism, Underdetermination, and a Causal Theory of Evidence," *Nous*, 7, 1–12.

Brandon, Robert. (1978). "Adaptation and Evolutionary Theory," *Studies in the History and Philosophy of Science*, 9, 181–206.

———, and Burian, Richard. (eds) (1984). *The Units of Selection*. Cambridge, MA: MIT Press.

Brandt, Richard. (1959). *Ethical Theory*. Englewood Cliffs, NJ: Prentice-Hall.

Bromberger, Sylvain. (1963). "A Theory about the Theory of Theories and a Theory about the Theory of Theory," in B. Baumrin (ed), *Proceedings of the Delaware Seminar*. New York: Interscience.

———. (1966). "Questions," *Journal of Philosophy*, 63, 597–606.

Buchwald, Jed. (1981). *From Maxwell to Microphysics*. Chicago: University of Chicago Press.

———. (1989). *The Rise of the Wave Theory of Light*. Chicago: University of Chicago Press.

Burian, Richard. (1985). "Adaptation," in M. Grene (ed), *Dimensions of Darwinism*. Cambridge: Cambridge University Press, 287–314.

Burkhardt, R. W. (1977). *The Spirit of System: Lamarck and Evolutionary Biology*. Cambridge: Cambridge University Press.

Cantor, Geoffrey. (1983). *Optics after Newton: Theories of Light in Britain and Ireland, 1704–1840*. Manchester: Manchester University Press.

———, and Hodge, M. J. S. (eds) (1981). *Conceptions of Ether: Studies in the History of Ether Theories*. Cambridge: Cambridge University Press.

Carlson, E. A. (1966). *The Gene: A Critical History*. Philadelphia: Saunders.

Carnap, Rudolf. (1951). *Logical Foundations of Probability*. Chicago: University of Chicago Press.

Carroll, Lewis. (1896). "What the Tortoise Said to Achilles," *Mind*, 4, 278–280.

Cartwright, Nancy. (1983). *How the Laws of Physics Lie*. Oxford: Oxford University Press.

Cavendish, Henry. (1783/1961). *The Composition of Water*. Manchester: Alembic Club Reprint.

Cherniak, Christopher. (1986). *Minimal Rationality*. Cambridge, MA: MIT Press.

Chisholm, Roderick. (1966). *Theory of Knowledge*. Englewood Cliffs, NJ: Prentice-Hall.

Churchland, Patricia S. (1986). *Neurophilosophy*. Cambridge, MA: MIT Press.

Churchland, Paul M. (1979). *Scientific Realism and the Plasticity of Mind*. Cambridge: Cambridge University Press.

———. (1988). "Perceptual Plasticity and Theoretical Neutrality: A Reply to Jerry Fodor," *Philosophy of Science*, 55, 167–187.

———. (1989). *A Neurocomputational Perspective*. Cambridge, MA: MIT Press.

———, and Hooker, Cliff. (eds) (1982). *Images of Science*. Chicago: University of Chicago Press.

Cole, Jonathan, and Cole, Stephen. (1973). *Social Stratification in Science*. Chicago: University of Chicago Press.

Coleman, William. (1964). *Georges Cuvier. Zoologist*. Cambridge, MA: Harvard University Press.

Collins, H. M. (1985). *Changing Order*. London: Sage.

———. (1987). "Pumps, Rock and Reality," *Sociological Review*, 21, 819–828.

————, and Pinch, Trevor. (1982). *Frames of Meaning*. London: Routledge.

Conant, James Bryant. (1957). *Harvard Case Histories in the Experimental Sciences*. Cambridge, MA: Harvard University Press.

Cosmides, Leda, and Tooby, John. (1988). "From Evolution to Behavior: Psychology as the Missing Link," in (Dupre 1988), 277–306.

Crow, James. (1979). "Genes That Violate Mendel's Rules," *Scientific American*, 240, no. 2, 134–146.

Culp, Sylvia, and Kitcher, Philip. (1989). "Theory Structure and Theory Change in Contemporary Molecular Biology," *British Journal for the Philosophy of Science*. 40, 459–483.

Cuvier, Georges. (1813). *Essay on the Theory of the Earth*. New York: Kirk and Mercein.

Darwin, Charles. (1859). *On The Origin of Species by Natural Selection*. (first edition). London: John Murray; facsimile reprint (1967), with introduction by Ernst Mayr. Cambridge, MA: Harvard University Press.

————. (1862). *On the Various Contrivances by Which Orchids Are Fertilized by Insects*. London: John Murray.

————. (1868). *The Variation of Plants and Animals under Domestication*. London: John Murray.

Darwin, Francis. (1888). *Life and Letters of Charles Darwin* (three volumes). London: John Murray; Reprint New York: Johnson Reprint, (1969).

————. (1903). *More Letters of Charles Darwin*, (two volumes). London: John Murray.

Davidson, Donald. (1973). "On the Very Idea of a Conceptual Scheme," in *Proceedings and Addresses of the American Philosophical Association*, reprint (1984) *Essays on Truth and Interpretation*. Oxford: Oxford University Press, 183–198.

Dawkins, Richard. (1976). *The Selfish Gene*. Oxford: Oxford University Press. (second edition, with additional material) (1990). Oxford: Oxford University Press.

————. (1982). *The Extended Phenotype*. San Francisco: Freeman.

————. (1986). *The Blind Watchmaker*. London: Longmans.

De Candolle, Alphonse. (1855). *Geographie Botanique Raisonée*. Paris: Victor Masson.

De Santillana, Georgio. (1955). *The Crime of Galileo*. Chicago: University of Chicago Press.

Dennett, Daniel. (1988). *The Intentional Stance*. Cambridge, MA: MIT Press.

Devitt, Michael. (1981). *Designation*. New York: Columbia University Press.

————. (1984). *Realism and Truth*. Princeton: Princeton University Press.

————, and Sterelny, Kim. (1987). *Language and Reality*. Cambridge, MA: MIT Press.

Dobzhansky, Theodosius. (1937). *Genetics and the Origin of Species* (first edition); (1982). reprint with introduction by Stephen Jay Gould. New York: Columbia University Press.

Donnellan, Keith. (1974). "Speaking of Nothing," *Philosophical Review*, 83, 3–31.

Doppelt, Gerald. (1978). "Kuhn's Epistemological Relativism: An Interpretation and Defense," *Inquiry*, 21, 33–86.

————. (1986). "Relativism and the Reticulational Model of Scientific Rationality," *Synthese*, 69, 225–252.

————. (1988). "The Philosophical Requirements for an Adequate Conception of Scientific Rationality," *Philosophy of Science*, 55, 104–133.

Drake, Stillman. (1978). *Galileo at Work*. Chicago: University of Chicago Press.

Duhem, Pierre. (1906). *The Aim and Structure of Physical Theory*, (1951) translation by Philip P. Wiener, Princeton: Princeton University Press.

Dummett, Michael. (1977). *Elements of Intuitionism*. Oxford: Oxford University Press.

Dupre, John. (1981). "Natural Kinds and Biological Taxa," *Philosophical Review,* 90, 66–90.

———. (1988). *The Latest on the Best: Essays on Optimality and Evolution*. Cambridge, MA: MIT Press.

Earman, John. (1986). *Determinism: A Primer*. Dordrecht: Kluwer.

———. (1992). *Bayes, or Bust*. Cambridge, MA: MIT Press.

———, and Fine, Arthur. (1977). "Against Indeterminacy," *Journal of Philosophy,* 74, 535–538.

Eldredge, Niles. (1985). *Unfinished Synthesis*. New York: Oxford University Press.

———, and Cracraft, Joel. (1980). *Phylogenetic Patterns and the Evolutionary Process*. New York: Columbia University Press.

———, and Gould, Stephen Jay. (1972). "Punctuated Equilibria: An Alternative to Phyletic Gradualism," in T. J. Schopf (ed), *Models in Paleobiology*. San Francisco: Freeman.

Endler, John. (1986). *Natural Selection in the Wild*. Princeton: Princeton University Press.

Evans, Gareth. (1975). "The Causal Theory of Names," *Proceedings of the Aristotelian Society,* 47, 187–208.

Faust, David. (1985). *The Limits of Scientific Reasoning*. Minneapolis: University of Minnesota Press.

Feigl, Herbert. (1950). *"De Principiis Non Est Disputandum. . . ?"* in Max Black (ed), *Philosophical Analysis* (1963), reprint. Englewood Cliffs, NJ: Prentice-Hall, 113–131.

Feyerabend, Paul. (1963a). "Explanation, Reduction, and Empiricism," in *Minnesota Studies in the Philosophy of Science,* volume III. Minneapolis: University of Minnesota, 28–97.

———. (1963b). "How to Be a Good Empiricist," *Proceedings of the Delaware Seminar,* volume 2. New York: Interscience, 3–39.

———. (1965). "Problems of Empiricism," in R. Colodny (ed), *Beyond the Edge of Certainty*. Englewood Cliffs, NJ: Prentice-Hall, 145–260.

———. (1970). "Against Method," in *Minnesota Studies in the Philosophy of Science,* volume IV. Minneapolis: University of Minnesota, 17–130.

———. (1975). *Against Method*. London: Verso.

———. (1979). *Science in a Free Society*. London: New Left Books.

———. (1987). *Farewell to Reason*. London: Verso.

Feynman, Richard et al. (1963). *The Feynman Lectures on Physics*. Reading, MA: Addison-Wesley.

Field, Hartry. (1972). "Tarski's Theory of Truth," *Journal of Philosophy,* 69, 347–375.

———. (1973). "Theory Change and the Indeterminacy of Reference," *Journal of Philosophy,* 70, 462–481.

———. (1978). "A Note on Jeffrey Conditionalization," *Philosophy of Science.* 45, 361–367.

Fine, Arthur. (1986). *The Shaky Game*. Chicago: University of Chicago Press.

Finocchiaro, Maurice. (1989). *The Galileo Affair*. Berkeley: University of California Press.

Fisher, R. A. (1918). "The Correlation between Relatives on the Supposition of Mendelian Inheritance," *Transactions of the Royal Society of Edinburgh,* 52, 399–433.

———. (1930). *The Genetical Theory of Natural Selection*. Oxford: Oxford University Press. (1956) reprint New York: Dover.

Fodor, Jerry. (1984). "Observation Reconsidered," *Philosophy of Science*, 51, 23–43.

Foley, Richard. (1988a). *Epistemic Rationality*. Cambridge, MA: Harvard University Press.

———. (1988b). "Some Different Conceptions of Rationality," in E. McMullin (ed), *Construction and Constraint*. Notre Dame: University of Notre Dame Press, 123–152.

Frege, Gottlob. (1892). "On Sense and Reference," in P. Geach and M. Black (eds), (1952) *Translations from the Writings of Gottlob Frege*. Oxford, Blackwell, 56–78.

Fresnel, Augustin. (1818). "Memoire sur la Diffraction de la Lumière (Coronné par l'Academie des Sciences)," in (Fresnel 1865/1965), volume I, 247–382.

———. (1865/1965). *Oeuvres*. (three volumes). Paris: Imprimerie Imperiale.

Friedman, Michael. (1974). "Explanation and Scientific Understanding," *Journal of Philosophy*, 71, 5–19.

Fuller, Steven. (1988). *Social Epistemology*. Bloomington: Indiana University Press.

Galileo Galilei. *Dialogue Concerning the Two Great Systems of the World*. Berkeley: University of California Press.

———. *The Starry Messenger* (trans Albert van Helden). Chicago: University of Chicago Press.

Galileo/Favaro. (1890–1909). [Galileo Galilei, edited by Antonio Favaro] *Opere*. Florence: Barbera.

Galison, Peter. (1987). *How Experiments End*. Chicago: University of Chicago Press.

Garber, Daniel. (1982). "Old Evidence and Logical Omniscience," in *Minnesota Studies in the Philosophy of Science,* volume X. Minneapolis: University of Minnesota, 99–131.

Ghiselin, Michael. (1969). *The Triumph of the Darwinian Method*. Berkeley: University of California Press.

———. (1974). "A Radical Solution to the Species Problem," *Systematic Zoology*. 23, 536–544.

———. (1989). *Intellectual Compromise: The Bottom Line*. New York: Paragon.

Giere, Ronald. (1985). "Philosophy of Science Naturalized," *Philosophy of Science*. 52, 331–356.

———. (1988). *Explaining Science*. Chicago: University of Chicago Press.

Gingerich, Owen, and Westman, Robert. (1991). "The Wittich Connection" *Transactions of the American Philosophical Society*. 78, part 7.

Glick, Thomas. (ed) (1972). *The Comparative Reception of Darwinism*. Austin: University of Texas Press.

Glymour, Clark. (1980). *Theory and Evidence*. Princeton: Princeton University Press.

Goldman, Alvin. (1979). "What Is Justified Belief?" in G. Pappas (ed), *Justification and Knowledge*. Dordrecht: Reidel, 1–23.

———. (1986). *Epistemology and Cognition*. Cambridge, MA: Harvard University Press.

———. (1987). "Foundations of Social Epistemics," *Synthese*, 73, 109–144.

———. (1989). "Epistemic Folkways and Justification and Scientific Epistemology," in A. Goldman (ed), *Liaisons*. Cambridge, MA: MIT Press, 1992, 155–175.

Goodman, Nelson. (1955). *Fact, Fiction, and Forecast*. Indianapolis: Bobbs-Merrill.

Gould, Stephen Jay. (1977). *Ontogeny and Phylogeny*. Cambridge, MA: Harvard University Press.

———. (1977a). *Ever Since Darwin*. New York: Norton.

———. (1980). *The Panda's Thumb*. New York: Norton.

———. (1980a). "The Return of the Hopeful Monster," in (Gould 1980), 186–193.

———. (1980b). "G. G. Simpson, Paleontology, and the Modern Synthesis," in (Mayr and Provine 1980).

———. (1980c). "Is a New and General Theory of Evolution Emerging?" *Paleobiology*, 6, 119–130.

———. (1981). *The Mismeasure of Man*. New York: Norton.

———. (1982). "Darwinism and the Expansion of Evolutionary Theory," *Science*, 216, 380–387.

———. (1983). "The Hardening of the Modern Synthesis," in M. Grene, *Dimensions of Darwinism*. Cambridge: Cambridge University Press, 71–93.

———. (1989). *Wonderful Life*. New York: Norton.

———, and Lewontin, Richard C. (1979). "The Spandrels of San Marco and the Panglossian Paradigm: A Critique of the Adaptationist Programme," *Proceedings of the Royal Society*, B 205, 581–598; reprint in E. Sober (ed) (1984). *Conceptual Issues in Evolutionary Biology*. Cambridge, MA: MIT Press, 252–270.

Grandy, Richard. (1973). "Reference, Meaning, and Belief," *Journal of Philosophy*, 70, 439–452.

Gray, Asa. (1876). *Darwiniana*. New York: Appleton.

Hacking, Ian. (1983). *Representing and Intervening*. Cambridge: Cambridge University Press.

Hagstrom, Warren. (1965). *The Scientific Community*. New York: Basic Books.

Hall, Mary Boas. (1966). *Robert Boyle on Natural Science*. Bloomington: Indiana University Press.

Hamilton, W. D. (1966). "The Genetical Evolution of Social Behavior," (parts I and II), *Journal of Theoretical Biology*, 7, 1–52.

Hanna, Joseph. (1985). "Sociobiology and the Information Metaphor," in James Fetzer (ed), *Sociobiology and Explanation*. Dordrecht: Reidel, 31–56.

Hanson, Norwood Russell. (1958). *Patterns of Discovery*. Cambridge: Cambridge University Press.

Hardwig, John. (1985). "Epistemic Dependence," *Journal of Philosophy*, 82, 335–49.

Hardy, G.H. (1940). *A Mathematician's Apology*. Oxford: Oxford University Press.

Harman, Gilbert. (1971). *Thought*. Princeton: Princeton University Press.

———. (1987). *Change in View*. Cambridge, MA: MIT Press.

Haugeland, John. (1985). *Artificial Intelligence: The Very Idea*. Cambridge, MA: MIT Press.

Heinrich, Bernd. (1981). *Bumblebee Economics*. Cambridge, MA: Harvard University Press.

Hempel, Carl G. (1945). "Studies in the Logic of Confirmation," in (Hempel 1965), 3–46.

———. (1950). "Problems and Changes in the Empiricist Criterion of Meaning," *Revue Internationale de Philosophie*, 11, 41–63.

———. (1951). "The Concept of Cognitive Significance: A Reconsideration," *Proceedings of the American Academy of Arts and Sciences*, 80, no. 1, 61–77.

———. (1965). *Aspects of Scientific Explanation*. New York: Free Press.

———. (1966). *Philosophy of Natural Science*. Englewood Cliffs, NJ: Prentice-Hall.

Hesse, Mary. (1962). *Forces and Fields*. Westport, CT: Greenwood.

——. (1970). "Is There an Independent Observation-Language?" in R. Colodny (ed), *The Nature of Scientific Theories*. Pittsburgh: University of Pittsburgh Press, 35–77.

Ho, Mae-Wan, and Fox, Sydney. (1987). *Metaphors in The New Evolutionary Paradigm*. New York: Wiley.

Ho, Mae-Wan, and Saunders, Peter. (1984). *Beyond Neo-Darwinism: An Introduction to the New Evolutionary Paradigm*. New York: Academic Press.

Hodge, M. J. S. (1971). "Species in Lamarck," in J. Schuller (ed), *Colloque Internationale "Lamarck."* Paris: A Blanchard, 31–46.

——. (1977). "The Structure and Strategy of Darwin's 'Long Argument'," *British Journal for the History of Science*, 10, 237–246.

Holland, John; Holyoak, Keith; Nisbett, Richard; and Thagard, Paul. (1986). *Induction*. Cambridge, MA: MIT Press.

Holmes, Frederic L. (1985). *Lavoisier and the Chemistry of Life*. Madison: University of Wisconsin.

——. (1988). "Lavoisier's Conceptual Passage," *Osiris*, 4, 82–92.

Horwich, Paul. (1982). *Probability and Evidence*. Cambridge: Cambridge University Press.

Hosiasson-Lindenbaum, Janina. (1940). "On Confirmation," *Journal of Symbolic Logic*, 5, 133–148.

Howson, Colin, and Urbach, Peter. (1989). *Scientific Reasoning*. LaSalle, IL: Open Court.

Hull, David. (1973a). *Philosophy of Biological Science*. Englewood Cliffs, NJ: Prentice-Hall.

——. (ed) (1973b). *Darwin and His Critics*. Cambridge, MA: Harvard University Press.

——. (1978). "A Matter of Individuality," *Philosophy of Science*. 45, 335–360.

——. (1982). "Darwinism as a Historical Entity," in P. Asquith and T. Nickles (eds), *PSA 1982*. East Lansing, MI: Philosophy of Science Association.

——. (1988). *Science as a Process*. Chicago: University of Chicago Press.

Huxley, Leonard. (1913). *The Life and Letters of Thomas Henry Huxley*. New York: Appleton.

Huxley, Thomas Henry. (1896). *Darwiniana*. New York: Appleton.

Hyde, T. S., and Jenkins J. J. (1973). "Recall for Words as a Function of Semantic, Graphic, and Syntactic Orienting Tasks," *Journal of Verbal Learning and Verbal Behavior*, 12, 471–480.

Jeffrey, Richard. (1983). *The Logic of Decision* (second edition). Chicago: University of Chicago Press.

Kant, Immanuel. (1781/1968). *Critique of Pure Reason*, trans. Norman Kemp Smith. London: MacMillan.

Kim, Jaegwon. (1988). "What Is Naturalistic Epistemology?," *Philosophical Perspectives*, 2, 381–405.

Kirwan, Richard. (1783). *An Essay on Phlogiston* (first edition). London.

——. (1789). *An Essay on Phlogiston;* reprint (Kirwan 1782) with translations of comments by Lavoisier and other French chemists, which had been included in a French translation of (Kirwan 1782), and replies by Kirwan. London: J. Johnson (both editions).

Kitcher, Patricia (1985). "Narrow Taxonomy and Wide Functionalism," *Philosophy of Science*, 52, 78–97.

———. (1988). "Marr's Computational Theory of Vision," *Philosophy of Science*, 55, 1–24.

Kitcher, Patricia. (1990). *Kant's Transcendental Psychology*. New York: Oxford University Press.

Kitcher, Philip. (1978). "Theories, Theorists, and Theoretical Change," *Philosophical Review*. 87, 519–547.

———. (1979). "Frege's Epistemology," *Philosophical Review*, 88, 235–262.

———. (1981). "Explanatory Unification," *Philosophy of Science*, 48, 507–531.

———. (1982). "Genes," *British Journal for the Philosophy of Science*, 33, 337–359.

———. (1982b). *Abusing Science: The Case against Creationism*. Cambridge, MA: MIT Press.

———. (1983a). *The Nature of Mathematical Knowledge*. New York: Oxford University Press.

———. (1983b). "Kant's Philosophy of Science," *Midwest Studies in Philosophy*, 8, 387–408.

———. (1983c). "Implications of Incommensurability," in P. Asquith and T. Nickles (eds), *PSA 1982*, volume II. East Lansing, MI: PSA, 689–703.

———. (1984). "1953 and All That: A Tale of Two Sciences," *Philosophical Review*, 93, 335–373.

———. (1985a). "Darwin's Achievement," in Nicholas Rescher (ed), *Reason and Rationality in Science*. Washington, DC: University Press of America, 123–185.

———. (1985b). *Vaulting Ambition: Sociobiology and the Quest for Human Nature*. Cambridge, MA: MIT Press.

———. (1986). "Projecting the Order of Nature," in Robert Butts (ed), *Kant's Philosophy of Physical Science*. Dordrecht: Reidel, 201–235.

———. (1988). "Why Not the Best?," in (Dupre 1988), 77–102.

———. (1989). "Explanatory Unification and the Causal Structure of the World," in P. Kitcher and W. Salmon (eds), *Scientific Explanation*. Minneapolis: University of Minnesota, 410–505.

———. (1990). "Developmental Decomposition and the Future of Human Behavioral Ecology," *Philosophy of Science*, 57, 96–117.

———. (1992). "The Naturalists Return," *Philosophical Review*, 101, 53–114.

———, and Salmon, Wesley. (eds) (1989). *Scientific Explanation*. Minneapolis: University of Minnesota.

Kohn, David. (ed) (1985). *The Darwinian Heritage*. Princeton: Princeton University Press.

Kordig, Carl. (1971). *The Justification of Scientific Change*. Dordrecht: Reidel.

Kornblith, Hilary. (1980). "Beyond Foundationalism and the Coherence Theory," *Journal of Philosophy*, 72, 597–612.

———. (1985). "Editorial Introduction" in *Naturalistic Epistemology*. Cambridge, MA: MIT Press.

Kosslyn, Stephen. (1981). *Image and Mind*. Cambridge, MA: Harvard University Press.

Kripke, Saul. (1972). "Naming and Necessity," in D. Davidson and G. Harman (eds), *Semantics of Natural Language*. Dordrecht: Reidel; reprinted as (Kripke 1980), 253–355.

———. (1980). *Naming and Necessity*. Cambridge, MA: Harvard University Press.

Kuhn, Thomas S. (1962). *The Structure of Scientific Revolutions*. (first edition). Chicago: University of Chicago Press.

―――. (1962/70). *The Structure of Scientific Revolutions*. (first or second). Chicago: University of Chicago Press.

―――. (1970). *The Structure of Scientific Revolutions*. (second edition). Chicago: University of Chicago Press.

―――. (1977). *The Essential Tension*. Chicago: University of Chicago Press.

―――. (1983). "Commensurability, Comparability, Communicability," in P. Asquith and T. Nickles (eds), *PSA 1982*. East Lansing, MI: PSA, 669–688.

Lakatos, Imre. (1970). "Falsification and the Methodology of Scientific Research Programmes," in I. Lakatos and A. Musgrave (eds), *Criticism and the Growth of Knowledge*. Cambridge: Cambridge University Press, 91–196.

―――. (1971). "History of Science and its Rational Reconstructions," in I. Lakatos (ed), *Philosophical Papers,* volume I. Cambridge: Cambridge University Press, 102–139.

―――, and Zahar, Elie. (1976). "Why Did Copernicus' Research Programme Succeed Ptolemy's?," in (Westman 1976), 354–383.

Landé, R. (1981). "Models of speciation by sexual selection on polygenic traits," *Proceedings of the National Academy of Sciences,* 78, 3721–3725.

Latour, Bruno. (1987). *Science in Action*. Cambridge, MA: Harvard University Press.

―――. (1988). *The Pasteurization of France*. Cambridge, MA: Harvard University Press.

―――. (1989a). "Postmodern? No, Simply AModern : Steps towards an Anthropology of Science," *Studies in the History and Philosophy of Science,* 21, 145–171.

―――. (1992). "One More Turn after the Social Turn . . .," in E. McMullin, *The Social Dimensions of Scientific Knowledge*. Notre Dame: University of Notre Dame Press, 272–294.

―――, and Woolgar, Steve. (1979). *Laboratory Life*. London: Sage, (1986) reprint Princeton: Princeton University Press.

Laudan, Larry. (1977). *Progress and Its Problems*. Berkeley: University of California Press.

―――. (1981). "A Confutation of Convergent Realism," *Philosophy of Science,* 48.

―――. (1984). *Science and Values*. Berkeley: University of California Press.

―――. (1987). "Progress or Rationality: The Prospects for Normative Naturalism," *American Philosophical Quarterly,* 24, 19–31.

―――. (1991). *Relativism: A Dialogue*. Chicago: University of Chicago Press.

Lavoisier, Antoine-Laurent. (1774). "Analyse du Mémoire sur l'Augmentation du Poids des Métaux par la Calcination," in (Lavoisier 1862), II, 97–99.

―――. (1777). "Memoire sur la Combustion en Général," in (Lavoisier 1862), II, 225–233.

―――. (1782). "Memoire sur l'Union du Principe Oxygine avec le Fer," in (Lavoisier 1862), II, 557–574.

―――. (1783). "Réflexions sur le Phlogistique," in (Lavoisier 1862), II, 623–655.

―――. (1789). *Traité Elementaire de Chimie*. Paris: Cuchet.

―――. (1792). "Détails Historiques sur la Cause de l'Augmentation de Poids qu'acquièrent les substances Métalliques, lorsqu'on les chauffe pendant leur exposition à l'air," in (Lavoisier 1862), II, 99–103.

―――. (1862). *Oeuvres*. Paris: Imprimerie Impériale.

Leplin, Jarrett. (ed) (1982). *Scientific Realism*. Berkeley: University of California Press.

Levi, Isaac. (1982). *The Enterprise of Knowledge*. Cambridge, MA: MIT Press.

Levins, Richard. (1966). *Evolution in Changing Environments*. Princeton: Princeton University Press.

———, and Lewontin, Richard C. (1985). *The Dialectical Biologist*. Cambridge, MA: Harvard University Press.

Lewis, David. (1973a). *Counterfactuals*. Oxford: Blackwell.

———. (1973b). "Causation," *Journal of Philosophy*, 70, 556–567.

Lewontin, Richard C. (1968). " 'Honest Jim' Watson's Big Thriller about DNA," in Gunnar Stent (ed), *The Double Helix*. New York: Norton, 185–187.

———. (1970). "The Units of Selection," *Annual Review of Ecology and Systematics*, 1, 1–18.

———. (1974). *The Genetic Basis of Evolutionary Change*. New York: Columbia University Press.

———. (1978). "Adaptation," *Scientific American*, 239, no. 3, 212–230.

———. Rose, Stephen, and Kamin, Leon. (1984). *Not in Our Genes*. New York: Pantheon.

Lloyd, Elisabeth. (1983). "The Nature of Darwin's Support for the Theory of Natural Selection," *Philosophy of Science*, 50, 112–129.

———. (1984). "A Semantic Approach to the Structure of Population Genetics," *Philosophy of Science*, 51, 242–264.

———. (1988). *The Structure and Confirmation of Evolutionary Theory*. Westport, CT: Greenwood.

Loftus, Elizabeth. (1979). *Eyewitness Testimony*. Cambridge, MA: Harvard University Press.

Longino, Helen. (1990). *Science as Social Knowledge*. Princeton: Princeton University Press.

MacArthur, Robert, and Wilson, E. O. (1967). *The Theory of Island Biogeography*. Princeton: Princeton University Press.

Machamer, Peter. (1973). "Feyerabend and Galileo: The Interaction of Theories and the Reinterpretation of Experience," *Studies in the History and Philosophy of Science*, 4, 1–46.

Mackie, J. L. (1963). "The Paradox of Confirmation," *British Journal for the Philosophy of Science*, 13, 265–277.

———. (1973). *The Cement of the Universe*. Oxford: Oxford University Press.

Manier, Edward. (1978). *The Young Darwin and His Cultural Circle*. Dordrecht: Reidel.

Margolis, Howard. (1982). *Selfishness, Altruism, and Rationality: A Theory of Social Choice*. Cambridge: Cambridge University Press.

———. (1987). *Patterns, Judgment, and Cognition*. Chicago: University of Chicago Press.

Marr, David. (1982). *Vision*. San Francisco: Freeman.

Martin, Paul, and Bateson, Patrick. (1986). *Measuring Behavior: An Introductory Guide*. Cambridge: Cambridge University Press.

Maxwell, Grover. (1963). "The Ontological Status of Theoretical Entities," in *Minnesota Studies in the Philosophy of Science*, volume III. Minneapolis: University of Minnesota, 3–37.

Maynard Smith, John. (1982). *Evolution and the Theory of Games*. Cambridge: Cambridge University Press.

———. (1988). "How to Model Evolution," in (Dupre 1988), 119–132.

———. (1989). *Did Darwin Get It Right?.* London: Chapman and Hall.

Mayr, Ernst. (1942). *Systematics and the Origin of Species.* New York: Columbia University Press.

———. (1963). *Animal Species and Evolution.* Cambridge, MA: Harvard University Press.

———. (1976). *Evolution and the Diversity of Life.* Cambridge, MA: Harvard University Press.

———. (1982). *The Growth of Biological Thought.* Cambridge, MA: Harvard University Press.

———, and Provine, William. (eds) (1980). *The Evolutionary Synthesis.* Cambridge, MA: Harvard University Press.

McClelland, Jay, and Rumelhart, David. (1986). *Parallel Distributed Processing.* Cambridge, MA: MIT Press.

McEvoy, John. (1988). "Continuity and Discontinuity in the Chemical Revolution," *Osiris,* 4, 195–213.

McMullin, Ernan. (ed) (1988). *Construction and Constraint.* Notre Dame: University of Notre Dame Press.

Merton, Robert K. (1973). *The Sociology of Science.* Chicago: University of Chicago Press.

Mill, John Stuart. (1859). *On Liberty.* Indianapolis: Bobbs-Merrill.

Miller, David. (1974). "On the Comparison of False Theories by Their Bases," *British Journal for the Philosophy of Science,* 25, 166–177.

———. (1979). "The Accuracy of Predictions," *Synthese,* 30, 159–191.

Miller, George. (1956). "The Magic Number Seven Plus or Minus Two," *Psychological Review,* 63, 81–97.

Milne-Edwards, Henri. (1844). "Considérations sur quelques principes relatifs à la classification naturelle des animaux," *Annales des Sciences Naturelles,* 3rd Series, 1, 65–99.

Mondadori, Fabrizio, and Morton, Adam. (1976). "Modal Realism: The Poisoned Pawn," *Philosophical Review,* 85 3–20.

Moore, James. (1985). "Darwin of Down: The Evolutionist as Squarson-Naturalist," in (Kohn 1985), 435–482.

Newton-Smith, William. (1981). *The Rationality of Science.* London: Routledge.

Nisbett, Richard, and Ross, Lee. (1980). *Human Inference: Strategies and Shortcomings of Social Judgment.* Englewood Cliffs, NJ: Prentice-Hall.

Olby, R. C. (1974). *The Path to the Double Helix.* London: Macmillan.

Ospovat, Dov. (1981). *The Development of Darwin's Theory: Natural History, Natural Theology, and Natural Selection 1838–1859.* Cambridge: Cambridge University Press.

Owen, Richard. (1848). *On the Archetype and Homologies of the Vertebrate Skeleton.* London: Van Voorst.

———. (1849). *On the Nature of Limbs.* London: Van Voorst.

Papineau, David. (1988). *Reality and Representation.* Oxford: Blackwell.

Partington, J. R. (1965). *A Short History of Chemistry.* London: MacMillan.

———, and McKie, Douglas. (1937–38). "Historical Studies on the Phlogiston Theory," *Annals of Science,* 2, 361–404, 4, 1–58, 4, 337–371.

Peacocke, Christopher. (1987). *Thoughts: An Essay on Content.* Oxford: Blackwell.

Peirce, C. S. (1958). *Collected Works.* Cambridge, MA: Harvard University Press.

Perrin, Carl B. (1988). "Research Traditions, Lavoisier, and the Chemical Revolution," *Osiris,* 4, 53–81.

Pickering, Andrew. (1984). *Constructing Quarks*. Chicago: University of Chicago Press.

Pinch, Trevor. (1985). "Theory Testing in Science—the Case of the Solar Neutrinos," *Philosophy of Social Science*, 15, 167–188.

———. (1986). "Strata Various," *Social Studies of Science*, 16, 705–713.

Popper, Karl R. (1959). *The Logic of Scientific Discovery*. London: Hutchinson.

Powell, Baden. (1855). *Essays on the Spirit of the Inductive Philosophy, the Unity of Worlds, and the Philosophy of Creation*. London: Longmans.

Priestley, Joseph (1775/1970). *Experiments and Observations on Different Kinds of Air*. New York: Johnson Reprint.

Provine, William. (1971). *The Origins of Theoretical Population Genetics*. Chicago: University of Chicago Press.

———. (1986). *Sewall Wright and Evolutionary Biology*. Chicago: University of Chicago Press.

Putnam, Hilary. (1973). "Meaning and Reference," *Journal of Philosophy*, 70, 699–711.

———. (1978). *Meaning and the Moral Sciences*. London: Routledge.

———. (1981). *Reason, Truth, and History*. Cambridge: Cambridge University Press.

Quine, W. V. (1951). "Two Dogmas of Empiricism," in *From a Logical Point of View*. New York: Harper, 20–46.

———. (1960). *Word and Object*. Cambridge, MA MIT Press.

———. (1966). "Carnap and Logical Truth," in *The Ways of Paradox*. New York: Random House, 107–132.

———. (1970). *Ontological Relativity and Other Essays*. New York: Columbia University Press.

———. (1970a). "Epistemology Naturalized," in (Quine 1970), 69–90.

———. (1970b). "Natural Kinds" in (Quine 1970), 114–138.

———. (1974). *The Roots of Reference*. LaSalle, IL: Open Court.

Redondi, Pietro. (1987). *Galileo, Heretic*. Princeton: Princeton University Press.

Rescher, Nicholas. (1989a). *A Useful Inheritance*. Totowa, NJ: Rowman and Littlefield.

———. (1989b). *Cognitive Economy*. Pittsburgh: University of Pittsburgh Press.

Rosenberg, Alexander. (1978). "The Supervenience of Biological Concepts," *Philosophy of Science*, 45, 368–386.

———. (1983). "Fitness," *Journal of Philosophy*, 80, 457–473.

———. (1985). *The Structure of Biological Science*. New York: Cambridge University Press.

Rosenkrantz, R. D. (1977). *Inference, Method and Decision: Towards a Bayesian Philosophy of Science*. Dordrecht: Reidel.

Roughgarden, Jonathan. (1979). *Theory of Population Genetics and Evolutionary Ecology: An Introduction*. New York: Macmillan.

Rouse, Joseph. (1987). *Knowledge and Power*. Ithaca: Cornell University Press.

Rudwick, Martin J. S. (1972/85). *The Meaning of Fossils*. Chicago: University of Chicago; (1985) reprint.

———. (1970). "The Strategy of Lyell's *Principles of Geology*," *Isis*, 61, 5–33.

———. (1985). *The Great Devonian Controversy*. Chicago: University of Chicago Press.

Ruse, Michael. (1973). *Philosophy of Biology*. London: Hutchinson.

———. (1979). *The Darwinian Revolution*. Chicago: University of Chicago Press.

———. (1986). *Taking Darwin Seriously*. Oxford: Blackwell.

Russell, E. B. (1916). *Form and Function.* Chicago: University of Chicago Press.

Salmon, Wesley. (1957). "Should We Attempt to Justify Induction?," *Philosophical Studies,* 8, 33–48.

———. (1967). *Foundations of Scientific Inference.* Pittsburgh: University of Pittsburgh Press.

———. (1968). "The Justification of Inductive Rules of Inference," in I. Lakatos (ed), *The Problem of Inductive Logic.* Amsterdam: North-Holland, 24–43.

———. (1970). "Bayes' Theorem and the History of Science," *Minnesota Studies in the Philosophy of Science,* volume V. Minneapolis: University of Minnesota Press, 68–86.

———. (1984). *Scientific Explanation and the Causal Structure of the World.* Princeton: Princeton University Press.

———. (1989). *Four Decades of Scientific Explanation.* Minneapolis: University of Minnesota; also in (Kitcher and Salmon 1989).

———. (1990). "Rationality and Objectivity in Science *or* Tom Kuhn Meets Tom Bayes," *Minnesota Studies in the Philosophy of Science,* volume XIV. Minneapolis: University of Minnesota Press, 175–204.

Sarkar, Husain. (1982). *A Theory of Method.* Berkeley: University of California Press.

Schaffner, Kenneth. (1972). *Nineteenth-Century Ether Theories.* Oxford: Pergamon.

Scheffler, Israel. (1967). *Science and Subjectivity.* Indianapolis: Bobbs-Merrill.

Schelling, Thomas. (1984). *Choice and Consequence.* Cambridge, MA: Harvard University Press.

Schlick, Moritz (1934/1959). "The Foundation of Knowledge," in A. J. Ayer (ed), *Logical Positivism.* New York: Free Press, 209–227.

Schofield, Robert. (1967). *A Scientific Autobiography of Joseph Priestley.* Cambridge, MA: MIT Press.

———. (1969). *Mechanism and Materialism.* Princeton: Princeton University Press.

Secord, James. (1985). "Darwin and the Breeders," in (Kohn 1985), 519–542.

Sellars, Wilfrid. (1956). "Empiricism and the Philosophy of Mind," in (Sellars 1963), chapter 5.

———. (1963). *Science, Perception, and Reality.* London: Routledge.

———. (1967). *Science and Metaphysics.* London: Routledge.

Shapere, Dudley. (1982). "The Concept of Observation in Science and in Philosophy," *Philosophy of Science,* 49, 485–525.

———. (1984). *Reason and the Growth of Knowledge.* Dordrecht: Reidel.

Shapin, Steven. (1982). "History of Science and Its Sociological Reconstructions," *History of Science.* 20, 157–211.

———, and Schaffer, Simon. (1985). *Leviathan and the Air-Pump.* Princeton: Princeton University Press.

Shepard, R., and Chipman, S. (1970). "Second-order Isomorphism of Internal Representations: Shapes of States," *Cognitive Psychology,* 1, 1–17.

———, and Metzler, J. (1971). "Mental Rotation of Three-Dimensional Objects," *Science.* 171, 701–703.

Simon, Herbert. (1957). *Models of Man.* New York: Wiley.

Simpson, George Gaylord. (1944). *Tempo and Mode in Evolution.* New York: Columbia University Press.

———. (1953). *The Major Features of Evolution.* New York: Columbia University Press.

Skyrms, Brian. (1986). *Choice and Chance* (third edition). Belmont, CA: Wadsworth.

Sneed, Joseph. (1971). *The Logical Structure of Mathematical Physics*. Dordrecht: Reidel.

Sober, Elliott. (1984). *The Nature of Selection*. Cambridge, MA: MIT Press.

———. (1988). Review of (Salmon 1984), *British Journal for the Philosophy of Science*, 38, 243–257.

———. (1988). *Reconstructing the Past*. Cambridge, MA: MIT Press.

Sokal, Robert, and Rohlf, James. (1974). *Biometry*. San Francisco: Freeman.

Solomon, Miriam. (1992). "Scientific Rationality and Human Reasoning," *Philosophy of Science* (forthcoming).

Stanley, Steven. (1979). *Macroevolution, Pattern and Process*. San Francisco: Freeman.

Sterelny, Kim, and Kitcher, Philip. (1988). "The Return of the Gene," *Journal of Philosophy*, 85, 339–361.

Stich, Stephen. (1983). *The Case against Belief: From Folk Psychology to Cognitive Science*. Cambridge, MA: MIT Press.

———. (1985). "Could Man Be an Irrational Animal?," *Synthese*, 64, 115–134.

———. (1990). *The Fragmentation of Reason*. Cambridge, MA: MIT Press.

Strawson, Peter. (1952). *Introduction to Logical Theory*. London: Methuen.

Strum, Shirley. (1987). *Almost Human*. New York: Random House.

Sulloway, Frank. (1982). "Darwin's Finches, the Evolution of a Legend," *Journal for the History of Biology*, 15, 1–53.

———. (forthcoming). "Orthodoxy and Innovation in Science," manuscript.

Suppe, Fred. (1972). "What's Wrong with the Received View of Scientific Theories?," *Philosophy of Science*, 39, 1–19.

———. (ed) (1977). *The Structure of Scientific Theories* (second edition). Urbana: University of Illinois Press.

Suppes, Patrick. (1967). "What Is a Scientific Theory?" in A. Danto and S. Morgenbesser (eds), *Philosophy of Science Today*. New York: Basic Books, 55–67.

Tarski, Alfred. (1936). "The Concept of Truth in Formalized Languages," in J. H. Woodger (ed. and trans.), *Logic, Semantics, Metamathematics*. Oxford: Oxford University Press (1956), 152–278.

———. (1944). "The Semantic Conception of Truth," in H. Feigl and W. Sellars (eds), *Readings in Philosophical Analysis*. New York: Appleton (1949), 341–374.

Toulmin, Stephen. (1961). *Foresight and Understanding*. New York: Harper and Row.

———. (1972). *Human Understanding*. Princeton: Princeton University Press.

Tversky, Amos, and Kahneman, Daniel. (1973). "Availability: A Heuristic for Judging Frequency and Probability," *Cognitive Psychology*, 5, 207–232.

———. (1974). "Judgment under Uncertainty: Heuristics and Biases," *Science*. 185, 1124–1131.

Van Fraassen, Bas. (1980). *The Scientific Image*. Oxford: Oxford University Press.

———. (1985). "Salmon on Explanation," *Journal of Philosophy*, 82, 639–651.

Van Helden, Albert. (1977). *The Invention of the Telescope*. Philadelphia: American Philosophical Society Monograph.

———. (1989). Introduction to translation of Galileo's *Sidereus Nuncius*. Chicago: University of Chicago Press.

Watson, James. (1967). *The Double Helix*. New York: Norton.

Westfall, R. S. (1982). *Never at Rest*. Cambridge: Cambridge University Press.

———. (1985). "Science and Patronage: Galileo and the Telescope," *Isis*, 76, 11–30.

Westman, Robert. (1975). "The Melanchthon Circle, Rheticus, and the Wittenberg Interpretation of Copernican Theory," *Isis*, 165–193.

———. (ed) (1976). *The Copernican Achievement.* Los Angeles: University of California Press.

———. (1980). "The Astronomer's Role in the Sixteenth Century: A Preliminary Study," *History of Science,* 18, 105–142.

———. (1990). "Proof, Poetics, and Patronage: Copernicus' Preface to *De Revolutionibus,*" in Westman and David Lindberg (eds), *Reappraisals of the Scientific Revolution.* Cambridge: Cambridge University Press, 167–205.

Williams, George C. (1966). *Adaptation and Natural Selection.* Princeton: Princeton University Press.

Williams, Mary. (1970). "Deducing the Consequences of Evolution," *Journal of Theoretical Biology,* 29, 343–385.

Wilson, David S. (1980). *The Natural Selection of Populations and Communities.* Menlo Park: Benjamin/Cummings.

———. (1983). "The Group Selection Controversy: History and Current Status," *Annual Review of Ecology and Systematics,* 14, 159–187.

Wilson, E. O.; Carpenter, F. M.; and Brown, W. L. (1967). "The First Mesozoic Ants," *Science,* 157, 1038–1040.

Wittgenstein, Ludwig. (1953). *Philosophical Investigations.* Oxford: Blackwell.

Woodward, James. (1989). "The Causal-Mechanical Model of Explanation," in (Kitcher and Salmon 1989), 357–383.

Woolfenden, G., and Fitzpatrick, J. (1984). *The Florida Scrubjay.* Princeton: Princeton University Press.

Worrall, John. (1976). "Thomas Young and the 'Refutation' of Newtonian Optics: A Case Study in the Interaction of Philosophy of Science and the History of Science," in Colin Howson (ed), *Method and Appraisal in the Physical Sciences.* Cambridge: Cambridge University Press, 107–180.

———. (1978). "The Ways in Which the Methodology of Scientific Research Programmes Improves on Popper's Methodology," in G. Radnitzky and G. Anderson (eds), *Progress and Rationality in Science.* Dordrecht: Reidel, 45–70.

———. (1985). "Scientific Discovery and Theory-Confirmation," in J. Pitt (ed), *Change and Progress in Modern Science.* Dordrecht: Reidel, 301–332.

———. (1988). "The Value of a Fixed Methodology," *British Journal for the Philosophy of Science,* 39, 263–275.

———. (1989). "Fresnel, Poisson, and the White Spot," in D. Gooding et al. (eds), *The Uses of Experiment.* Cambridge: Cambridge University Press, 135–157.

Wright, Sewall. (1931). "Evolution in Mendelian Populations," *Genetics,* 16, 97–159.

Index